Stratospheric Ozone Reduction,
Solar Ultraviolet Radiation and Plant Life

NATO ASI Series

Advanced Science Institutes Series

A series presenting the results of activities sponsored by the NATO Science Committee, which aims at the dissemination of advanced scientific and technological knowledge, with a view to strengthening links between scientific communities.

The Series is published by an international board of publishers in conjunction with the NATO Scientific Affairs Division

A **Life Sciences**	Plenum Publishing Corporation
B **Physics**	London and New York
C **Mathematical and Physical Sciences**	D. Reidel Publishing Company Dordrecht, Boston and Lancaster
D **Behavioural and Social Sciences**	Martinus Nijhoff Publishers Boston, The Hague, Dordrecht and Lancaster
E **Applied Sciences**	
F **Computer and Systems Sciences**	Springer-Verlag Berlin Heidelberg New York Tokyo
G **Ecological Sciences**	

Series G: Ecological Sciences Vol. 8

Stratospheric Ozone Reduction, Solar Ultraviolet Radiation and Plant Life

Edited by

Robert C. Worrest

Department of General Science, Oregon State University, Corvallis, Oregon, USA
and
U.S. Environmental Protection Agency, Environmental Research Laboratory, Corvallis, Oregon, USA

Martyn M. Caldwell

Department of Range Science and the Ecology Center Utah State University, Logan, Utah, USA

Springer-Verlag Berlin Heidelberg New York Tokyo
Published in cooperation with NATO Scientific Affairs Division

Proceedings of the NATO Advanced Research Workshop on The Impact of Solar Ultraviolet Radiation upon Terrestrial Ecosystems: I. Agricultural Crops held at Bad Windsheim, Germany, September 27–30, 1983

QK
757
.N28
1986

ISBN 3-540-13875-7 Springer-Verlag Berlin Heidelberg New York Tokyo
ISBN 0-387-13875-7 Springer-Verlag New York Heidelberg Berlin Tokyo

Library of Congress Cataloging in Publication Data. NATO Advanced Research Workshop on the Impact of Solar Ultraviolet Radiation upon Terrestrial Ecosystems: I. Agricultural Crops (1983 : Bad Windsheim, Germany) Stratospheric ozone reduction, solar ultraviolet radiation, and plant life. (NATO ASI series. Series G, Ecological sciences ; vol. 8) "Proceedings of the NATO Advanced Workshop on the Impact of Solar Ultraviolet Radiation upon Terrestrial Ecosystems: I. Agricultural Crops held at Bad Windsheim, September 27–30, 1983"—T.p. verso. "Published in cooperation with NATO Scientific Affairs Division." Includes index. 1. Plants Effect of ultraviolet radiation on—Congresses. 2. Plants, Effect of solar radiation on—Congresses. 3. Atmospheric ozone—Environmental aspects—Congresses. 4. Ultraviolet radiation—Congresses. 5. Solar radiation—Congresses. 6. Ultraviolet radiation—Physiological effect—Congresses. 7. Solar radiation—Physiological effect—Congresses. 8. Stratosphere—Congresses. I. Worrest, Robert C., 1935-. II. Caldwell, Martyn M., 1941-. III. North Atlantic Treaty Organization. Scientific Affairs Division. IV. Title. V. Series: NATO ASI series. Series G, Ecological sciences ; no. 8. QK757.N28 1983 581.5'222 86-3791
ISBN 0-387-13875-7 (U.S.)

© Springer-Verlag Berlin Heidelberg 1986
Printed in Germany

Printing: Beltz Offsetdruck, Hemsbach; Bookbinding: J. Schäffer OHG, Grünstadt
2131/3140-543210

Preface

Inadvertent alterations of the earth's atmosphere by man's activities are now of regional and even global proportion. Increasing concern has been focused in the last decade on consequences of acid rain, carbon dioxide enrichment of the atmosphere and reduction of ozone in the upper atmosphere. The latter two problems are of truly global scale. This book focuses on the atmospheric ozone reduction problem and the potential consequences for plant life.

Unlike carbon dioxide enrichment, reduction of the total atmospheric ozone column has not yet taken place to a noticeable degree -- it is a problem of the future. The processes leading to ozone reduction involve time periods on the scale of decades. However, by the same token, if society finds ozone reduction to be unacceptable it will take even longer for the process to be reversed. Thus, anticipation of the consequences of ozone reduction is of obvious importance.

Speculation of the possibility of ozone reduction first appeared in the early 1970's and was focused on the consequences of the injection of large quantities of nitrogen oxides into the upper atmosphere by supersonic aircraft flying at high altitudes. Other sources of nitrogen oxides originating from the earth's surface were also considered. With further refinement, the concerns of nitrogen oxide pollution of the upper atmosphere were diminished since the quantities likely to be involved were insufficient to cause a serious threat to the ozone layer. However, in the mid-1970's, the concern for halogen pollution of the upper atmosphere surfaced. The halogen of immediate concern is chlorine, and the source is primarily from chlorofluorocarbons that are released worldwide from aerosol spray cans, certain plastic foams and refrigerative air conditioners. The atmospheric models that predict future ozone depletion are in a continual process of refinement. Through the years the predicted scenarios for resultant changes in the ozone column have ranged from decreases in excess of 20 percent to slight increases in the stratospheric ozone content. The magnitude and sign of the predicted change in total ozone over the next century depend critically on the multiple trace-gas scenarios that are assumed.

Although ozone constitutes a very small proportion of the stratosphere, it plays a major role in the life on this planet. The result of changes in the density of the total ozone column could be far-reaching. Not only could these changes alter the earth's climate due to an altered energy balance for the earth, but it could also have important effects on the nature of sunlight penetrating the earth's atmosphere. By absorbing solar radiation, primarily in the 200- to 310-nanometer waveband, atmospheric ozone serves as a protective shield surrounding the earth. If a decrease in total atmospheric ozone were to occur, exposure to solar radiation in the 290- to 320-nanometer waveband (ultraviolet-B or UV-B radiation) would increase at the surface of the earth. This would have an effect on humans, other animals and plants. Most of the known biological effects of UV-B radiation are damaging; therefore, the possibility of increased exposure to solar UV-B radiation gives particular cause for concern.

This book treats many aspects of the question concerning possible decreases in the amount of stratospheric ozone -- themes ranging from the current status of atmospheric model predictions about the ozone layer and the altered optical properties of the atmosphere to the biological implications, with particular reference to plants. Biological considerations covered in this volume include how photobiological damage is caused by ultraviolet radiation of various wavelengths and how this relates to potential changes in spectral transmission for the atmosphere, as well as the basic types of biological damage that can occur and how this can affect different stages of plant development. These are represented both from theoretical considerations and the results of laboratory and field experiments. Implications of stratospheric ozone reduction for agricultural and nonagricultural ecosystems are also explored.

This book is a result of the efforts of many scientists who share in the concern about the possible threat to the biosphere resulting from anthropogenic changes in the total ozone column. Each chapter represents a contribution to a NATO-sponsored Advanced Research Workshop held in Bad Windsheim, Federal Republic of Germany, 27-30 September, 1983. The workshop was dedicated to discussions about the potential causes and effects of a decrease in the amount of stratospheric ozone. The Director of the workshop was R. C. Worrest, and the following scientists served both as an Organizing Committee and as an Editorial Board:

R. C. Worrest, Co-Editor
    Oregon State University, Corvallis, Oregon USA
M. M. Caldwell, Co-Editor
    Utah State University, Logan, Utah USA
H. Bauer
    Gesellschaft für Strahlen- und Umweltforschung, Munich, FRG
M. Tevini
    University of Karlsruhe, FRG

The Board acknowledges the outstanding support provided for the Workshop and for the production of this monograph by the North Atlantic Treaty Organization, the Bundesministerium für Forschung und Technologie and the U.S. Environmental Protection Agency.

R. C. Worrest
M. M. Caldwell

Corvallis, Oregon USA
Logan, Utah USA
December, 1984

TABLE OF CONTENTS

Agents and Effects of Ozone Trends in the Atmosphere

G. Brasseur* and A. De Rudder

Institut d'Aéronomie Spatiale de Belgique, Bruxelles, Belgium

*Aspirant au Fonds National de la Recherche Scientifique. Also at the Institut d'Astronomie, d'Astrophysique et de Géophysique - Université Libre de Bruxelles.

ABSTRACT

Ozone is produced from the photodissociation of molecular oxygen by ultra-violet radiation ($\lambda < 242.4$ nm). It is destroyed by direct recombination with oxygen atoms. This recombination can be catalyzed by hydroxyl radicals, nitrogen and chlorine oxides. In other words, the ozone layer can be altered by species produced either by natural processes in the biosphere or by human activity in relation with agriculture or industry. As an example, man-made chlorofluorocarbons should reduce the ozone amount by a few percent if the present release into the atmosphere continues. On the other hand, an increase of carbon dioxide should cool the upper strato-sphere and consequently increase the ozone concentration. This chapter presents results from an interactive chemical/radiative/dynamic one-dimensional model with several coupled perturbation scenarios.

INTRODUCTION

Although its relative amount is only of the order of 5×10^{-7}, ozone is of great importance in the atmosphere. This gas, which is photochemically produced, is indeed a strong absorber of ultraviolet radiation with wave-lengths shorter than 310 nm. It therefore protects the earth's surface from a series of potentially harmful biological effects of solar UV radia-tion. Moreover, ozone, which is confined essentially to the stratosphere, is responsible for the heating of the middle atmosphere and in particular for the temperature maximum at 50 km altitude. Ozone therefore is not only a protection for life on earth, but it is also linked to the dynamic state of the atmosphere and finally to the terrestrial climate.

As indicated by Fig. 1 which shows an average vertical distribution of ozone below 100 km, the maximum O_3 concentration is located at an altitude of 20 to 25 km with a corresponding value of about 4×10^{12} cm^{-3}. In the troposphere the ozone concentration is less than 10^{12} cm^{-3} and at the stratopause it is of the order of 5×10^{10} cm^{-3}. The integrated concentra-tion, also called the ozone column, is equal to about 9×10^{18} cm^{-2}, corre-sponding to a layer at STP conditions of 0.35 ± 0.05 cm. In fact, continuous observations show that the ozone column is rather constant in the equatorial regions (about 0.29 cm), but exhibits a great deal of diurnal and seasonal variability at higher latitudes. An ozone maximum (> 0.4 cm) occurs in the polar region at the end of the winter, suggesting

NATO ASI Series, Vol. G8
Stratospheric Ozone Reduction, Solar Ultraviolet Radiation
and Plant Life. Edited by R. C. Worrest and M. M. Caldwell
© Springer-Verlag Berlin Heidelberg 1986

a strong control of the ozone layer by dynamics (general circulation and waves). Figure 2 shows a time-latitude variation of total ozone resulting from the analysis of numerous observations made from ground level instruments.

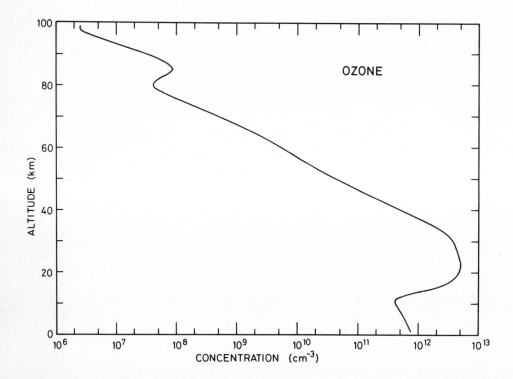

Fig. 1. Calculated vertical distribution of the ozone concentration between 0 and 100 km altitude.

The stability of the ozone layer has been a subject of great concern in the last 15 years since the destruction of the O_3 molecule is catalyzed by several radicals which are produced in fairly large amounts as a result of human activity. For example, nitrogen oxides can be injected in the atmosphere directly from aircraft engines or produced from the N_2O molecules which are released when nitrogen fertilizers are used in agricultural practices. These NO and NO_2 molecules are involved in the catalytic destruction of ozone (Crutzen 1970) and introduce a possible effect of aircraft emissions (Johnston 1971; Crutzen 1972) or nitrogen fertilizers (McElroy et al. 1976) on the ozone layer. Moreover, ozone can be depleted by halogen compounds such as chlorine oxides (Stolarski and Cicerone 1974) which are produced from the man-made chlorofluorocarbons (Molina and Rowland 1974a,b).

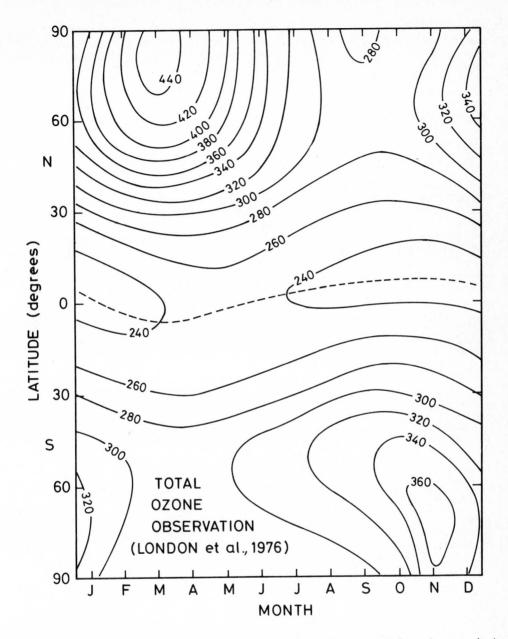

Fig. 2. Seasonal variation of the ozone column (expressed in Dobson units) as a function of latitude (London et al. 1976).

The estimation of the possible ozone depletion and the related increase in the UV penetration due to agricultural, industrial and domestic activity is usually based on theoretical models. These models take into account the most important chemical and photochemical reactions and are usually coupled with a radiative transfer code. The transport of the long-lived species and of heat has to be somewhat parameterized since a full three-dimensional calculation of the mean circulation and the waves requires prohibitive computer resources for most applications. Two-dimensional models assume a zonal symmetry and they parameterize the meridional transport either by the eulerian mean circulation and by eddy diffusion processes (Rao-Vupputuri 1973, 1979; Harwood and Pyle 1975; Crutzen 1975; Brasseur 1978; Miller et al. 1981) or by eulerian residual meridional wind components (Dunkerton 1978; Pyle and Rogers 1980; Garcia and Solomon 1983; Ka Kit Tung 1983).

One-dimensional models assume complete horizontal mixing and derive the globally averaged vertical distributions of the species concentrations and in some cases the temperature. The vertical exchanges are parameterized by a phenomenological eddy diffusion coefficient (K). Although these models are highly simplified from the dynamic point of view, they usually contain a rather detailed chemical and photochemical scheme and they are appropriate to study the direct couplings between chemistry, radiation and temperature. These models are commonly used to estimate the global impact of several types of chemical perturbations (Logan et al. 1978; Wuebbles et al. 1983; Nicoli and Visconti 1983).

OZONE PHOTOCHEMISTRY IN A PURE OXYGEN ATMOSPHERE

A simple theory of ozone photochemistry has been given by Chapman, considering a pure oxygen atmosphere in which the following reactions take place:

(J_2): $O_2 + h\nu$ ($\lambda < 242.4$ nm) $\longrightarrow$ $O + O$

(k_1): $O + O + M \longrightarrow O_2 + M$

(k_2): $O + O_2 + M \longrightarrow O_3 + M$

(k_3): $O + O_3 \longrightarrow O_2 + O_2$

(J_3): $O_3 + h\nu$ ($\lambda < 1180$ nm) $\longrightarrow$ $O_2 + O$

The recombination of atomic oxygen by reaction (k_1) can, however, be neglected below the mesopause (85 km) so that, in the stratosphere and in the mesosphere, the kinetic equation which can be derived for odd oxygen $(O_x = O_3 + O)$ is given by

$$\frac{dn(O_x)}{dt} + 2k_3\, n(O)\, n(O_3) = 2J_2\, n(O_2), \tag{1}$$

if $n(X)$ represents the concentration of species X (see Nicolet 1971 or Brasseur 1982). The ratio between the atomic oxygen and the ozone concentrations can be derived from the photoequilibrium equation

$$\frac{n(O)}{n(O_3)} = \frac{J_3}{k_2 n(M)\, n(O_2)} \tag{2}$$

The lifetime of odd oxygen, which is given by

$$\tau(O_x) = \frac{0.275 \; n(O_x)}{J_2 \; n(O_2)} \; , \qquad (3)$$

is of the order of 1 hour at the stratopause (50 km) but increases gradually with decreasing altitude to reach more than 5 years below 20 km. Since the dynamic time constant is of the order of 2 years in the stratosphere, ozone is controlled by dynamics only (general circulation and waves) below say 25 km and by photochemistry only above 30 km or so. In the intermediate layer both chemistry and transport have to be considered to explain the behavior of ozone. It should be noted that in the troposphere (z < 12 km) the chemistry again becomes more active with regard to O_3 and could even play the major role at low altitude.

When photochemical equilibrium conditions apply, the ozone concentration in a pure oxygen atmosphere is given by

$$n(O_3) = \left\{ \frac{J_2}{J_3} \; n(M) \; n^2(O_2) \; \frac{k_2(T)}{k_3(T)} \right\}^{1/2} \qquad (4)$$

The value of the rate constants k_1, k_2 and k_3 are given in Table 1 and the averaged physical parameters which enter into the calculation of O_3 are specified in Table 2. The exact determination of the photodissociation frequencies of O_2 and O_3 is not straightforward since it requires a detailed knowledge of the solar irradiance spectrum (see Brasseur and Simon 1981) and of the absorption cross section of O_2 and O_3 [see Nicolet (1981) or Simon and Brasseur (1983) for a discussion]. In particular, one has to determine quantitatively the penetration of the solar radiation taking into account absorption and scattering processes. Indeed the solar irradiance Φ at level z for a zenith angle of the sun χ and at wavelength λ is given by

$$\Phi(\lambda;z,\chi) = \Phi_\infty(\lambda) \; \exp - [\tau_a(\lambda;z,\chi) + \tau_s(\lambda;z,\chi)] + \int_{4\pi}^{\infty} L(\lambda;z,\omega) \; d\omega \qquad (5)$$

if $\Phi_\infty(\lambda)$ is the solar irradiance at the top of the atmosphere, $L(\lambda;z,\omega)$ the radiance in direction ω due to the multiple scattering of the light, $\tau_a(\lambda;z,\chi)$ the optical depth due to the absorption, essentially by O_2 and O_3:

$$\tau_a(\lambda;z,\chi) = [\sigma(O_2;\lambda) \; N(O_2;z) + \sigma(O_3;\lambda) \; N(O_3;z)] \; \sec \chi \qquad (6)$$

where σ is the absorption cross section and N the vertically integrated concentration of the absorber (O_2 and O_3, respectively) above altitude z.

The optical depth due to the Rayleigh scattering is given by

$$\tau_S = \sigma_R \, n(M)$$

where $n(M)$ is the atmosphere concentration,

$$\sigma_R = \frac{8\pi^3}{3} \frac{(\mu^2 - 1)}{\lambda^4 n^2(M)} D \sec \chi \qquad (7)$$

the corresponding cross section, μ the air refractive index, $D \cong 1.06$ the depolarization factor and χ the solar zenith angle. Moreover, the amplification factor of the solar irradiance at ground level due to scattering and the earth's albedo in the 310-400 nm range, that is the factor by which the direct solar flux has to be multiplied to account for a given surface albedo and the multiple scattering processes, is shown in Fig. 3 (Nicolet et al. 1983). The photodissociation coefficient of a species X is then derived from

$$J(X;z,\chi) = \int_\lambda \sigma(X;\lambda) \, q(\lambda;z,\chi) \, d\lambda \qquad (8)$$

if $q = \Phi/h\nu$ is the photon flux and $h\nu$ the photon energy at frequency ν. A variation in the ozone column $N(O_3;z)$ consequently modifies the solar irradiance value in the atmosphere and at ground level and changes the photodissociation rates of the molecules which absorb in the same spectral range as O_3. The thermal structure of the atmosphere is also modified.

The Chapman scheme, which has been described previously, does not explicitly consider the electronic state of the oxygen atoms which are produced. When the wavelength of the incident radiation is shorter than 310 nm, the photodissociation of ozone leads to excited atomic oxygen $O(^1D)$. The atoms are rapidly quenched by the major gases N_2 and O_2. However a small fraction of them react with several trace species and initiate several reaction chains which will be discussed hereafter.

Table 1a. Reaction rates adopted in the model.

Reactions	Rate constants (cm^3 s^{-1})
$O + O + M \longrightarrow O_2 + M$	$k_1 = 4.7 \times 10^{-33} (300/T)^2 n(M)$
$O + O_2 + M \longrightarrow O_3 + M$	$k_2 = 6.0 \times 10^{-34} (300/T)^{2.3} n(M)$
$O + O_3 \longrightarrow 2O_2$	$k_3 = 1.5 \times 10^{-11} e^{-2218/T}$
$O(^1D) + N_2 \longrightarrow O(^3P) + N_2$	$k_4 = 1.8 \times 10^{-11} e^{107/T}$
$O(^1D) + O_2 \longrightarrow O(^3P) + O_2$	$k_5 = 3.2 \times 10^{-11} e^{67/T}$
$H + O_2 + M \longrightarrow HO_2 + M$	$a_1 = 5.5 \times 10^{-32} (T/300)^{-1.4} n(M)$
$H + O_3 \longrightarrow O_2 + OH(v\ 9)$	$a_2 = 1.4 \times 10^{-10} e^{-470/T}$
$OH + O \longrightarrow H + O_2$	$a_5 = 2.2 \times 10^{-11} e^{117/T}$
$OH + O_3 \longrightarrow HO_2 + O_2$	$a_6 = 1.6 \times 10^{-12} e^{-940/T}$
$HO_2 + O_3 \longrightarrow OH + 2O_2$	$a_{6b} = 1.4 \times 10^{-14} e^{-580/T}$
$HO_2 + O \longrightarrow O_2 + OH(v\ 6)$	$a_7 = 3.0 \times 10^{-11} e^{200/T}$
$OH + OH \longrightarrow H_2O + O$	$a_{16} = 4.2 \times 10^{-11} e^{-242/T}$
$OH + HO_2 \longrightarrow H_2O + O2$	$a_{17} = 1.6 \times 10^{-11} e^{436/T}$
$OH + H_2 \longrightarrow H_2 + H$	$a_{19} = 6.1 \times 10^{-12} e^{-2030/T}$
$H + HO_2 \longrightarrow 2OH$	$a_{23a} = 4.2 \times 10^{-10} e^{-950/T}$
$H + HO_2 \longrightarrow H_2 + O_2$	$a_{23b} = 4.2 \times 10^{-11} e^{-350/T}$
$H + HO_2 \longrightarrow H_2O + O$	$a_{23c} = 8.3 \times 10^{-11} e^{-500/T}$
$H_2 + O \longrightarrow OH + H$	$a_{24} = 8.8 \times 10^{-12} e^{-4200/T}$
$HO_2 + NO \longrightarrow NO_2 + OH$	$a_{26} = b_{29} = 3.7 \times 10^{-12} e^{240/T}$
$HO_2 + HO_2 \longrightarrow H_2O_2 + O_2$	$a_{27} = 2.4 \times 10^{-13} e^{560/T}$
$OH + H_2O_2 \longrightarrow H_2O + HO_2$	$a_{30} = 3.1 \times 10^{-12} e^{-187/T}$
$OH + CO \longrightarrow CO_2 + H$	$a_{36} = 1.35 \times 10^{-13} (1 + P_{atm})$
$O(^1D) + H_2O \longrightarrow 2OH$	$a_1^* = 2.2 \times 10^{-10}$
$O(^1D) + CH_4 \longrightarrow CH_3 + OH$	$a_2^* = 1.4 \times 10^{-10}$
$O(^1D) + H_2 \longrightarrow OH + H$	$a_3^* = 1.0 \times 10^{-10}$

Table 1a (cont.)

Reactions	Rate constants ($cm^3 \ s^{-1}$)
$O(^3p) + NO_2 \longrightarrow NO + O_2$	$b_3 = 9.3 \times 10^{-12}$
$O_3 + NO \longrightarrow NO_2 + O_2$	$b_4 = 2.2 \times 10^{-12} \ e^{-1430/T}$
$N(^4S) + NO \longrightarrow N_2 + O$	$b_6 = 3.4 \times 10^{-11}$
$N(^4S) + O_2 \longrightarrow NO + O$	$b_7 = 4.4 \times 10^{-12} \ e^{-3220/T}$
$NO_2 + O_3 \longrightarrow NO_3 + O_2$	$b_9 = 1.2 \times 10^{-13} \ e^{-2450/T}$
$NO_3 + NO_2 + M \longrightarrow N_2O_5 + M$	b_{12}
$NO_2 + OH + M \longrightarrow HNO_3 + M$	b_{22} } see Table 1b
$HO_2 + NO_2 + M \longrightarrow HO_2NO_2 + M$	b_{23}
$HNO_3 + OH \longrightarrow H_2O \ NO_3$	$b_{27} = 9.4 \times 10^{-15} \ e^{778/T}$
$HO_2NO_2 + OH \longrightarrow H_2O + NO_2 + O_2$	$b_{28} = 20 \times 10^{-12}(1 - 0.9e^{-0.0587P_{Torr}})$
$N_2O_5 + M \longrightarrow NO_3 + NO_2 + M$	$b_{32} = 2.2 \times 10 \times 10^{-5} \ e^{-9700/T}$
$N_2O + O(^1D) \longrightarrow N_2 + O_2$	$b_{38} = 4.9 \times 10^{-11}$
$N_2O + O(^1D) \longrightarrow 2NO$	$b_{39} = 6.7 \times 10^{-11}$
$CH_4 + OH \longrightarrow CH_3 + H_2O$	$c_2 = 2.4 \times 10^{-12} \ e^{-1710/T}$
$CH_3 + O_2 + M \longrightarrow CH_3O_2 + M$	c_4 : see Table 1b
$CH_3O_2 + NO \longrightarrow CH_3O + NO_2$	$c_5 = 4.2 \times 10^{-12} \ e^{180/T}$
$CH_3O_2 + HO_2 \longrightarrow CH_3OOH + O_2$	$c_7 = 7.7 \times 10^{-14} \ e^{1300/T}$
$CH_3O_2 + CH_3O_2 \longrightarrow 2CH_3O + O_2$	$c_{14} = 1.6 \times 10^{-13} \ e^{220/T}$
$CH_3OOH + OH \longrightarrow CH_3O_2 + H_2O$	$c_{17} = 2.6 \times 10^{-12} \ e^{-190/T}$
$C_2H_2 + Cl \longrightarrow products$	$c_{20} = 1.0 \times 10^{-12}$
$C_2H_2 + OH \longrightarrow products$	$c_{21} = 6.5 \times 10^{-12} \ e^{-650/T}$
$C_2H_6 + Cl \longrightarrow products$	$c_{22} = 7.7 \times 10^{-11} \ e^{-90/T}$
$C_2H_6 + OH \longrightarrow products$	$c_{23} = 1.9 \times 10^{-11} \ e^{-1260/T}$
$C_3H_8 + Cl \longrightarrow products$	$c_{24} = 1.4 \times 10^{-10} \ e^{40/T}$
$C_3H_8 + OH \longrightarrow products$	$c_{25} = 1.6 \times 10^{-11} \ e^{-800/T}$
$CH_3HN + OH \longrightarrow products$	$c_{29} = 5.86 \times 10^{-13} \ e^{-750/T}$

Table 1a (cont.)

Reactions	Rate constants ($cm^3 s^{-1}$)
$CH_3 + OH \longrightarrow CH_2Cl + H_2O$	$d_1 = 1.8 \times 10^{-12} e^{-1112/T}$
$Cl + O_3 \longrightarrow ClO + O_2$	$d_2 = 2.8 \times 10^{-11} e^{-257/T}$
$ClO + NO \longrightarrow Cl + O_2$	$d_3 = 7.7 \times 10^{-11} e^{-130/T}$
$ClO + NO \longrightarrow Cl + NO_2$	$d_4 = 6.2 \times 10^{-12} e^{294/T}$
$Cl + CH_4 \longrightarrow HCl + CH_3$	$d_5 = 9.6 \times 10^{-12} e^{-1350/T}$
$Cl + H_2 \longrightarrow HCl + H$	$d_6 = 3.7 \times 10^{-11} e^{-2300/T}$
$Cl + HO_2 \longrightarrow HCl + O_2$	$d_7 = 1.8 \times 10^{-11} e^{170/T}$
$Cl + H_2O_2 \longrightarrow HCl + HO_2$	$d_8 = 1.1 \times 10^{-11} e^{-980/T}$
$Cl + CH_2O \longrightarrow HCl + CHO$	$d_{10} = 8.2 \times 10^{-11} e^{-34/T}$
$HCl + OH \longrightarrow Cl + H_2O$	$d_{11} = 2.8 \times 10^{-12} e^{-425/T}$
$HCl + O(^3P) \longrightarrow Cl + OH$	$d_{12} = 1.0 \times 10^{-11} e^{-3340/T}$
$ClO + NO_2 + M \longrightarrow ClONO_2 + M$	$d_{22} = d_{31}$: see Table 1b
$ClONO_2 + O \longrightarrow ClO + NO_3$	$d_{32} = 3.0 \times 10^{-12} e^{-808/T}$
$ClO + HO_2 \longrightarrow HOCl + O_2$	$d_{33} = d_{35} = 4.6 \times 10^{-13} e^{710/T}$
$CH_3CCl_3 + OH \longrightarrow CH_2CCl_3 + H_2O$	$d_{50} = 5.4 \times 10^{-12} e^{-1820/T}$

Table 1b. Three-body reactions. The rate constants of these reactions are expressed as follows:

$$\frac{k_O \, n(M)}{1 + n(M) \, k_O/k_\infty} \times 0.6 \left[1 + \log_{10}^2 \left(n(M) k_O/k_\infty\right)\right]^{-1}$$

with
$$k_O = k_O^{300}(T/300)^{-n} \text{ cm}^6 \text{ s}^{-1}$$
$$k_\infty = k_\infty^{300}(T/300)^{-m} \text{ cm}^3 \text{ s}^{-1}$$

Reactions	Name of the constant	Values of the parameters
$NO_3 + NO_2 + M \longrightarrow N_2O_5 + M$	b_{12}	$k_O^{300} = 2.2 \times 10^{-30}$ $n = 2.8$ $k_\infty^{300} = 1.0 \times 10^{-12}$ $m = 0$
$NO_2 + OH \, M \longrightarrow HNO_3 + M$	b_{22}	$k_O^{300} = 2.6 \times 10^{-30}$ $n = 2.9$ $k_\infty^{300} = 2.4 \times 10^{-11}$ $m = 1.3$
$HO_2 + NO_2 + M \longrightarrow HO_2NO_2 + M$	b_{23}	$k_O^{300} = 2.3 \times 10^{-31}$ $n = 4.6$ $k_\infty^{300} = 4.2 \times 10^{-12}$ $m = 0$
$CH_3 + O_2 + M \longrightarrow CH_3O_2 + M$	c_4	$k_O^{300} = 1.8 \times 10^{-31}$ $n = 2.2$ $k_\infty^{300} = 2.0 \times 10^{-12}$ $m = 1.7$
$ClO + NO_2 + M \longrightarrow ClONO_2 + M$	$d_{22} = d_{31}$	$k_O^{300} = 1.8 \times 10^{-31}$ $n = 3.4$ $k_\infty^{300} = 1.5 \times 10^{-11}$ $m = 1.9$

Table 1c. Photochemical reactions considered in the model.

O_2 + hv --> $2O(^3P)$

O_2 + hv --> $O(^3P)$ + $O(^1D)$

O_3 + hv --> O_2 + $O(^3P)$

O_3 + hv --> O_2 $^1\Delta g$ + $O(^1D)$

NO + hv --> NO + $O(^3P)$

NO_2 + hv --> NO + $O(^3P)$

NO_3 + hv --> NO + O_2

NO_3 + hv --> NO_2 + O

N_2O + hv --> N_2 + $O(^1D)$

N_2O_5 + hv --> NO_2 + NO_3

HNO_3 + hv --> NO_2 + OH

HNO_4 + hv --> NO_2 + HO_2

$ClONO_2$ + hv --> NO_2 + ClO

$ClONO_2$ + hv --> NO_3 + Cl

ClO + hv --> Cl + $O(^3P)$

HCl + hv --> H + Cl

HOCl + hv --> ClO + H

HOCl + hv --> Cl + OH

H_2O + hv --> H + OH

H_2O_2 + hv --> 2OH

CO_2 + hv --> CO + O

CH_4 + hv --> products

CCl_4 + hv --> products

$CFCl_3$ + hv --> products

CF_2Cl_2 + hv --> products

CH_3Cl + hv --> products

CH_3CCl_3 + hv --> products

Table 2. Temperature, pressure, atmospheric and molecular oxygen as a function of altitude between 0 and 120 km (US Standard Atmosphere 1976).

Altitude (km)	Temperature (K)	Pressure (mb)	Total concentration (cm^{-3})	Molecular oxygen concentration (cm^{-3})
0	288	1013	2.55×10^{19}	5.33×10^{18}
5	256	540	1.53×10^{19}	3.20×10^{18}
10	223	265	8.60×10^{18}	1.80×10^{18}
15	217	121	4.05×10^{18}	8.47×10^{17}
20	217	55.3	1.85×10^{18}	3.87×10^{17}
25	222	25.5	8.33×10^{17}	1.74×10^{17}
30	227	12.0	3.83×10^{17}	8.01×10^{16}
35	237	5.75	1.76×10^{17}	3.68×10^{16}
40	250	2.87	8.31×10^{16}	1.74×10^{16}
45	264	1.49	4.09×10^{16}	8.55×10^{15}
50	271	7.98×10^{-1}	2.14×10^{16}	4.47×10^{15}
55	261	4.25×10^{-1}	1.18×10^{16}	2.47×10^{15}
60	247	2.20×10^{-1}	6.44×10^{15}	1.35×10^{15}
65	233	1.09×10^{-1}	3.39×10^{15}	7.09×10^{14}
70	220	5.22×10^{-2}	1.72×10^{15}	3.60×10^{14}
75	208	2.39×10^{-2}	8.30×10^{14}	1.73×10^{14}
80	199	1.05×10^{-2}	3.84×10^{14}	8.02×10^{13}
85	189	4.46×10^{-3}	1.71×10^{14}	3.57×10^{13}
90	187	1.84×10^{-3}	7.12×10^{13}	1.48×10^{13}
95	188	7.60×10^{-3}	2.92×10^{13}	5.83×10^{12}
100	195	3.20×10^{-4}	1.19×10^{13}	2.15×10^{12}
105	209	1.45×10^{-4}	5.02×10^{12}	7.65×10^{11}
110	240	7.10×10^{-5}	2.14×10^{12}	2.62×10^{11}
115	300	4.01×10^{-5}	9.68×10^{11}	9.65×10^{10}
120	360	2.54×10^{-5}	5.11×10^{11}	4.40×10^{10}

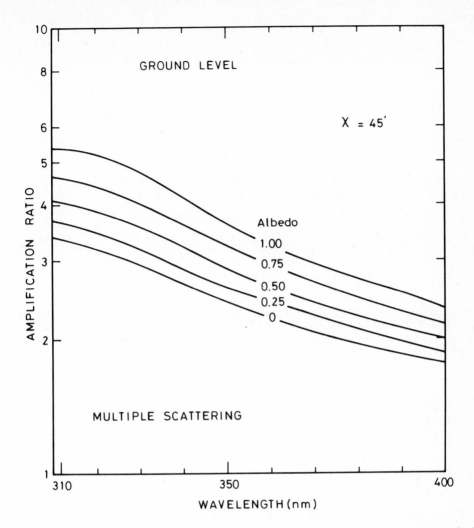

Fig. 3. Amplification ratio of the solar radiation at ground level for wavelength between 310 and 400 nm and for different values of the surface albedo. The amplification ratio is defined by q_1/q_0 where q_1 is the irradiance calculated including multiple scattering and albedo and q_0 is the irradiance of the direct solar radiation only.

OZONE AND TEMPERATURE

Assuming again a pure oxygen atmosphere, a relation between the ozone
concentration and the temperature can be deduced from expression (4) by
considering the numerical values of k_2 and k_3 as a function of the temperature, namely

$$\frac{\Delta n(O_3)}{n(O_3)} = \frac{-1400}{T^2} \Delta T \tag{9}$$

If the temperature at the stratopause (270°K) were increased by 10°K, the
ozone concentration would be reduced by about 20 percent. This value
however has to be considered as an upper limit since the introduction of a
more detailed chemical scheme reduces the sensitivity of ozone to temperature. Nevertheless, when considering the action of ozone-depleting agents,
the temperature feedback should be introduced in the computation. Since
the heating rate is directly proportional to the amount of ozone (Fig. 4)
expression (9) introduces a negative feedback mechanism which stabilizes
ozone against several chemical agents. However, it introduces a relation
between O_3 and CO_2 amount. This latter gas, which emits infrared radiation
due to its 15 µm band, is responsible for most of the cooling in the
vicinity of the stratopause (Fig. 4). The observed enhancement in the
carbon dioxide concentration should lead therefore to an increase in the
amount of ozone.

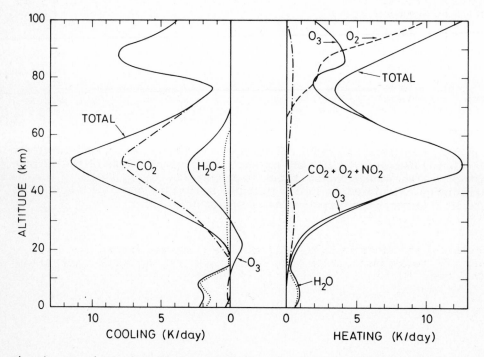

Fig. 4. Vertical distribution of the heating and the cooling rates due to
several atmospheric species (London 1980).

AGENTS OF OZONE DEPLETION

The loss of odd oxygen ($O + O_3$), namely through the recombination of ozone and atomic oxygen ($O + O_3 \rightarrow 2O_2$), can be catalyzed in the presence of hydroxyl radicals (OH), nitric oxide (NO) and chlorine oxide (ClO). Different cycles have to be considered with specific efficiencies depending on the altitude. In the mesosphere the main destruction of odd oxygen is due to the hydrogen species:

$$OH + O \rightarrow H + O_2$$
$$H + O_2 + M \rightarrow HO_2 + M$$
$$HO_2 + O \rightarrow OH + O_2$$

net: $\quad O + O \rightarrow O_2$

or

$$OH + O \rightarrow H + O_2$$
$$H + O_3 \rightarrow OH + O_2$$

net: $\quad O + O_3 \rightarrow 2O_2$

(Bates and Nicolet 1950; Nicolet 1971) while, in the stratosphere, the most efficient processes are

$$NO + O_3 \rightarrow NO_2 + O_2$$
$$NO_2 + O \rightarrow NO + O_2$$

net: $\quad O + O_3 \rightarrow 2O_2$

(Crutzen 1970; Johnston 1971; Nicolet 1971) or

$$Cl + O_3 \rightarrow ClO + O_2$$
$$ClO + O \rightarrow Cl + O_2$$

net: $\quad O + O_3 \rightarrow 2O_2$

(Stolarski and Cicerone 1974). In the lower stratosphere, and particularly near the tropopause, other cycles involving odd hydrogen have to be considered (Nicolet 1975), for example:

$$OH + O_3 \longrightarrow HO_2 + O_2$$
$$HO_2 + O_3 \longrightarrow OH + 2O_2$$

net:
$$2O_3 \longrightarrow 3O_2$$

Supplementary reaction chains involving species such as NO_3, N_2O_5, $HOCl$, $ClONO_2$, etc. can also be introduced (see Brasseur 1982 for a comprehensive review of these cycles), showing the complexity of the ozone chemistry in the stratosphere. Furthermore, in the troposphere and in the lower strato- sphere, additional ozone resulting from the oxidation of methane in the presence of nitric oxide has to be considered. The reaction chain is

$$CH_4 + OH \longrightarrow CH_3 + H_2O$$
$$CH_3 + O_2 + M \longrightarrow CH_3O_2 + M$$
$$CH_3O_2 + NO \longrightarrow CH_3O + NO_2$$
$$NO_2 + h\nu \longrightarrow NO + O$$
$$O + O_2 + M \longrightarrow O_3 + M$$
$$CH_3O + O_2 \longrightarrow CH_2O + HO_2$$
$$HO_2 + NO \longrightarrow NO_2 + OH$$
$$NO_2 + h\nu \longrightarrow NO + O$$
$$O + O_2 + M \longrightarrow O_3 + M$$
$$CH_2O + h\nu \longrightarrow CO + H_2$$

net:
$$CH_4 + 4O_2 + 3h\nu \longrightarrow H_2O + CO + H_2 + 2O_3$$

(Crutzen 1974; Nicolet 1975). The equilibrium concentration of ozone depends on the amount of catalytic agents which are present in the atmo- sphere. Above 30 km expression (4) can still be used if the k_3 coefficient is replaced by an effective recombination rate

$$k_{3A} = k_3 [1 + A] \tag{10}$$

where A is a correction factor which is given to a first approximation by

$$A = \frac{a_5 \, n(OH) + b_3 \, n(NO_2) + d_3 \, n(ClO)}{k_3 \, n(O)} \tag{11}$$

a_5, b_3, and d_3 being the rate constants of reaction $O + OH$, $O + NO_2$ and $O + ClO$ (Nicolet 1971). In the middle and lower stratosphere where the transport can no longer be neglected, a general continuity equation must be used, namely

$$\frac{\delta n(O_3)}{\delta t} + \vec{\nabla}\vec{\Phi}(O_3) = 2J_2\, n(O_2) + a_{26}\, n(HO_2)\, n(NO) +$$
$$c_5\, n(CH_3O_2)\, n(NO) -$$
$$[2k_3\, n(O)\, n(O_3) + a_5\, n(OH)\, n(O) +$$
$$a_6\, n(OH)\, n(O_3) + a_{6b}\, n(HO_2)\, n(O_3) +$$
$$a_7\, n(HO_2)\, n(O) + 2b_3\, n(NO_2)\, n(O) +$$
$$2d_3\, n(ClO)\, n(O)] \tag{12}$$

assuming that $n(O_x) \cong n(O_3)$ in this altitude range. The value of the individual rate constants is given in Table 1.

As shown by these latter expressions, a detailed understanding of the ozone balance requires a study of the processes which are responsible for the formation and the destruction of the radicals which are involved (Cariolle 1983).

This problem is not yet completely solved since, as indicated by Schmailzl and Crutzen (in this volume), some inconsistencies in the description of the ozone budget remain.

The major mesospheric source of odd hydrogen (H, OH, HO_2) is the photo-dissociation of water vapor while, in the stratosphere, the main production mechanism of HO_x is the oxidation of water vapor, methane and molecular hydrogen by the excited oxygen atom $O(^1D)$. Consequently an increase in the methane amount as a function of time should lead to an ozone reduction in the mesosphere and upper stratosphere where OH is the main depleting agent of O_3, and to an ozone increase in the troposphere where the reaction chain involving methane and nitric oxide (see previous description) plays a dominant role.

Nitric oxide in the stratosphere is also produced by the $O(^1D)$ atoms which dissociate nitrous oxide as follows:

$$N_2O + O(^1D) \longrightarrow 2NO$$
$$\longrightarrow N_2 + O_2$$

The N_2O molecules are produced at ground level by bacterial processes associated with complex nitrification and denitrification mechanisms in soils. Since the industrial fixation of nitrogen can no longer be neglected in comparison to the corresponding natural processes, it is now believed that the future anthropogenic production of N_2O, especially through nitrogenous fertilizers, combustion and organic wastes, could perturb the ozonosphere. Direct production of odd nitrogen (N, NO, NO_2, NO_3, N_2O_5, HNO_3, HO_2NO_2) from aircraft engine emissions should also be considered in the future ozone balance.

Finally the atmospheric release of chlorofluorocarbons is responsible for an increase in the amount of odd chlorine (essentially Cl, ClO, HCl, $ClONO_2$, HOCl) and consequently for a depletion in the stratospheric ozone (Molina and Rowland 1974a). The most important sources of ClX are the dissociation of CCl_4 (or F-10), $CFCl_3$ (or F-11), CF_2Cl_2 (or F-12) and

CH_3CCl_3, which are anthropogenically produced, and the destruction of the natural CH_3Cl by OH radicals and by sunlight.

MODEL CALCULATIONS OF THE OZONE RESPONSE TO CHEMICAL PERTURBATIONS

Theoretical models are currently used to predict the changes in ozone concentration and in temperature due to the intrusion into the atmosphere of several chemical constituents. These coupled chemical/radiative models have to be validated by comparing the calculated, current concentrations of the different species with the corresponding observed values (see e.g., Brasseur et al. 1982). Although a certain number of discrepancies, one should point out the difference in the calculated and observed vertical distribution of several source species such as $CFCl_3$, CF_2Cl_2, N_2O, CH_4 and H_2O, especially in the upper stratosphere. An incorrect treatment of the solar penetration in the O_2 Schumann-Runge bands region could partly explain these discrepancies. Moreover, models do not correctly reproduce the ClO and HNO_3 distribution in the upper stratosphere and results in ozone concentrations above 40 km which are somewhat smaller than the known observed values. These problems among others indicate that the predictions made by models, and especially by one-dimensional approaches, should be considered with care and are subject to quantitative changes as long as some important chemical or photochemical parameters are still inaccurately known.

In order to study the sensitivity of the ozone layer to several chemical agents, a series of individual and coupled scenarios has been applied and the new equilibrium ozone concentration has been calculated. The CFC perturbation (P) corresponds to a constant emission of $CFCl_3$ (3.4×10^5 T/yr), CF_2Cl_2 (4.1×10^5 T/yr), CCl_4 (1.0×10^5 T/yr) and CH_3CCl_3 (3.6×10^5 T/yr). For the other species (CO_2, N_2O and CH_4), the current concentration has been uniformly multiplied by two. Table 3 shows the resulting variation in the total ozone column. To estimate the importance of the temperature feedback, which is introduced in most model runs, a number of cases have been computed using a fixed temperature profile. The results show that, with the presently adopted chemistry, the ozone depletion for a CFC perturbation lies between 3 and 3.5 percent. On the contrary, a doubling in the CO_2 amount leads to an ozone increase of 3.14 percent which is explained by a cooling of the stratosphere. Since the processes which are involved are highly nonlinear, the two perturbations applied simultaneously do not cancel. The combined effects lead to an ozone depletion of 1.3 percent. The reduction in the O_3 column when the amount of nitrous oxide is doubled, is of the order of 8 percent. When the methane concentration is multiplied by two, the total ozone amount is increased by one percent. Finally a combined perturbation of CFC's, CO_2, N_2O and CH_4 leads to an ozone reduction of 3 percent which is comparable to the corresponding depletion when the sole action of CFC's is considered.

Table 3. Perturbation of the ozone column.

Case	CO_2	N_2O	CH_4	CFC's	Temperature feedback	$\dfrac{\Delta O_3}{O_3}$ %
A				P[a]	Yes	− 3.50
A'				P	No	− 3.37
B	x 2				Yes	+ 3.14
C		x 2			Yes	− 7.79
C'		x 2			No	− 8.92
D			x 2		Yes	+ 0.97
D'			x 2		No	+ 0.87
E	x 2			P	Yes	− 1.30
F		x 2	x 2	P	Yes	− 5.56
F'		x 2	x 2	P	No	− 6.84
G	x 2	x 2	x 2	P	Yes	− 3.06

[a]Emission of CFC's.

These results, when compared with values obtained previously (NASA 1979; National Research Council 1979), indicate a larger effect of nitrogen compounds and a smaller influence of the chlorofluorocarbons. These changes can be explained by smaller calculated concentration values of the OH radicals as a result of recent revisions in the reaction rates involved in the odd hydrogen destruction. Smaller OH amount lead to a smaller $n(ClO)/n(HCl)$ ratio and to a larger $n(NO_2)/n(HNO_3)$ value.

The perturbations which have been applied are useful to test the model but are rather arbitrary. To predict the future state of the atmosphere would require the introduction of a realistic scenario into the model. Because of the large uncertainties in the projected emissions of the pollutants, different scenarios have been considered. In Case 1 only the CFC contribution to the pollution has been considered. Figure 5 shows the historical data and the projected values which have been used for the industrial sources of the several halocarbons which are released in the atmosphere (Logan et al. 1978). The corresponding ClX mixing ratio at the stratopause from year 1950 to year 2080 is reproduced in Fig. 6. The next two cases (2 and 3) consider, together with the CFC emission, a gradual increase in the CO_2 and N_2O mixing ratio (Fig. 7).

A number of scenarios for the emission of carbon dioxide have been developed (see e.g., Niehaus 1976; Rotty 1977; Council on Environmental Quality 1981) and are based on the projected fossil fuel usage. The projection adopted in this work is based on the expression suggested by Wuebbles et al. (1983):

$$f(CO_2) = 335.0 \exp [0.0056(t - 1979)]$$

where $f(CO_2)$ is the CO_2 mixing ratio expressed in ppmv and t the year. Such an expression leads to a doubling of the CO_2 amount at the end of the

21st century. The past evolution of carbon dioxide is based on historical records (World Meteorological Organization 1977).

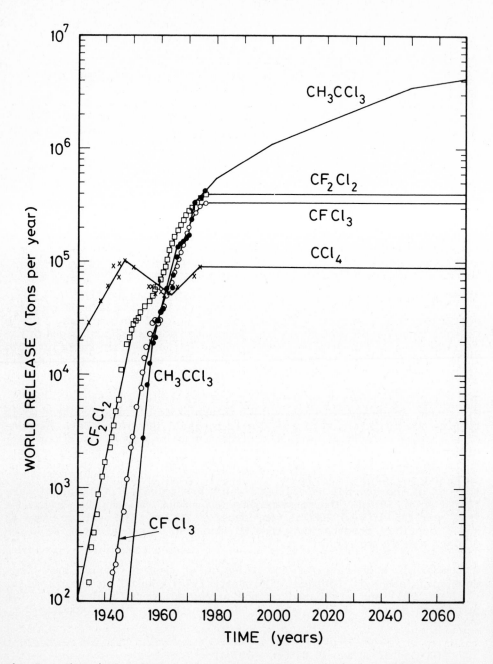

Fig. 5. Historical release and assumed, future emission of different halo-carbons adopted as scenario for the perturbation calculations (Logan et al. 1978).

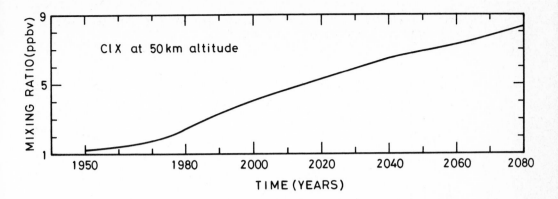

Fig. 6. Variation of the ClX mixing ratio in the upper stratosphere derived for the halocarbon release shown in Fig. 5.

Atmospheric measurements carried out in the last decade have shown a systematic increase in the N_2O concentration of about 0.25 percent per year (see e.g., Weiss 1981; Khalil and Rasmussen 1981, 1983). It has been assumed that such a growth rate will remain constant until 2080.

The existence of a systematic trend in the methane amount is subject to discussion. Analyzing a series of measurements made in the past decade and taking into account several corrections made for older data, Ehhalt et al. (1983) does not derive any significant change in the CH_4 mixing ratio. However, continuous observations made in recent years by Rasmussen and Khalil (1981), Blake et al. (1982) and Rowland (personal communication) definitely show an increase of 1 to 2 percent per year. Because of the present difficulty in proposing a future trend in the CH_4 concentration, two scenarios have been considered: in Case 2 the amount of atmospheric methane remains constant (mixing ratio of 1.5 ppmv) while in Case 3 a continuous growth of 1.5 percent/year has been applied after 1970.

The response of ozone and temperature for these three scenarios has been determined by using a coupled chemical and radiative one-dimensional model. This model includes a detailed chemical and photochemical scheme taking into account the action of the oxygen, hydrogen, nitrogen, chlorine and carbon species (Brasseur et al. 1982). The vertical temperature profile is derived from a thermal scheme in which heating is calculated using the parameterization suggested by Schoeberl and Strobel (1978). The radiative transfer in the infrared is treated by determining the transmission function in four large spectral intervals. The upward and downward infrared flux, as well as the corresponding cooling rates, are obtained by solving the radiative transfer equation. An energy balance is achieved at the top of the atmosphere assuming a global earth-atmosphere albedo of 0.3. The model extends from 0 to 70 km. The vertical transport of the trace species and of heat (especially in the troposphere where convective insta-bility appears) is parameterized by means of an eddy diffusion coefficient.

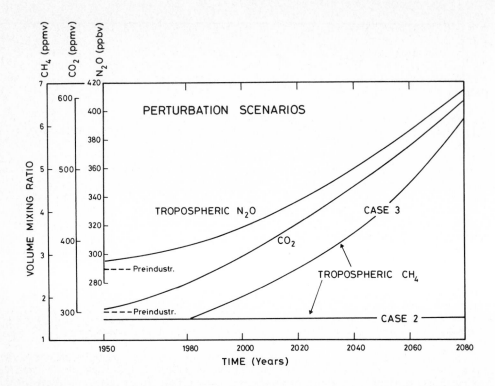

Fig. 7. Assumed variations in the mixing ratio of N_2O, CO_2 and CH_4 from 1950 to 2080. Case 2 refers to a constant methane content and Case 3 to an increase of 1.5%/yr.

The change in the ozone column relative to a preindustrial atmosphere is shown in Fig. 8. When only a continuous injection of CFC's is considered (Case 1) the ozone amount decreases gradually and the corresponding depletion reaches 3 percent in the year 2080. When the increase in CO_2 and N_2O is added to the CFC's emission (Case 2), the decrease in the ozone column is somewhat reduced (1.5 percent in year 2080). However, as shown by Fig. 9a and b, this enhancement results from an increase in the troposphere and a simultaneous decrease in the upper stratosphere. It should be noted however that, at these levels, the presence of a large quantity of methane prevents the chlorine species from depleting the ozone. This is explained by the fact that methane, when sufficiently abundant, converts the active chlorine atoms into inactive HCl molecules. Moreover, the simultaneous introduction of nitrogen and chlorine oxides leads to the formation of $ClONO_2$ in the middle and lower stratosphere. The $ClONO_2$ molecule acts as a temporary reservoir for both ClX and NO_x and its presence therefore introduces a nonlinear effect in the ozone depletion problem. Finally, the model calculations show that, in the presence of a large ClX concentration, the efficiency of CO_2 in increasing the ozone concentration by cooling the stratosphere is considerably reduced since the sensitivity of O_3 to temperature becomes very small when the loss rate of O_3 is reaction (d_3) instead of (k_3). Equation (9) must be modified when correction factor A (Eq. 11) becomes equal to or greater than 1 (Haigh and Pyle 1982).

The calculated change in the stratospheric temperature corresponding to Case 3 is shown in Fig. 9c. It results from both the reduction in the ozone concentration (less UV absorption and less heating) and the increase in the CO_2 amount (more IR emission and more cooling). These quantitative results depend on the scenarios which are adopted and on the relative contribution of O_3 and CO_2 in the cooling processes.

In the troposphere the introduction of molecular species such as CO_2, N_2O, CH_4 and CFC's, which are optically active in the infrared, enhances the greenhouse effect and should somewhat increase the averaged temperature at ground level. These climatic changes, although very important for the world's future, will not be treated here. Most radiative-convective models suggest, for a doubling of CO_2, a temperature increase at the earth's surface of 1 to 3°K (see Ramanathan 1981). It is usually believed that in the future the combined effect of the other trace gases (N_2O, CH_4 and CFC's) could modify the temperature by a value which is comparable to that inferred by the CO_2 increase.

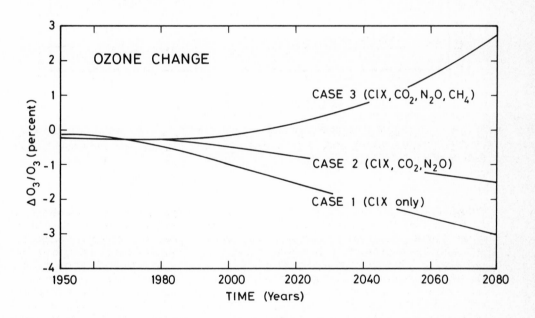

Fig. 8. Calculated changes (expressed in percent) of the ozone column for three different scenarios. Case 1 refers to a perturbation by halocarbons only, Case 2 takes into account also the increase in the CO_2 and N_2O amount (Fig. 7) and Case 3 considers the combined effects of halocarbons, carbon dioxide, nitrous oxide and methane.

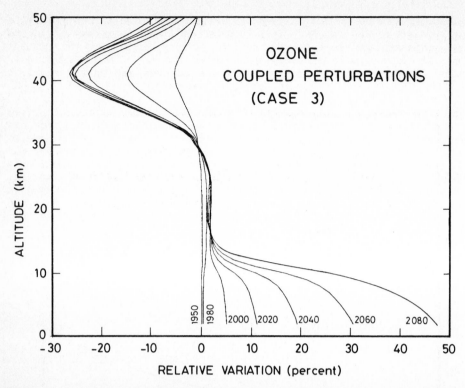

Fig. 9a. Relative variation of the ozone concentration as a function of altitude corresponding to Case 3 (ClX, CO_2, N_2O and CH_4 increase--see Figs. 6 and 7) between years 1950 and 1980.

SUMMARY

Ozone drives a number of important mechanisms occurring in the atmosphere and involving chemical, radiative, thermal and dynamic processes. Since this gas is a strong absorber of UV radiation, its depletion by several chemical agents leads to an increase in the UV radiation level at the earth's surface. It is, for example, generally assumed that the percentage change of the biologically effective ultraviolet at ground level would be greater than the percentage ozone decrease (see Caldwell et al., in this volume). Agricultural and industrial activity is responsible for the emission of several gases which modify the morphology of the ozone layer and consequently the penetration of ultraviolet radiation. Determination of the quantitative ozone change requires complex models taking into account many interactive processes and feedback mechanisms. The models are based on laboratory data which are measured with a certain degree of uncertainty and on phenomenological parameters which are sometimes poorly known.

Consequently the model results should be associated with an uncertainty which is difficult to determine because of the large number of processes which are involved.

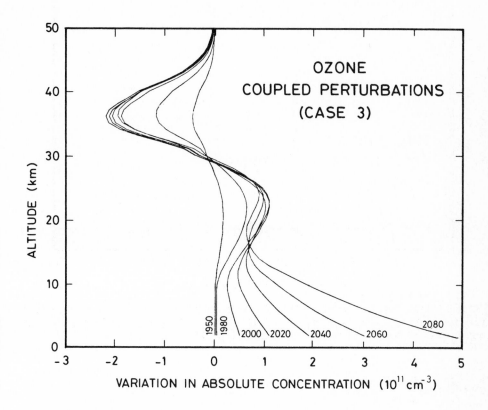

Fig. 9b. Same as in Fig. 9a but for the absolute variation in the ozone concentration.

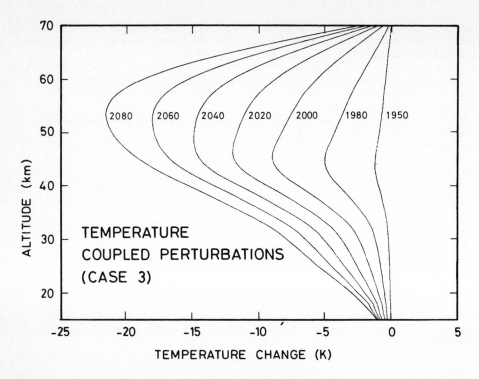

Fig. 9c. Same as in Fig. 9a but for temperature.

ACKNOWLEDGMENTS

The authors wish to thank Mr. Chr. Tricot for useful discussion during the preparation of this chapter. This work has been supported by the Chemical Manufacturers Association under contract 82-396.

REFERENCES

Bates DR, Nicolet M (1950) The photochemistry of atmospheric water vapor. J Geophys Res 55:301-327
Blake DR, Mayer EW, Tyler SC, Makida Y, Montagne DC, Rowland FS (1982) Global increase in atmospheric methane concentrations between 1978 and 1980. Geophys Res Lett 9:477-480
Brasseur G (1978) Un modèle bidimensionnel du comportement de l'ozone dans la stratosphère. Planet Space Sci 26:139-159
Brasseur G (1982) Physique et chimie de l'atmosphere moyenne. Masson ed, Paris
Brasseur G, Simon PC (1981) Stratospheric chemical and thermal response to long-term variability in solar UV irradiance. J Geophys Res 86:7343-7362

Brasseur G, De Rudder A, Roucour A (1982) The natural and perturbed ozonosphere. In: Proceeding of the international conference on environmental pollution, Thessaloniki, p 840

Cariolle D (1983) The ozone budget in the stratosphere: results of a one-dimensional photochemical model. Planet Space Sci 31:1033-1052

Council on Environmental Quality (1981) Global energy futures and the carbon dioxide problem.

Crutzen PJ (1970) The influence of nitrogen oxides on the atmospheric ozone content. Quart J Roy Met Soc 96:320-325

Crutzen PJ (1971) Ozone production rates in an oxygen-hydrogen-nitrogen oxide atmosphere. J Geophys Res 76:7311-7327

Crutzen PJ (1972) SST's - A threat to the earth's ozone shield. Ambio 1:41-51

Crutzen PJ (1974) Photochemical reactions initiated by and influencing ozone in unpolluted tropospheric air. Tellus 26:47-57

Crutzen PJ (1975) A two-dimensional photochemical model of the atmosphere below 55 km: Estimates of natural and man-caused ozone perturbations due to NO_x, Proceedings of the 4th CIAP conference US DOT, Washington, DC

Dunkerton T (1978) On the mean meridional mass motions of the stratosphere and mesosphere. J Atm Sci 35:2325-2333

Ehhalt D, Roeth EP, Schmidt U (1983) On the temporal variance of stratospheric trace gas concentrations. J Atmos Chem 1:27-51

Garcia RR, Solomon S (1983) A numerical model of the zonally averaged dynamical and chemical structure of the middle atmosphere. J Geophys Res 88:1379-1400

Haigh JD, Pyle JA (1982) Ozone perturbation experiments in a two-dimensional circulation model. Quart J Roy Met Soc 108:551-574

Harwood RS, Pyle JA (1975) A two-dimensional mean circulation model for the atmosphere below 80 km. Quart J Roy Met Soc 101:723-747

Johnston H (1971) Reduction of stratospheric ozone by nitrogen oxide catalysts from supersonic transport exhaust. Science 173:517-522

Ka Kit Tung (1982) On the two-dimensional transport of stratospheric trace gases in isentropic coordinates. J Atm Sci 39:2330-2355

Khalil MAK, Rasmussen RA (1981) Increases in atmospheric concentrations of halocarbons and N_2O. Geophys Monit Clim Change 9:134-138

Khalil MAK, Rasmussen RA (1983) Increase and seasonal cycles of nitrous oxide in the Earth's atmosphere. Tellus 35:161-169

Logan JA, Prather MJ, Wofsy SC, McElroy MB (1978) Atmospheric chemistry: response to human influence. Phil Trans Roy Soc Lond 290:187-234

London J (1980) Radiative energy sources and sinks in the stratosphere and mesosphere. In: Aikin AC (ed) Proceedings in the NATO Advanced Study Institute on atmospheric ozone. Report FAA-EE-80-20, p 703

London J, Bojkov RD, Oltmans S, Kelley JI (1976) Atlas of the global distribution of total ozone, July 1957-June 1967, National Center for Atmospheric Research

McElroy MB, Elkins JW, Wofsy SC, Yung YL (1976) Sources and sinks for atmospheric N_2O. Rev Geophys Space Phys 14:143-150

Miller C, Filkin DL, Owens AJ, Steed JM, Jesson PJ (1981) A two-dimensional model of stratospheric chemistry and transport. J Geophys Res 86:12039-12065

Molina MJ, Rowland FS (1974a) Stratospheric sink for chlorofluoromethanes: chlorine atom catalyzed destruction of ozone. Nature 249:810-812

Molina MJ, Rowland FS (1974b) Predicted present stratospheric abundances of chlorine species from photodissociation of carbon tetrachloride. Geophys Res Lett 1:309-312

NASA (1979) The stratosphere: present and future. NASA Reference publication 1049

National Research Council (1979) Stratospheric ozone depletion by halocarbons: Chemistry and transport. National Academy of Sciences, Washington, DC

Nicolet M (1971) Aeronomic reactions of hydrogen and ozone. In: Fiocco G (ed) Mesospheric models and related experiments. p 1

Nicolet M (1975) Stratospheric ozone: An introduction to its study. Rev Geophys Space Phys 13:593-636

Nicolet M (1981) The solar spectral irradiance and its action in the atmospheric photodissociation processes. Planet Space Sci 29:951-974

Nicolet M, Meier RR, Anderson DE Jr (1983) Radiation field in the troposphere and stratosphere - II Numerical analysis. Planet Space Sci 30:935-983

Nicoli MP, Visconti G (1982) Impact of coupled perturbations of atmospheric trace gases on Earth's climate and ozone. Pure Appl Geophys 120:626-641

Niehaus F (1976) A non-linear eight level tandem model to calculate future CO_2 and C-14 burden to the atmosphere. Int Inst for Applied Systems Analysis Research Memorandum, RM 76-35

Pyle JA, Rogers CF (1980) A modified diabatic circulation model for stratospheric tracer transport. Nature 287:711-714

Ramanathan V (1981) The role of ocean-atmospheric interactions in the CO_2 climate problem. J Atm Sci 38:918-930

Rao-Vupputuri RK (1973) Numerical experiments on the steady state meridional structure and ozone distribution in the stratosphere. Mon Weather Rev 101:510-527

Rao-Vupputuri RK (1979) The structure of the natural stratosphere and the impact of chlorofluoromethanes on the ozone layer investigated in a 2-D time dependent model. Pure Appl Geophys 117:448-485

Rasmussen RA, Khalil MAK (1981) Atmospheric methane (CH_4): Trends and seasonal cycles. J Geophys Res 86:9826-9832

Rotty RM (1977) The atmospheric CO_2 consequences of heavy dependence on coal. Oak Ridge National Laboratory Report ORAU/IEA-77-27

Schoeberl MR, Strobel DF (1978) The zonally averaged circulation of the middle atmosphere. J Atm Sci 35:577-591

Simon PC, Brasseur G (1983) Photodissociation effects of solar UV radiation. Planet Space Sci 31:987-999

Schmailzl U, Crutzen PJ (this volume) Inconsistencies in current photochemical models deduced from considerations of the ozone budget. p 29

Stolarski RS, Cicerone RJ (1974) Stratospheric chlorine: A possible sink for ozone. Can J Chem 52:1610-1615

US Standard Atmosphere (1976) NOAA--S/T 76-1562, NOAA/NASA/US Air Force, Washington, DC

Weiss RW (1981) The temporal and spatial distribution of tropospheric nitrous oxide. J Geophys Res 86:7185-7195

World Meteorological Organization (1977) Effects of human activities on global climate, note 156

Wuebbles DJ, Luther FM, Penner JE (1983) Effect of coupled anthropogenic perturbations on stratospheric ozone. J Geophys Res 88:1444-1456

Inconsistencies in Current Photochemical Models Deduced from Considerations of the Ozone Budget

U. Schmailzl and P. J. Crutzen

Max-Planck-Institut für Chemie (Otto-Hahn-Institut)
Mainz, Federal Republic of Germany

ABSTRACT

During the last decade there have been varying predictions for an expected decrease of total atmospheric ozone through catalytic photochemical processes caused by human activities such as release of various chlorinated hydrocarbons, operation of high-altitude aircraft or use of mineral fertilizers. In the last years the picture has been broadened by consideration of photochemical ozone production in the troposphere, which can compensate for some of the stratospheric ozone depletion. This process is sensitive to injection of methane and nitrogen oxides into the troposphere. Recently, it has also been pointed out that an increase of CO_2 can increase stratospheric ozone by decreasing the temperature in the stratosphere. While model scenarios predict changes for total ozone between +0.5% and -1.5% until the year 1980, a statistical analysis of total ozone measurements gives a slightly positive trend with most probable values below 0.5%.

Since the knowledge of kinetic parameters and trace gas concentrations has improved considerably in the last decade, we can now give a much better assessment of the importance of different ozone sources and sinks. However, substantial uncertainties may still exist. The budget of stratospheric ozone, as calculated adopting the ozone, nitrous oxide and methane distributions from Nimbus-4 and Nimbus-7 satellite data, allows us to estimate the uncertainties in current one- and two-dimensional models. A major discrepancy in the stratosphere below 30 km suggests inconsistencies between observations and model schemes presently used for ozone depletion estimates.

INTRODUCTION

Depletion of ozone has been the subject of many studies which simulate the photochemistry of the atmosphere assuming different scenarios for the release of anthropogenic trace gases. These estimates changed frequently and substantially with updates in rate coefficients and the inclusion of additional gases like $ClONO_2$, $HOCl$ or HNO_4. Maximum predictions of asymptotic ozone depletion by fluorocarbons of 17% were calculated in 1980. Since more recent studies include additional factors that either add to the ozone depletion (release of additional chlorinated hydrocarbons, increase of nitrous oxide) or compensate for it (increase of tropospheric odd nitrogen and methane, stratospheric cooling due to CO_2 increase or ozone decrease), and since their effects add in a nonlinear manner, the predic-

NATO ASI Series, Vol. G8
Stratospheric Ozone Reduction, Solar Ultraviolet Radiation
and Plant Life. Edited by R. C. Worrest and M. M. Caldwell
© Springer-Verlag Berlin Heidelberg 1986

tions have become very dependent on the assumptions made for future anthropogenic releases. Scenario studies (Fig. 1), which use 1981 rate coefficients (Callis et al. 1983; Derwent 1982) or 1982 rate coefficients (Wuebbles et al. 1983; Cicerone et al. 1983; Brasseur et al. 1983; Sze et al. 1982) show changes of total ozone ranging between +0.5% and -1.5% for the year 1980, compared to calculated values in the pre-industrial atmosphere. This is in reasonable agreement with the results from time series analysis studies of the total ozone records, which give changes in total ozone by 0.49% $\pm$1.35% (Reinsel 1981) or 0.1% $\pm$ 0.55% (Bloomfield et al. 1983a,b) for the time period 1970-1979. Local ozone depletions due to chlorocarbons should first be detected in the stratosphere around 40 km. A significant negative trend at that altitude could not yet be confirmed (Reinsel et al. 1983). However, before accepting the results from the ozone simulations we should ask how well photochemical models describe today's atmosphere. A critical test for the completeness of an ozone model is whether it satisfies the requirement of a global balance between ozone production and destruction in the stratosphere when using the best available experimental data and current photochemical schemes and models. Recently rather complete data for the important source gases N_2O and CH_4 have become available from the Nimbus-7 satellite SAMS experiment (Jones and Pyle 1984). Also, excellent data on the latitudinal and seasonal distribution of ozone were obtained from the Nimbus-4 BUV instrument (Frederick et al. 1983). We used our 2-D photochemical model to calculate the production of ozone as a function of latitude and season on the one hand, and the distribution of ozone-destroying catalysts and resulting ozone loss terms on the other. The comparison of ozone sources and sinks, which includes considerations of transport processes and the uncertainty in the input data, shows the existence of a major discrepancy which cannot be resolved with presently accepted photochemical schemes.

THE MODEL

The model used in this study has been described in Hidalgo and Crutzen (1977), Gidel et al. (1983) and Crutzen and Schmailzl (1983). We use the rate constants from DeMore et al. (1982) with the higher constant for the reaction

$$NO_2 + ClO \longrightarrow ClONO_2. \tag{1}$$

For the absorption cross section of O_2 we take the lower values in the Herzberg continuum region, as published recently by Herman and Mentall (1982) and also lower cross sections around 195 nm (Bucchia et al. 1983). We define "odd oxygen" as

$$O_x = O + O(^1D) + O_3 + ClO + HOCl + NO_2 + 2NO_3 +$$
$$3NO_2O_5 + HNO_3 + HNO_4 + 2ClONO_2, \tag{2}$$

so that the balance of odd oxygen (in practice, ozone) may be described by the equation

$$d/dt\ O_x = P - D_O - D_N - D_H - D_{Cl} \tag{3}$$

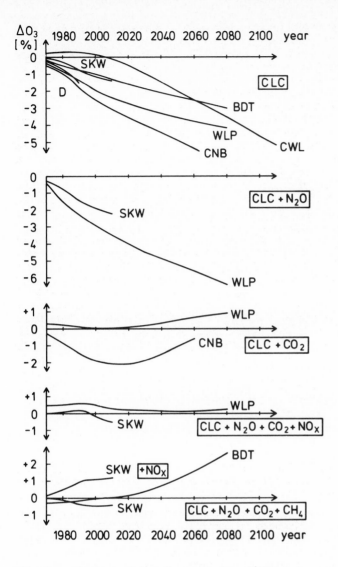

Fig. 1. Change of total ozone for different scenarios with increases of chlorocarbons (CLC), nitrous oxide (N_2O), carbon dioxide (CO_2), odd nitrogen (NO_x) and/or methane (CH_4) from Derwent (1982), average (D); Callis et al. (1983) (CNB); Wuebbles et al. (1983) (WLP); Cicerone et al. (1983) (CWL); Sze et al. (1982), scenario C (SKW); Brasseur et al. (1983) (BDT).

where P denotes the production of odd oxygen by photodissociation of O_2 and the various D terms denote the loss through cycles involving the Chapman reactions or noncatalytic reactions (D_O) and reactions with odd nitrogen (D_N), odd hydrogen (D_H) or odd chlorine (D_{Cl}) catalysts. The full expressions for the terms are:

$$P = 2J_1(O_2) \tag{4}$$

$$D_O = 2k_4(O)(O_3) + 2k_6(O^1D)(O_3) + \\ k_{14}(O^1D)(H_2O) + k_{17}(O^1D)(CH_4) - k_{38}(OH)(HNO_3) \tag{5}$$

$$D_N = 2k_{23}(O)(NO_2) + 2J_{36}(NO_3) \tag{6}$$

$$D_{Cl} = 2k_{42}(O)(ClO) + k_{45}(ClO)(HO_2) - (ClOH)(OH) \tag{7}$$
$$= 2k_{42}(O)(ClO) + J_{48}(ClOH) \tag{8}$$

$$D_H = k_8(O)(OH) + k_7(O)(HO_2) + k_{10}(O_3)(H) + \\ k_9(O_3)(OH) + k_{11}(O_3)(HO_2) - \\ k_{19}(NO)(HO_2) - k_{18}(NO)(CH_3O_2) \tag{9}$$

The rate coefficients (k) and the J values are given in Crutzen and Schmailzl (1983). For the ozone-budget calculations the distributions of CH_4, N_2O and O_3 are kept constant at their monthly values. Since the BUV ozone spectrometer showed an instrument drift from year 1970 to 1977 (Reinsel et al. 1983), we used an average of the 1970 and 1971 values. The data for CH_4 and N_2O were measured in 1979 on Nimbus-7. The different observational periods for these species should not cause a major problem for our analysis, because large changes in ozone did not occur during this period and likewise are not expected for N_2O and CH_4. H_2O is obtained from the assumption that the sum of hydrogen atoms, which is mainly defined by $2\,H_2O + 2\,H_2 + 4\,CH_4$ is constant throughout the stratosphere, with a tropopause value for H_2O of 4 ppmv, CH_4 of about 1.6 ppmv and H_2 of about 0.6 ppmv. The distribution of odd chlorine was calculated for 1979 conditions assuming historical release rates for $CFCl_3$, CF_2Cl_2 (CMA 1982), CH_3CCl_3 (Neely and Plonka 1978), CCl_4 (Galbally 1976), CH_2Cl_2, C_2HCl_3, C_2Cl_4 (Bauer 1978) and a ground value of 0.7 ppbv for CH_3Cl. The photochemical production and loss rates are averaged over 24 hours.

DISCUSSION OF THE OZONE BUDGET

Figures 2-6 show meridional cross sections for January and April of the terms P, D_O, D_N, D_{Cl} and D_H. We note that production of ozone is greatest around 42 km in the summer hemisphere or above the equator. The loss terms peak around 47.5 km (D_H), 43 km (D_O), 41 km (D_{Cl}) and 37 km (D_N). There is a negative maximum of D_H at northern latitudes near the ground, where ozone is produced by photochemical oxidation of methane and CO, which we have included in the term D_H (Crutzen and Schmailzl 1983). Figure 7 shows stratospheric production and loss terms averaged from 55°S to 55°N and over one year. Only small contributions to the ozone budget come from outside this latitude range. It can be seen that destruction of ozone is dominated by the term D_N between 25 and 40 km and by D_H above 45 km. Between 40 and 45 km D_{Cl} becomes important, while D_O is a less important loss term at all heights. Looking at the total loss term D it also becomes obvious that there is a discrepancy between ozone production and loss (P-D) of different sign depending on altitude. Based on the observed distributions of O_3, CH_4 and N_2O and the chosen model reaction scheme and rate coefficients, we obtain excess ozone destruction above and excess ozone production below 29 km. Figure 8 shows the latitudinal distribution of the term (P-D).

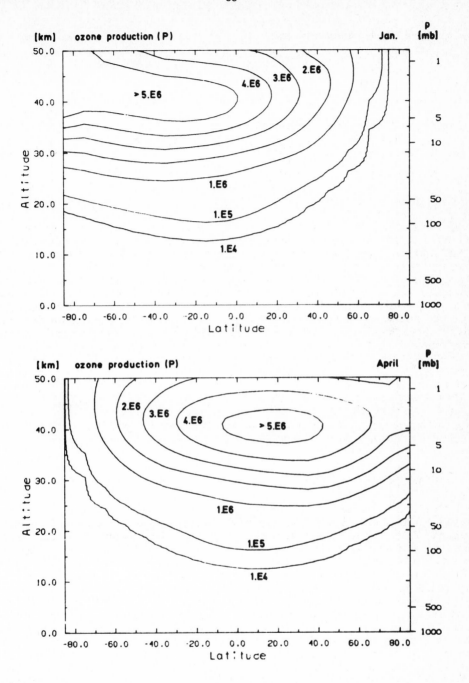

Fig. 2. Meridional cross-section for production of odd oxygen by photo-dissociation of O_2 (molecules cm^{-3} s^{-1}).

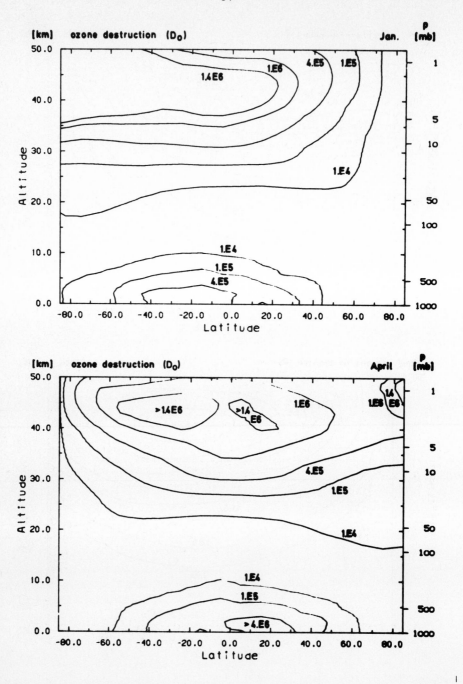

Fig. 3. Meridional cross-section for destruction of odd oxygen without catalysts (molecules cm^{-3} s^{-1}).

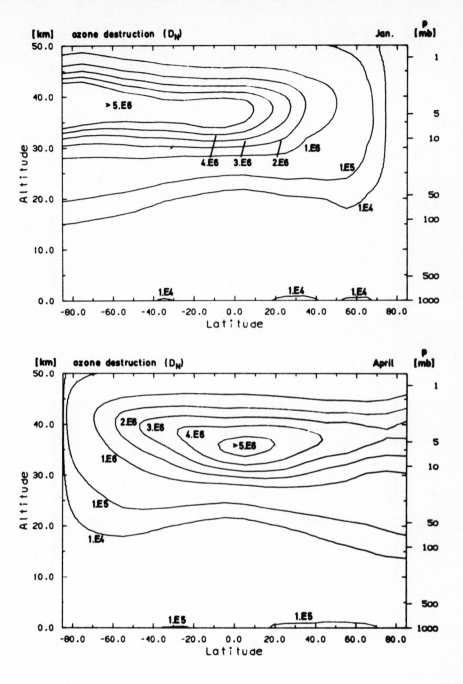

Fig. 4. Meridional cross-section for destruction of odd oxygen with odd nitrogen catalysts (molecules cm^{-3} s^{-1}).

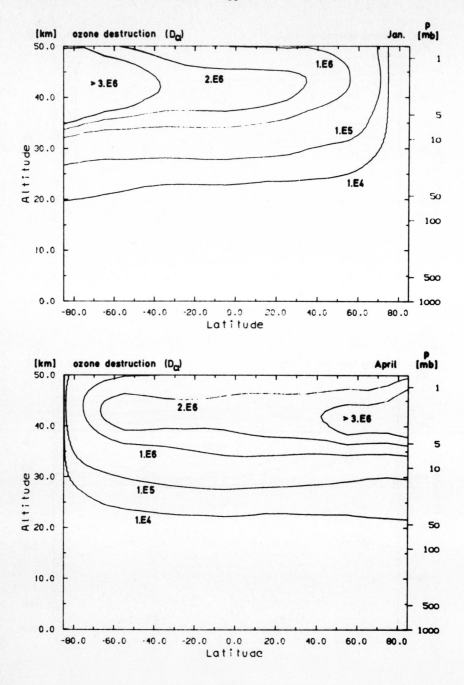

Fig. 5. Meridional cross-section for destruction of odd oxygen with chlorine catalysts (molecules cm^{-3} s^{-1}).

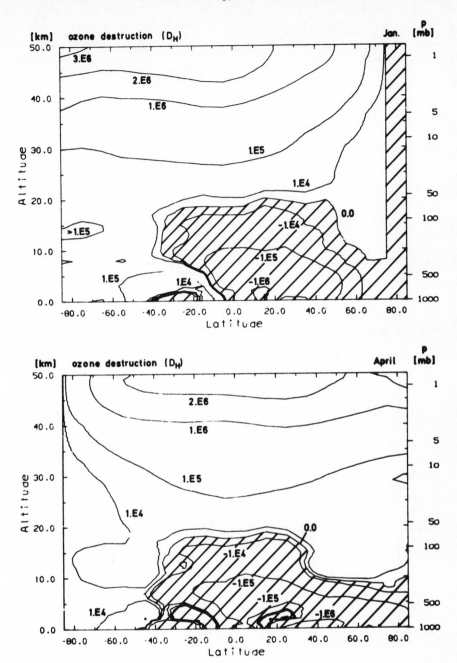

Fig. 6. Meridional cross-section for destruction of odd oxygen with odd hydrogen catalysts (molecules cm^{-3} s^{-1}). The area of negative values (i.e., photochemical ozone production) is hatched.

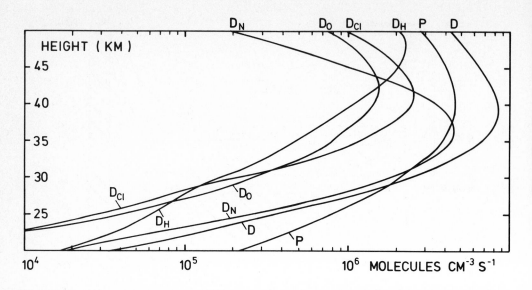

Fig. 7. Production and loss rates for odd oxygen averaged over latitude and over one year.

Since the chemical lifetime of ozone (Fig. 9) is much shorter than strato-spheric characteristic transport times, ozone is controlled by chemical processes in the upper stratosphere. This means that only part of the discrepancy found in the ozone balance can be accounted for by transport processes. Figure 10 shows (P-D)/P, the excess production of ozone in percentage of the total production terms as a function of altitude. For comparison we have also plotted the transport term in percentage of total ozone production, which shows that the ozone produced below 29 km cannot compensate for ozone loss at higher altitudes. This leaves us with an excess ozone loss of up to 88% in the upper stratosphere. But let us first discuss the lower regions where we observe excess ozone production. The total column excess ozone production in the stratosphere must be balanced by transport through the tropopause. Figure 11 gives the integral $\int$(P-D)dz from tropopause height to the height where ozone is balanced ($\cong$29 km) for latitudes $\lesssim$55° and every month of the year. It shows a surplus of ozone production for all months with maxima at subpolar latitudes between 4.5×10^{11} and 7×10^{11} molecules $cm^{-2} s^{-1}$; whereas the flux of ozone into the troposphere is known to be smaller than 10^{11} molecules $cm^{-2} s^{-1}$ (Junge 1962; Fabian and Pruchniewicz 1977; Danielsen and Mohnen 1977; Mahlman et al. 1980; Gidel and Shapiro 1980). Figure 12 gives the total excess ozone depletion below 41 km with maxima between 4×10^{12} and 7×10^{12} molecules $cm^{-2} s^{-1}$, the highest values occurring in summer at high latitudes.

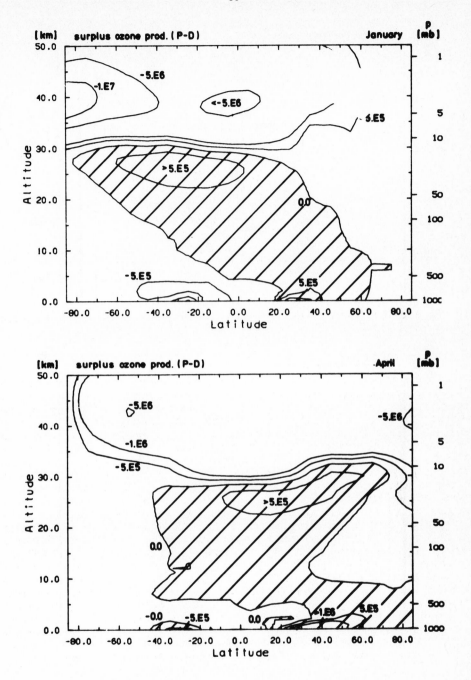

Fig. 8. Meridional cross-section for surplus production of odd oxygen. The area of positive values (net production) is hatched.

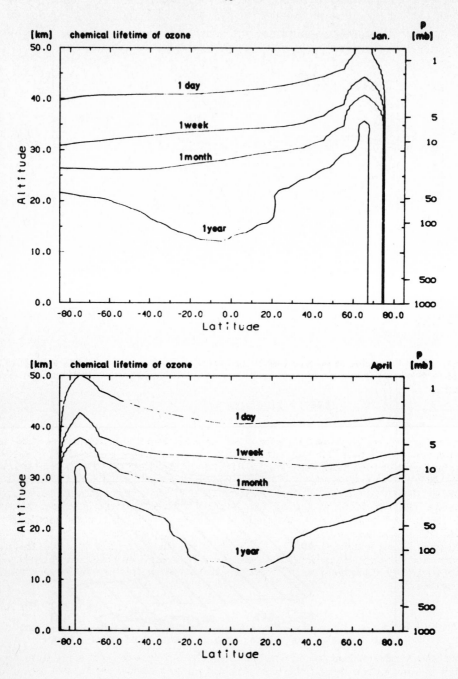

Fig. 9. Meridional cross-section for lifetime of odd oxygen (molecules cm^{-3} s^{-1}).

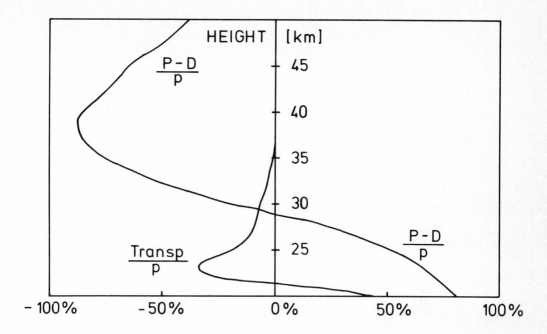

Fig. 10. Excess production and transport of odd oxygen versus altitude in percent of the total ozone production, averaged over latitude and over one year.

In our previous paper (Crutzen and Schmailzl 1983) we calculated a similar estimate using less complete data for N_2O and CH_4, two different ozone distributions from ozonesonde compilations (Dütsch 1969; Dütsch 1978) and two sets of O_2 absorption cross sections (Ackerman 1971; Herman and Mentall 1982). With the ozone distributions by Dütsch the excess ozone loss was up to 40% (Dütsch 1969) and up to 35% (Dütsch 1978), respectively, and shifted to altitudes above 35 km. With the higher O_2 absorption cross sections we did not calculate any excess ozone loss. It appears as if the ozone balance at altitudes above 30 km is very sensitive to the ozone distributions and absorption cross sections used; whereas the excess production in the lower stratosphere is at least as large as in this study for all cases considered.

CONCLUSIONS

Our study of the budget of ozone in the stratosphere shows an imbalance with excess ozone destruction above and excess ozone production below 29 km, which cannot easily be explained. Transport processes, which are a major factor of uncertainty in 2-D models, cannot account for this imbalance. While the excess ozone destruction above 29 km is highly dependent on the choice of absorption cross sections for O_2 and on the ozone distribution used, the excess ozone production below 29 km shows less sensitivity to these factors. If the discrepancy is due to errors in the chemical rate coefficients or in the source gas distributions used, it should also show in the concentrations calculated for ozone-destroying radicals, especially NO_2 and ClO, when we compare them with field measurement data. In Figs. 13 and 14 we show our model results with experimental values for $NO_x = NO+NO_2$ and ClO (WMO 1982), respectively. The agreement is good with the model curve lying on the higher side below 40 km in both cases. To compensate for the excess ozone production between 20 and 29 km we would have to multiply NO_2 by the factor 2.3 or ClO by a factor of 12.5 at 25 km, also taking into account the average contribution of transport. The imbalance of ozone production and destruction rates in the stratosphere below 30 km must be due to errors in reaction coefficients, in the source gas distributions, in the solar fluxes or other deficiencies in our understanding of stratospheric photochemistry. If this is the case, we do not know how the inclusion of missing processes would influence the ozone-depletion predictions for different scenarios. Until those inconsistencies are resolved, further updates in the predicted ozone reduction should be expected.

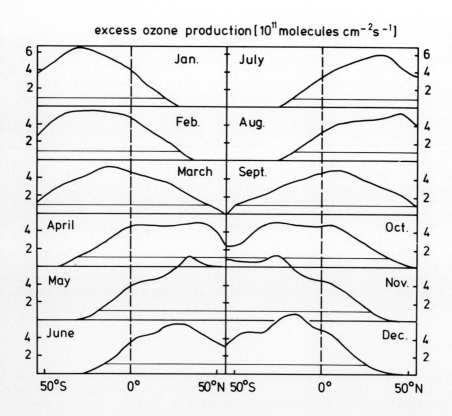

Fig. 11.   Excess column production of odd oxygen versus latitude.

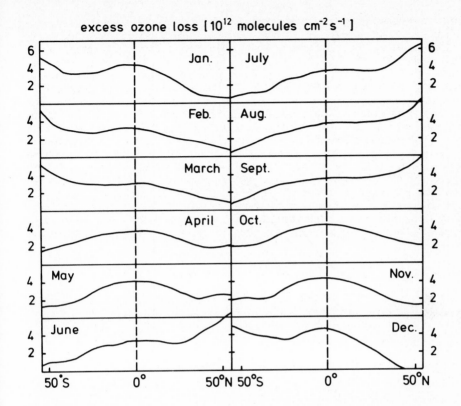

Fig. 12.   Excess column destruction of odd oxygen versus latitude.

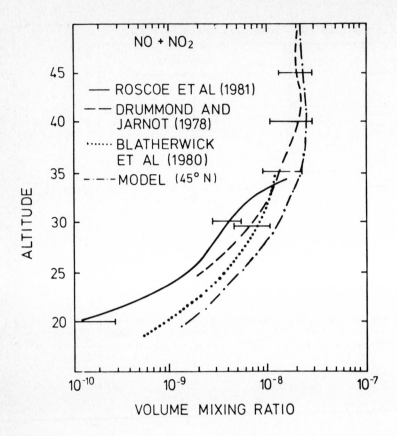

Fig. 13. Experimental data for NO+NO$_2$ and model result for 45°N.

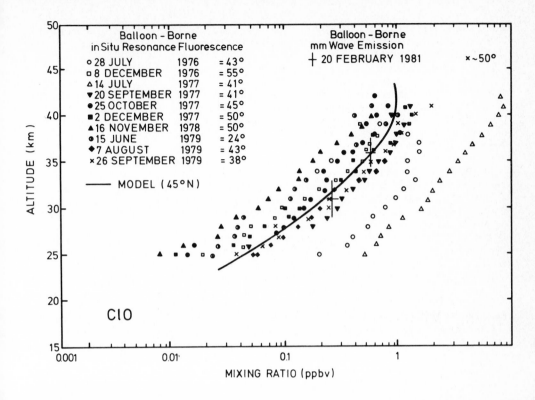

Fig. 14. Experimental data for ClO and model result for 45°N.

REFERENCES

Ackerman M (1971) Ultraviolet solar radiation related to mesospheric processes. In: Fiocco G (ed) Mesospheric models and related experiments. Reidel, Dordrecht, p 149

Bauer E (1978) A catalogue of perturbing influences on stratospheric ozone, 1955-1975. Dept of Transportation Report No. FAA-EQ-78-20, US DOT, Washington, DC

Bloomfield P, Oehlert G, Zeger S (1983a) Frequency domain trend analysis of stratospheric ozone. EOS 64:468

Bloomfield P, Oehlert G, Thompson ML, Zeger S (1983b) Frequency domain analysis of trends in Dobson total ozone records. J Geophys Res 98:8512-8522

Brasseur G, De Rudder A, Tricot C (1983) Chemical and thermal response of the middle atmosphere to coupled man-made perturbations. Presented at the IAMAP meeting, Hamburg, 15- 27 August 1983

Bucchia M, Megie G, Nicolet M (1983) Atmospheric transmittance in the 200 $\pm$ 20 nm spectral region and stratospheric predissociation of O_2. Presented at the IAMAP meeting, Hamburg, 15-27 August 1983

Callis LB, Natarajan M, Boughner RE (1983) On the relationship between the greenhouse effect, atmospheric photochemistry and species distribution. J Geophys Res 88:1401–1426

Chemical Manufacturers Association (1982) Van Horn JC (ed) 1925 Connecticut Ave NW, Washington DC 20009

Cicerone RJ, Walters S, Liu SC (1983) Nonlinear response of stratospheric ozone column to chlorine injections. J Geophys Res 88:3647–3661

Crutzen PJ, Schmailzl U (1983) Chemical budgets of the stratosphere. Planet Space Sci 31:1009–1032

Danielsen EF, Mohnen VA (1977) Project dustorm report: ozone transport, in situ measurements, and meteorological analyses of tropopause folding. J Geophys Res 82:5867–5877

DeMore WB, Watson RT, Golden DM, Hampson RF, Kurylo M, Howard CJ, Molina MJ, Ravishankara AR (1982) Chemical kinetics and photochemical data for use in stratospheric modeling. NASA Jet Propulsion Laboratory, JPL Publication 82–57

Derwent RG (1982) Two-dimensional model studies of the impact of aircraft exhaust emissions on tropospheric ozone. Atmos Environ 16:1997–2007

Dütsch HU (1969) Atmospheric ozone and ultraviolet radiation. In: Landsberg HE, Rex DF (eds) World survey of climatology, vol 4. p 383–432

Dütsch HU (1978) Vertical ozone distribution on a global scale. Pageoph 116:511–529

Fabian P, Pruchniewicz PG (1977) Meridional distribution of ozone in the troposphere and its seasonal variations. J Geophys Res 82:2063–2073

Frederick JE, Huang FT, Douglass AR, Reber CA (1983) The distributions and annual cycle of ozone in the upper stratosphere. J Geophys Res 88:3819–3828

Galbally IE (1976) Man-made carbon tetrachloride in the atmosphere. Science 193:573–576

Gidel LT, Shapiro MA (1980) General circulation model estimates of the net vertical flux of ozone in the lower stratosphere and the implication for the tropospheric ozone. J Geophys Res 85:4049–4058

Gidel LT, Crutzen PJ, Fishman J (1983) A two-dimensional photochemical model of the atmosphere I. Chlorocarbon emissions and their effect on stratospheric ozone. J Geophys Res 88:6622–6640

Herman JR, Mentall JE (1982) O_2 absorption cross section (187– 225 nm) from stratospheric solar flux measurements. J Geophys Res 87:8967–8975

Hidalgo H, Crutzen PJ (1977) The tropospheric and stratospheric composition perturbed by NO_x emissions of high-altitude aircraft. J Geophys Res 82:5833–5866

Jones RL, Pyle JA (1984) Observations of CH_4 and N_2O by the Nimbus 7 SAMS: A comparison with in situ data and two-dimensional numerical-model calculations. J Geophys Res 89:5263–5279

Junge CE (1962) Global ozone budget and exchange between stratosphere and troposphere. Tellus 14:363–377

Mahlman JD, Levy H II, Moxim WJ (1980) Three-dimensional tracer structure behavior as simulated in two ozone precursor experiments. J Atmos Sci 37:655–685

Neely WB, Plonka JH (1978) Estimation of time-averaged hydroxyl radical concentration in the troposphere. Environ Sci Tech 12:317–321

Reinsel GC (1981) Analysis of total ozone data for the detection of recent trends and the effects of nuclear testing during the 1960's. Geophys Res Lett 8:1227–1230

Reinsel GC, Tiao GC, Lewis R, Bobkoski M (1983) Analysis of upper stratospheric ozone profile data from the ground-based Umkehr method and the Nimbus-4 BUV satellite experiment. J Geophys Res 88:5393–5402

Sze ND, Ko MKW, Wang WC (1982) Atmospheric ozone: response to combined emissions of CFC's, N_2O, CH_4, NO_x and CO_2. Special report for distribution by CMA, Atmospheric and Environmental Research, Inc., 840 Memorial Drive, Cambridge, Massachusetts 02139 USA, December 1982

WMO Report No. 11 (1982) The stratosphere 1981. World Meteorological Organization, Geneva, Switzerland

Wuebbles DJ, Luther FM, Penner JE (1983) Effect of coupled anthropogenic perturbations on stratospheric ozone. J Geophys Res 88:1444–1456

Computation of Spectral Distribution and Intensity of Solar UV-B Radiation

Robert Rundel

Mississippi State University, Mississippi State, Mississippi USA

ABSTRACT

A computer model has been developed which is capable of calculating clear-sky UV-B spectral irradiance and spectral actinic flux, both upward and downward, as a function of ozone and aerosol concentrations, altitude, surface albedo, and latitude. The calculated irradiance can be integrated over a day, a season, or a year to provide a total dose, or weighted with any desired action spectrum to provide a total biologically effective dose. The basis of the model is a parameterization by the Schippnick and Green full radiative transfer calculations. The results of the present models are compared with those obtained using earlier parameterizations by Green and coworkers, and also with experimental measurements of clear-sky spectral irradiance.

COMPUTATION OF SPECTRAL IRRADIANCE

The computation of the spectral irradiance of solar ultraviolet radiation at the earth's surface has been of great importance since the inception of research into the effects of ozone depletion. Most potentially harmful effects of ozone depletion, such as increase in human skin cancer incidence or decrease in crop yields, occur due to the resulting increased fluence of biologically harmful UV in the wavelength range of 290-320 nm (UV-B).

In this wavelength range, radiation is strongly scattered in collisions with aerosol particles and with atmospheric molecules (Rayleigh scattering), as well as being absorbed by ozone. Thus at the earth's surface only about one half of the irradiance appears as direct unscattered radiation. The rest appears as diffuse skylight distributed over the entire hemisphere. Unlike the direct irradiance, this diffuse irradiance can be calculated accurately only by the numerical solution of complex radiative transfer equations. Although such radiative transfer calculations can, and have been, made (Braslau and Davé 1973a,b,c; Davé and Halpern 1976; Gerstl et al. in this volume), it is not feasible to carry out these calculations for all the possible combinations of input parameter values which are of interest.

An alternative approach to performing radiative transfer calculations has been made by Green and coworkers at the University of Florida. Using, at first, measured solar UV-B spectral irradiances (Bener 1972), and later the results of radiative transfer calculations (Braslau and Davé 1973a,b,c;

NATO ASI Series, Vol. G8
Stratospheric Ozone Reduction, Solar Ultraviolet Radiation
and Plant Life. Edited by R. C. Worrest and M. M. Caldwell
© Springer-Verlag Berlin Heidelberg 1986

Davé and Halpern 1976), as input data, they have developed a series of models, of increasing accuracy and versatility, to predict UV-B spectral irradiance under virtually any conditions (Green et al. 1974a,b, 1980; Schippnick and Green 1982). These models utilize multidimensional mathematical parameterizations which are optimized to reproduce the available input data and which provide the capability of interpolation or extrapolation to other sets of conditions.

The input parameters needed to calculate solar UV-B spectral irradiance are the extraterrestrial solar spectral irradiance, the solar zenith angle (which depends in turn on latitude, time of day, and date), and the optical depths for Rayleigh scattering, aerosol scattering and absorption, and ozone absorption. In addition, it is desirable to include effects due to altitude and reflection of UV radiation from the earth's surface. With accurate knowledge of these input parameter values, it is relatively simple to calculate the irradiance D for direct solar UV-B radiation striking a horizontal surface

$$D = \mu H \exp[-\Sigma(\tau_i/\mu)] \tag{1}$$

where μ is the cosine of the solar zenith angle, H is the extraterrestrial irradiance and the τ_i are the four optical depths referred to above.

The most recent approach taken by the Florida group to model the diffuse irradiance S is to consider the ratio of diffuse to direct irradiance. They define

$$M(\lambda) = S(\lambda,0°)/D(\lambda,0°) \tag{2}$$

as the ratio, as a function of wavelength, of diffuse to direct irradiance for overhead sun (solar zenith angle zero), and

$$\mathcal{S}(\lambda,\theta) = S(\lambda,\theta)/S(\lambda,0°) \tag{3}$$

as the ratio, as a function of wavelength, of diffuse irradiance at solar zenith angle θ to that for $\theta = 0°$. These auxiliary functions are represented by fairly complex polynomial functions of the input parameters discussed above containing adjustable parameters which reoptimized to fit the results of radiative transfer calculations. Once adequate representations of the auxiliary functions are obtained, the irradiance of diffuse solar UV-B radiation can be calculated from

$$S(\lambda,\theta) = \mathcal{S}(\lambda,\theta) \, M(\lambda) \, D(\lambda,0°) \tag{4}$$

The major advantage of using these auxiliary functions occurs because the diffuse irradiance S varies by nearly six orders of magnitude between 290 nm and 320 nm. When S is modeled using Eq. (4), most of this variation is contained in the accurately-known term $D(\lambda,0°)$, allowing the auxiliary functions S and M to optimize to a high degree of accuracy.

For any such models to produce results which represent the real world, it is of utmost importance that fixed physical input parameters such as extraterrestrial irradiance and the ozone absorption coefficient be accurately known. The most recent Florida model (Schippnick and Green 1982) used an improved parameterization of extraterrestrial irradiance based on recent satellite measurements of Heath and Park (1980). This parameterization includes Gaussian modifier terms which accurately reproduce solar Fraunhofer line structure convoluted with a slit function of 1-nm width. Figure 1 shows the parameterization compared with experimental measurement. The necessity of including Fraunhofer structure is evident. In Fig. 2 this new parameterization is compared with that used previously by Green et al. (1980). As demonstrated, this change leads to a substantial improvement in the agreement of model predictions with experimental measurements of surface irradiance.

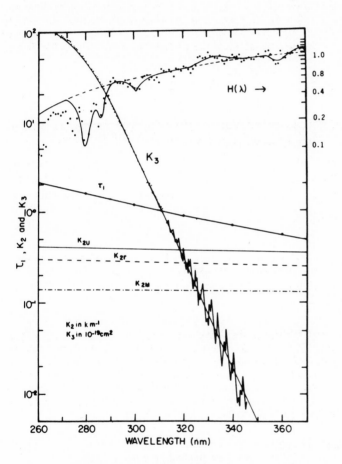

Fig. 1. Physical input quantities for model of Schippnick and Green (1982). Solid curve $H(\lambda)$ represents analytic fit to the extraterrestrial irradiance data of Heath and Park (1980) shown by points. The dashed line, $H(\lambda)$, is the parameterization used previously by Green et al. (1980). Curve K_3 represents ozone absorption coefficient, τ_1 represents optical depth for Rayleigh scattering, and K_{2m}, K_{2r}, and K_{2u} represent aerosol scattering coefficient for maritime, rural, and urban environments respectively.

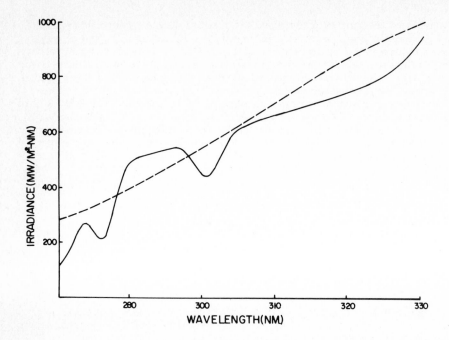

Fig. 2. Extraterrestrial irradiance used by Schippnick and Green (1982) (solid curve) compared with that used by Green et al. (1980) (broken curve).

There is at present still some uncertainty in the values to be used for the ozone absorption coefficient as a function of wavelength, and new measurements continue to appear (Klenk 1980; Bass and Paur 1981). The Florida group has used

$$k_{oz}(\lambda) = k_o(\beta + 1)/[\beta + \exp\{(\lambda - 300)/\delta\}] \qquad (5)$$

with several slightly different sets of values for the adjustable parameters k_o, β, and δ. Figure 1 shows Eq. (5) (with parameter values from Green, 1983) compared with experimental results of Vigroux (1967). Table 1 gives parameter values used recently in Eq. (5). Calculations reported here use the values suggested by Green (personal communication).

Although considerable effort has been devoted to model improvements as discussed previously, there have been few attempts to validate the model predictions by comparison with experimentally measured surface UV-B irradiances. The major purpose of this report is to present such comparisons, and to demonstrate that the model of Schippnick and Green (1982) is capable of making highly accurate predictions under a variety of conditions.

Data for these comparisons were provided by Martyn Caldwell (personal communication). All data were taken in the summer of 1983 at Logan, Utah (latitude 41.75°N, elevation 1.46 km) and consist of cosine-weighted spectral irradiance versus wavelength at 1-nm intervals from 290 nm to

Table 1.   Parameter Values for Equation 5

K_O (atm-cm^{-1})	ß	δ	References
9.517	0.0445	7.294	Green et al. (1980)
10.89	0.0355	7.150	Green (1983)
9.788	0.0556	6.798	Green (personal communication)

320 nm, with date and mean solar time for each measurement. The solar zenith angle θ for each measurement can be calculated from

$$\cos(\theta) = \cos(15t)\cos(L)\cos(D) + \sin(L)\sin(D) \qquad (6)$$

where t is the mean solar time in hours measured from noon (i.e., for 9 AM, t = -3 h), L is the latitude, and D is the declination of the earth's axis with respect to its orbital plane, given by

$$\sin(D) = \sin(23.5°)\sin(360T/365.25) \qquad (7)$$

with T the number of days after the vernal equinox (March 21).

Although ozone and aerosol column content at the time of the measurements are not known, the model predictions appear to be sufficiently accurate to allow these quantities to be deduced from the comparison, as demonstrated below.

Table 2 summarizes the various cases for which comparisons were made, and Figs. 3-9 show a series of measurements made at half-hour intervals from 10:00 AM to 11:30 AM on June 14. In each figure data are plotted both logarithmically and linearly to demonstrate clearly the comparison with model predictions both for very small irradiances (near 290 nm) and for very large irradiances (near 320 nm). The solid lines represent model predictions for assumed ozone column contents of 0.28 atm-cm (highest), 0.30 atm-cm (middle) and 0.32 atm-cm (lowest). The logarithmic plots show clearly that a column content of 0.30 atm-cm, which is typical at this latitude in summer (Bener 1972) fits the data extremely well. (Small deviations at 290-291 nm are probably due to scattered light in the spectrometer.) The series of half-hourly data shows the ability of the model to track accurately the variation of irradiance with solar zenith angle.

Figure 7 shows data for the much larger solar zenith angle of nearly 50°. Model predictions are shown for the same three ozone column contents. On this day the ozone column appears slightly higher--about 0.31 atm-cm.

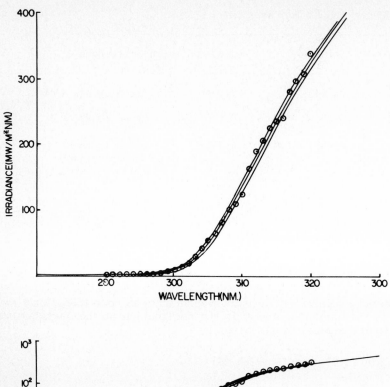

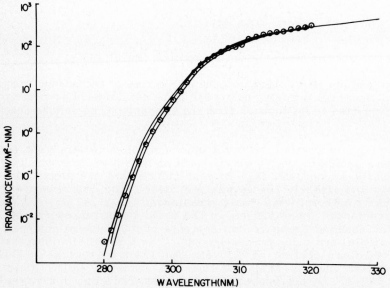

Fig. 3. Spectral irradiance versus wavelength, Logan, Utah, 10:00 AM, June 14, 1983. Points are measurements, solid lines are model predictions for ozone column contents of 0.28 atm-cm (upper), 0.30 atm-cm (middle), and 0.32 atm-cm (lower). Upper figure shows a linear plot and lower figure shows a logarithmic plot.

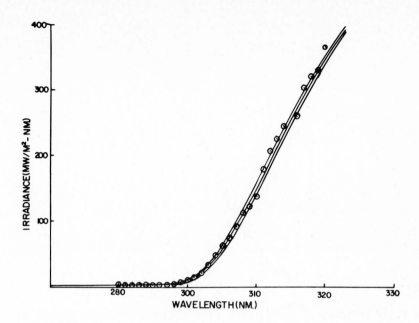

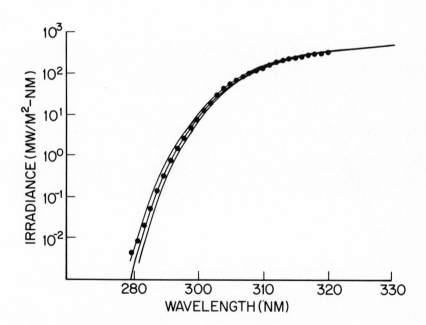

Fig. 4. Same for 10:30 AM, June 14, 1983.

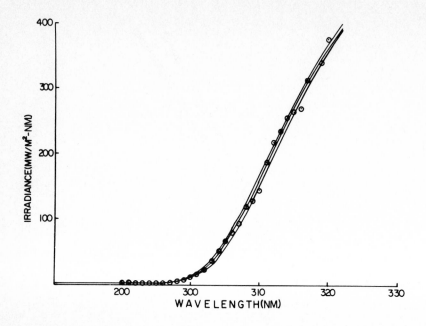

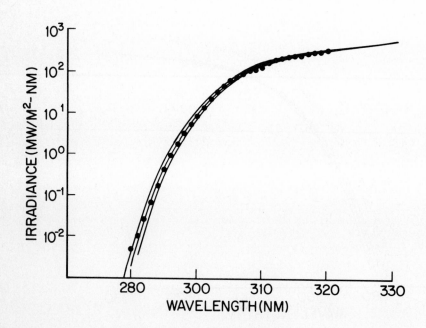

Fig. 5. Same for 11:00 AM, June 14, 1983.

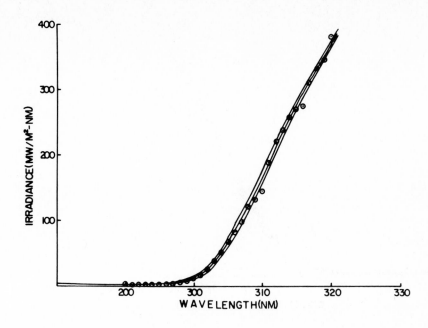

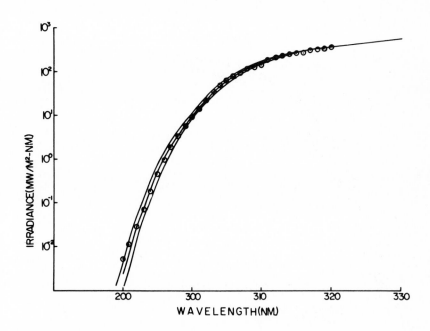

Fig. 6.   Same for 11:30 AM, June 14, 1983.

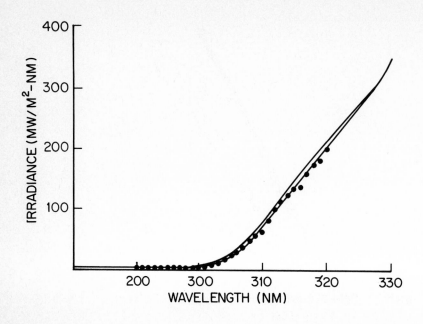

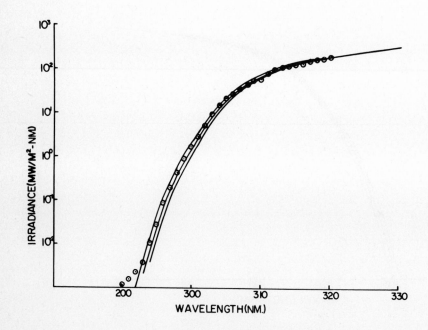

Fig. 7.   Same for 8:30 AM, July 18, 1983.

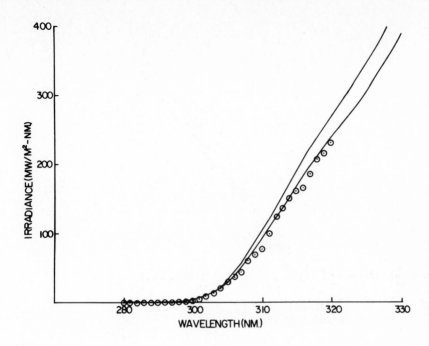

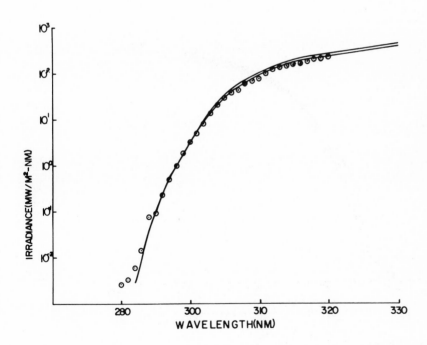

Fig. 8. Same for 10:25 AM, August 30, 1983. Both curves use 0.30 atm–cm of ozone. Upper curve for "normal" aerosol, lower curve for five times increase in aerosol optical depth.

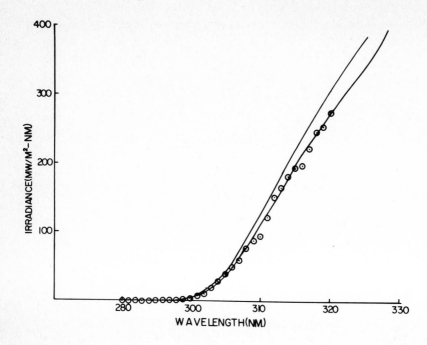

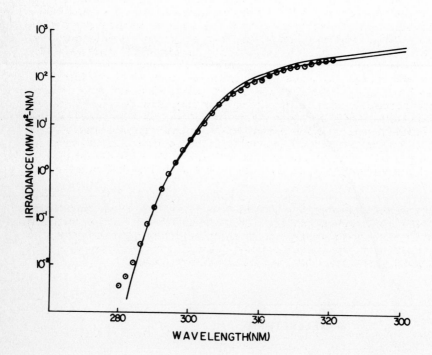

Fig. 9. Same for 12:07 PM, July 18, 1983.

Table 2. Summary of Conditions for Measured Spectral Irradiance (Latitude 41.75°N)

Date	T (days)	D	Solar Time	θ
June 14	85	23.35°	10:00 AM	31.00°
			10:30 AM	26.27°
			11:00 AM	22.25°
			11:30 AM	19.44°
July 18	118	20.94°	8:30 AM	48.55°
August 30	161	8.33°	10:25 AM	39.45°
			12:07 PM	33.46°

Figures 8 and 9 show data for two times on August 30. The higher solid line shows the model prediction for 0.30 atm-cm of ozone. Note on the linear plots that the model predicts too much irradiance at longer wavelengths, although it is quite accurate at the shorter wavelengths. This effect appears to be due to an unusually large aerosol optical depth on that day. Note in Fig. 1 that the aerosol optical depth is essentially independent of wavelength, so that an increase in aerosol optical depth will be most noticeable at longer wavelengths where the irradiance is higher. The lower solid curve is the model prediction with the aerosol optical depth increased by a factor of 5. This increase can be in both the aerosol absorption coefficient (change in particle size distribution) and the aerosol column content (more particles). An overall net increase of a factor of 5 is probably not unreasonable if the wind is blowing from an urban area toward the measurement site. Clearly it would be helpful if conditions of atmospheric visibility were noted when irradiance measurements are made.

In summary, the model of Schippnick and Green has been shown to predict measured UV-B irradiance extremely well. Comparison of model predictions with measured irradiance permits accurate estimation of the optical depths of both ozone and aerosols. See Appendix 1 for a listing of the code used for the Schippnick and Green UV spectral irradiance model.

REFERENCES

Bass AM, Paur RJ (1981) Ozone absorption coefficients. Natl Bur of Standards, Washington, DC. Tech Note No 800

Bener P (1972) Approximate values of intensity of natural ultraviolet radiation for different amounts of ozone. Final technical report. European Research Office, US Army, London. Contract No DAJA37-68-C-1017, p 59

Braslau N, Davé JV (1973a) Effects of aerosols on the transfer of solar energy through realistic model atmospheres. J Appl Meteorol 12:601–615

Braslau N, Davé JV (1973b) Effect of aerosols on the transfer of solar energy through realistic model atmospheres. Part II: Partly-absorbing aerosols. J Appl Meteorol 12:616–619

Braslau N, Davé JV (1973c) Effect of aerosols on the transfer of solar energy through realistic model atmospheres, part III: ground level fluxes in the biologically active bands .3850–.3700 microns. IBM Research Report, RC 4308

Davé JV, Halpern P (1976) Effects of changes in ozone amount on the ultraviolet radiation received at sea level of a model atmosphere. Atmos Environ 10:547–555

Gerstl SAW, Zardecki A, Wiser HL (this volume) A new UV-B handbook, vol 1.

Green AES (1983) The penetration of ultraviolet radiation to the ground. Physiol Plant 558:351–359

Green AES, Cross KR, Smith LA (1980) Improved analytic characterization of ultraviolet skylight. Photochem Photobiol 31:59–65

Green AES, Mo T, Miller JH (1974a) A study of solar erythema radiation doses. Photochem Photobiol 20:473–482

Green AES, Sawada T, Shettle EP (1974b) The middle ultraviolet reaching the ground. Photochem Photobiol 19:251–259

Heath DF, Park HW (1980) Ultraviolet extraterrestrial solar spectral irradiance. Geophys Union Meeting, Toronto

Klenk KF (1980) Absorption coefficients of ozone for the backscatter UV experiment. Appl Opt 19:236–242

Schippnick PF, Green AES (1982) Analytical characterization of spectral actinic flux and spectral irradiance in the middle ultraviolet. Photochem Photobiol 35:89–101

Vigroux E (1967) Determination des coefficients moyens d'absorption de l'ozone en vue des observations concernant l'ozone atmospherique á l'aide du spectrometre Dobson. Ann Phys 14:209–215

A New UV-B Handbook, Vol. 1

S. A. W. Gerstl, A. Zardecki and H. L. Wiser*

Los Alamos National Laboratory, Los Alamos, New Mexico USA

*U. S. Environmental Protection Agency, Washington DC USA

ABSTRACT

Results of detailed computations of the solar ultraviolet radiation reaching the earth's surface are reported in the form of 63 figures and 80 tables. This UV-B Handbook is intended for use by biologists and experimenters to find values of expected UV fluxes at various wavelength/latitude/season/time-of-day combinations for normal and depleted ozone amounts. The data are presented in tabular and graphical form for the seven wavelengths 290(5)320 nm, the eight northern latitudes 0(10)70°, the 39 solar zenith angles 0(2)76°, the five assumed ozone depletions 0(5)20%, and for measured summer and winter ozone profiles. Although only clear standard atmospheric conditions are considered in Volume I, the computations include the effects of all orders of multiple scattering (Rayleigh and Mie) plus molecular absorption. The effects of varying aerosol loads and cloud conditions are being studied separately and will be reported in Volume II. In addition to the UV flux data the Handbook also gives graphs and formulas to determine the local solar zenith angle at any location, for any date and any local time of day. Algorithms to estimate daily mean values of the UV flux and spectral integrals are also given.

PHYSICAL BASIS

This handbook is intended as a quick reference guide to estimate the quantity of solar ultraviolet radiation reaching the earth's surface at any given location and any specified time. We envision as the predominant users of the handbook the group of biological and physical researchers that have a day-by-day need to determine the spectral composition and intensity of solar radiation in the biologically active region of the UV spectrum at a specific open-air location. To use our tables and graphs it is not necessary to have a detailed knowledge of the physics or mathematics that governs the radiative transfer process through the atmosphere. We refrain therefore from giving any derivations of the equations that we used or to discuss detailed aspects of the physical processes of radiative transfer, such as line absorptions or multiple scattering. However, to understand precisely for which atmospheric and solar source conditions our calculations have been performed, we give in an appendix a complete description of all input data and cite the data sources. A preliminary version of Vol. I of this handbook is available from the principal authors, Gerstl et al. (1983).

NATO ASI Series, Vol. G8
Stratospheric Ozone Reduction, Solar Ultraviolet Radiation
and Plant Life. Edited by R. C. Worrest and M. M. Caldwell
© Springer-Verlag Berlin Heidelberg 1986

The amount and spectral distribution of the solar UV flux (irradiance) reaching the earth's surface depends on many diverse factors, compare e.g., Liou (1980):

1. the wavelength of the radiation, λ,
2. the solar zenith angle, θ_0,
3. the solar source spectrum incident at the top of the atmosphere,
4. the location of the detector,
5. molecular absorption and scattering,
6. aerosol absorption and scattering,
7. absorption, scattering and reflection by clouds,
8. the reflectance characteristics of the ground (albedo), and
9. the altitude of the detector above sea level.

The first six of these factors have been considered for our calculations explicitly as reported in this Volume I. The influences of items 7 through 9 will be evaluated and reported in a separate Volume II of this handbook. Some specific results for realistic atmospheres are reported by Zardecki and Gerstl (1982).

The solar zenith angle itself depends on latitude, date of the year, and the time of day. It is most accurately and easily determined by locally measuring the solar elevation through the course of a day. However, in order to be independent of any measurement requirements, we provide formulae and graphs to estimate θ_0 for any latitude, date and time.

Aerosol absorption and scattering (item 6 above), as well as a measured solar source spectrum (item 3 above) have been considered in our calculations but were held constant for all cases. For the aerosol composition, size distribution and vertical sensitivity profile, a reasonable "standard" was selected that can be described as a typical rural aerosol imbedded in a standard troposphere with 70% relative humidity, resulting in a visual range of 50 km along a horizontal path above the ground. In addition to this boundary layer aerosol model, assumed to be present only between 0 and 2 km height above the ground, a moderate background aerosol distribution is assumed for the upper troposphere and stratosphere; cf. Shettle and Fenn (1975 and 1979) for details. In general, our "standard" aerosol model represents therefore a very mild pollution scenario with fairly clear air over urban as well as rural regions.

The dependence of any recorded UV flux on the detector location (item 4 above) is due to varying atmospheric conditions (mainly ozone amount and distribution) and due to the declination of the sun against the equatorial plane. The solar declination gives rise to a latitude-dependent solar zenith angle for any fixed point in time. The location-dependent atmospheric conditions are included in our calculations to the degree that we employed five latitude- and season-dependent atmosphere models: Tropical, Midlatitude Summer, Midlatitude Winter, Subarctic Summer, and Subarctic Winter.

Molecular absorption and scattering, item 5, is explicitly contained in the cross-section data by which the different atmosphere models are defined. The largest contribution to the molecular absorption cross-section in the UV-B region is, of course, the absorption due to atmospheric ozone. The main contribution to molecular scattering is Rayleigh scattering of air molecules which is treated in full detail as described by the Rayleigh phase function. Scattering and absorption by our "standard" aerosol content is superimposed over the molecular interaction cross-sections.

Absorption, scattering and reflection of radiation by clouds (item 7) is not included in any calculations reported in this first volume. In fact, a cloudless atmosphere is assumed. In addition, the ground albedo (item 8) is set to zero for all calculations reported here. It is expected that realistic ground reflectance characteristics may enhance the total UV flux at ground level by up to 35%, cf. Sundararaman et al. (1975) and Schippnick and Green (1982). Also, no altitude-dependence of the UV flux (item 9) is considered; all values reported here refer to sea level. A flux enhancement of up to about 15% may occur if a detector is located at 2000 m above sea level, Sundararaman et al. (1975). A quantitative study on how the UV fluxes reported here must be adjusted for variable aerosol scenarios, cloudiness conditions, ground reflectance characteristics and different altitudes above sea level, is planned; its results will be reported in a separate Volume II of this handbook.

CALCULATED QUANTITIES

The major quantity reported in the Handbook is the monochromatic total downward solar UV flux at ground level, denoted simply as "downward flux" in our tables and graphs. This quantity is defined as the amount of radiant power which crosses a unit horizontal area in all downward directions, where the spectral distribution of the radiative energy flow is expressed per unit wavelength interval centered on a particular wavelength of interest. The units we chose are Watts per square meter surface area and per nanometer wavelength interval, $W\ m^{-2}\ nm^{-1}$. This "downward flux" is also called the spectral irradiance $F\downarrow(\lambda)$. Contrary to many calculations of $F\downarrow(\lambda)$ reported in the literature, our computations did not require the separation of direct and diffuse flux components. We report, therefore, only the total downward fluxes.

If a local measurement of the solar zenith angle θ_0 is undesirable, its value can also be estimated for any latitude, date and time of day from astronomical data. Analytic formulae and figures of θ_0 vs. time of the day for four representative dates and the northern latitudes 0, 10, ... , 70 degrees are given. An effective solar zenith angle Z_0 is defined as the daily zenith angle averaged over the duration of sunlight. A Figure gives a contour plot of the effective solar zenith angle Z_0 vs. latitude and season. This concept of an effective (daily averaged) solar zenith angle is helpful when estimates of daily averaged solar UV fluxes or dose rates are desired, cf. Gerstl et al. (1981).

USE OF THE HANDBOOK

The results of our radiative transfer calculations are presented in tabular as well as in graphical form. While the graphs contain the same information as the tables, the tables give the downward flux with greater detail and higher accuracy than the graphs. Three categories of graphical representations are chosen:

Category I: 18 figures plot the downward UV-flux versus wavelength, $F\downarrow(\lambda)$, for selected values of the ozone depletion ΔO_3, the solar zenith angle θ_o, and the northern latitude, for both summer and winter ozone distributions.

Category 2: 18 other figures plot the downward UV-flux versus solar zenith angle, $F(\theta_o)$, for selected values of the other parameters.

Category 3: 18 more figures plot the downward UV-flux versus latitude, $F\downarrow(\text{Lat.})$, for the same selected parameter values.

The 80 tables are all of the same category, namely the downward UV-flux at seven wavelengths (considered as the central values of 5-nanometer-wide wavelength bins) and 39 discrete solar zenith angles, $F\downarrow(\theta_o,\lambda)$, for specific values of the other parameters.

If it is necessary or desirable to obtain integral values over the entire spectral range of the biologically active UV-B radiation--like a damaging UV dose rate, DUV, or an integral UV flux rate--numerical or graphical integrations may be performed over the data. Analytical formulae and approximate expressions for numerical integrations over any wavelength interval are given in the Handbook. Similarly, practical algorithms are also given for temporal integrations: for example, to determine daily, monthly, or annual averages.

A SAMPLE APPLICATION

In the following we reproduce five typical graphs and a representative table from the Handbook in the context of how these data might be used. A simple four-step procedure should be followed to look-up the computed solar UV-flux for clear sky conditions:

Step I: Set the geographic location and local time for which the solar UV-flux is desired.

   Example: Bad Windsheim, W. Germany, 27 September, 11:30 h.

Step II: Determine the latitude for this location and the solar zenith (or elevation) angle for that time of day.

   Our example:

      Lat. = 49°N
      θ_o = 54°

Step III: Set the desired ozone depletion and chose summer or winter.

   Our example:

      ΔO_3 = 0 (standard)
      Summer

Step IV: Look up the solar UV-flux for the desired wavelength from the
tables or the figures

Our example:

$$\lambda \; = 305 \; nm$$
$$F\downarrow = 1.07 \times 10^{-2} \; W \, m^{-2} \, nm^{-1}$$

For our sample case we estimate the solar zenith angle θ_O for the given
latitude of 49°N and the local time of 11:30 h from one of the eight
figures included in the Handbook, which we reproduce as Fig. 1. We take
the latitude of 50°N and the October 1 date as an approximation. It is
likely that in practice a simple measurement of the actual solar elevation
against the horizon might produce an even more accurate value of θ_O.

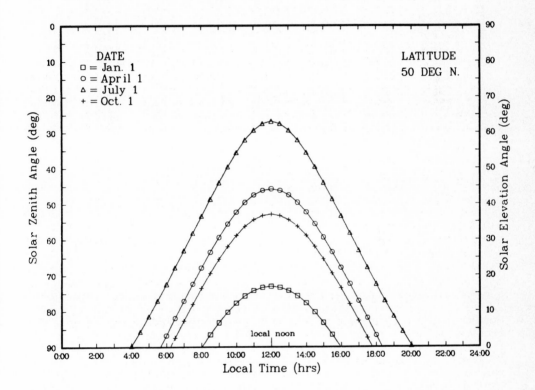

Fig. 1. Diurnal variation of solar zenith angle (Figure 6 from Gerstl et
al. 1983).

Table 1. Downward solar flux as a function of solar zenith angle and wavelength (Table 6 from Gerstl et al. 1983).

DOWNWARD SOLAR UV FLUX (W/M**2/NM) AS A FUNCTION OF SOLAR ZENITH ANGLE & WAVELENGTH

LATITUDE = 50.0 DEGREES.　　　　SUMMER
STANDARD OZONE AMOUNT　= 0.327 CM-ATM
ACTUAL OZONE AMOUNT　　= 0.327 CM-ATM　(0% DEPLETION)

ZENITH ANGLE	WAVELENGTH (NM)						
	290	295	300	305	310	315	320
0	5.209E-06	1.009E-03	1.524E-02	7.481E-02	1.932E-01	3.289E-01	4.407E-01
2	4.956E-06	9.776E-04	1.493E-02	7.374E-02	1.911E-01	3.260E-01	4.373E-01
4	4.703E-06	9.461E-04	1.461E-02	7.266E-02	1.890E-01	3.231E-01	4.339E-01
6	4.450E-06	9.146E-04	1.430E-02	7.159E-02	1.869E-01	3.202E-01	4.306E-01
8	4.197E-06	8.832E-04	1.399E-02	7.052E-02	1.848E-01	3.173E-01	4.272E-01
10	3.944E-06	8.517E-04	1.367E-02	6.944E-02	1.827E-01	3.144E-01	4.238E-01
12	3.632E-06	8.102E-04	1.325E-02	6.794E-02	1.798E-01	3.103E-01	4.190E-01
14	3.272E-06	7.606E-04	1.272E-02	6.610E-02	1.761E-01	3.053E-01	4.130E-01
16	2.912E-06	7.110E-04	1.220E-02	6.426E-02	1.725E-01	3.002E-01	4.071E-01
18	2.554E-06	6.571E-04	1.161E-02	6.211E-02	1.682E-01	2.941E-01	4.000E-01
20	2.197E-06	5.988E-04	1.094E-02	5.963E-02	1.632E-01	2.870E-01	3.916E-01
22	1.840E-06	5.405E-04	1.026E-02	5.715E-02	1.582E-01	2.799E-01	3.833E-01
24	1.519E-06	4.828E-04	9.562E-03	5.447E-02	1.527E-01	2.721E-01	3.740E-01
26	1.245E-06	4.260E-04	8.818E-03	5.152E-02	1.465E-01	2.632E-01	3.635E-01
28	9.714E-07	3.691E-04	8.074E-03	4.858E-02	1.404E-01	2.544E-01	3.529E-01
30	7.431E-07	3.162E-04	7.333E-03	4.554E-02	1.339E-01	2.449E-01	3.416E-01
32	5.786E-07	2.689E-04	6.597E-03	4.235E-02	1.269E-01	2.346E-01	3.292E-01
34	4.142E-07	2.216E-04	5.861E-03	3.916E-02	1.199E-01	2.243E-01	3.168E-01
36	2.888E-07	1.804E-04	5.161E-03	3.597E-02	1.127E-01	2.136E-01	3.038E-01
38	2.124E-07	1.469E-04	4.506E-03	3.277E-02	1.052E-01	2.022E-01	2.899E-01
40	1.360E-07	1.133E-04	3.851E-03	2.957E-02	9.775E-02	1.909E-01	2.761E-01
42	8.602E-08	8.698E-05	3.266E-03	2.649E-02	9.025E-02	1.792E-01	2.616E-01
44	5.940E-08	6.698E-05	2.744E-03	2.351E-02	8.271E-02	1.673E-01	2.468E-01
46	3.277E-08	4.699E-05	2.221E-03	2.053E-02	7.517E-02	1.553E-01	2.319E-01
48	1.986E-08	3.409E-05	1.806E-03	1.784E-02	6.787E-02	1.433E-01	2.166E-01
50	1.309E-08	2.437E-05	1.438E-03	1.527E-02	6.067E-02	1.313E-01	2.012E-01
52	6.321E-09	1.465E-05	1.070E-03	1.271E-02	5.348E-02	1.193E-01	1.858E-01
54	4.605E-09	1.054E-05	8.375E-04	1.066E-02	4.697E-02	1.077E-01	1.704E-01
56	3.237E-09	6.806E-06	6.143E-04	8.655E-03	4.051E-02	9.614E-02	1.551E-01
58	2.153E-09	3.816E-06	4.221E-04	6.816E-03	3.435E-02	8.486E-02	1.399E-01
60	1.794E-09	2.702E-06	3.090E-04	5.408E-03	2.895E-02	7.428E-02	1.252E-01
62	1.435E-09	1.589E-06	1.959E-04	4.001E-03	2.356E-02	6.370E-02	1.105E-01
64	1.208E-09	1.090E-06	1.329E-04	3.005E-03	1.912E-02	5.422E-02	9.672E-02
66	1.025E-09	7.975E-07	8.652E-05	2.148E-03	1.499E-02	4.511E-02	8.326E-02
68	8.637E-10	5.718E-07	5.118E-05	1.439E-03	1.133E-02	3.664E-02	7.043E-02
70	7.400E-10	4.675E-07	3.567E-05	9.998E-04	8.507E-03	2.933E-02	5.872E-02
72	6.201E-10	3.687E-07	2.147E-05	5.919E-04	5.825E-03	2.227E-02	4.729E-02
74	5.275E-10	3.080E-07	1.636E-05	4.060E-04	4.143E-03	1.690E-02	3.771E-02
76	4.364E-10	2.486E-07	1.146E-05	2.292E-04	2.525E-03	1.168E-02	2.830E-02

Table 1 is a reproduction of one of 80 tables in the Handbook from which the desired value for $F\downarrow$ can be looked up. Similar tables are given for summer and winter dates, and for different latitudes and ozone depletions. Figure 2 is a representative plot of $F\downarrow(\lambda)$ for Lat. = 60°N and θ_0 = 50°. Note that this latitude and solar zenith angle only crudely approximate the required values for our sample case, and that therefore the value for $F\downarrow$ from this figure does not precisely coincide with the value from the table. This note also applies to Figs. 3 and 4 which plot $F\downarrow(\theta_0)$ and $F\downarrow(Lat.)$, respectively. Figure 5 is a representative plot taken from the 18 similar figures given in the Handbook's appendix. The vertical optical depths, $\tau(\lambda)$, for the aerosol extinction, molecular absorption (mainly ozone), and molecular scattering (pure Rayleigh scattering) can be extracted from this figure, if this might be desired. Figure 6 gives a contour plot of the effective solar zenith angle that may be used to estimate the daily averaged solar UV flux. For our example we obtain $Z_0 \cong 67°$. Thus the average UV flux for Bad Windsheim on 27 September for clear sky is estimated from Table 1 to be $\overline{F} = 1.7 \times 10^{-3}$ W m^{-2} nm^{-1} at 305 nm. To obtain the total radiative energy deposition for a 5-nm wide wavelength bin centered at 305 nm $\overline{E}(305 \text{ nm})$, the average flux F must be multiplied by the duration of daylight τ and integrated over this wavelength integral $\Delta\lambda$ = 5 nm:

69

$$\overline{E}(305 \text{ nm}) = \overline{F} \tau \Delta\lambda.$$

An estimate of τ may be obtained from Fig. 1 which, for $\theta_O = 90°$, gives the local times for sunrise and sunset. Thus, for our example we obtain $\tau = 11$ h and $\overline{E}(305 \text{ nm}) = 337$ J m^{-2}.

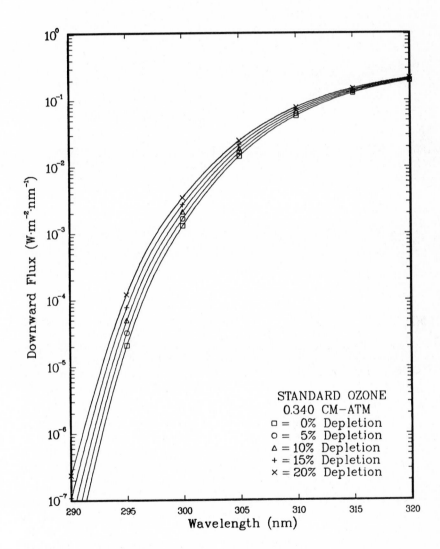

Fig. 2. Downward flux versus wavelength for several different ozone deple-tions at 60°N latitude in the summer with a solar zenith angle of 50° (Fig. 15 from Gerstl et al. 1983).

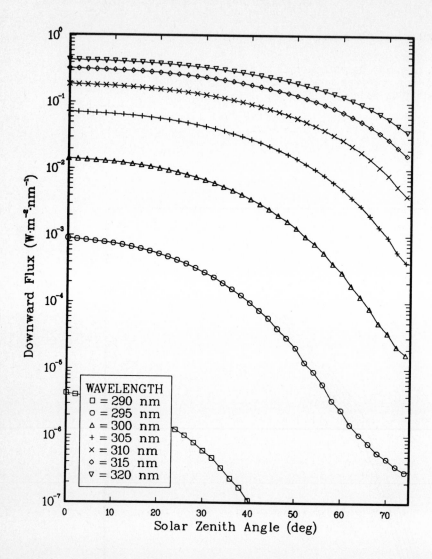

Fig. 3. Downward flux at several different wavelengths of UV-B radiation versus solar zenith angle for 60°N latitude in the summer with an ozone amount of 0.34 atm-cm (Fig. 30 from Gerstl et al. 1983).

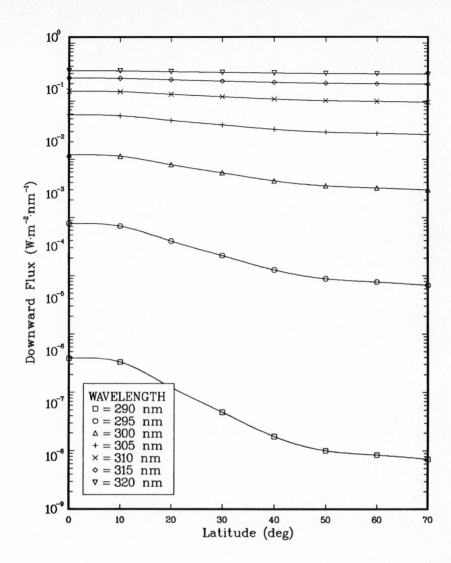

Fig. 4. Downward flux at several different wavelengths of UV-B radiation versus latitude in the summer with a solar zenith angle of 50° and an ozone amount of 0.34 atm-cm (Fig. 49 from Gerstl et al. 1983).

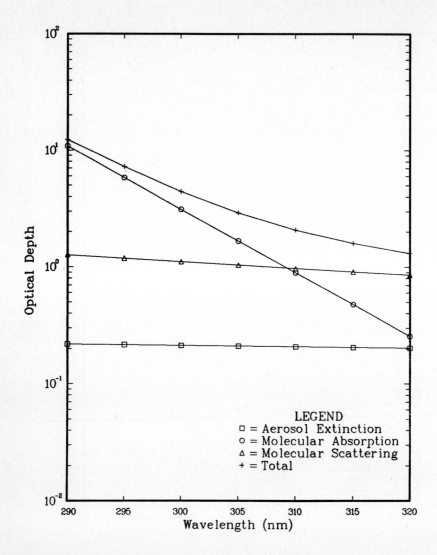

Fig. 5. Aerosol extinction, molecular absorption and scattering optical depth versus wavelength at 60°N latitude in the summer with an ozone amount of 0.34 atm-cm (Fig. 66 from Gerstl et al. 1983).

73

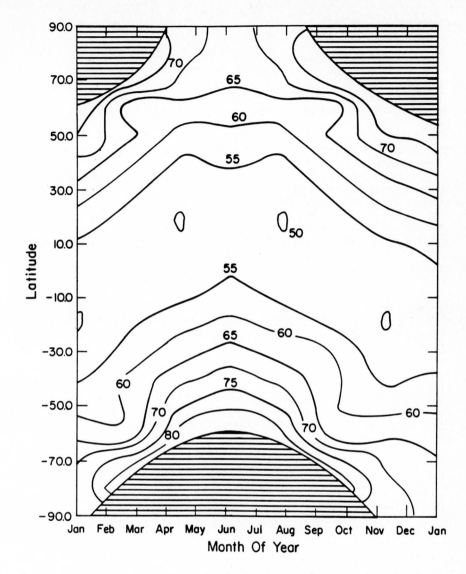

Fig. 6. Effective solar zenith angle for latitude versus month of the year
(Fig. 9 from Gerstl et al. 1983).

REFERENCES

Gerstl SAW, Zardecki A, Wiser HL (1981) Biologically damaging radiation amplified by ozone depletions. Nature 294:352–354

Gerstl SAW, Zardecki A, Wiser HL (1983) UV-B handbook vol I. Los Alamos National Laboratory report LA-UR-83-728 (Sept 1983)

Liou K-N (1980) An introduction to atmospheric radiation. Academic Press, New York

Schippnick PF, Green AES (1982) Analytical characterization of spectral actinic flux and spectral irradiance in the middle ultraviolet. Photochem Photobiol 35:89–101

Shettle EP, Fenn RW (1975) Models of the atmospheric aerosols and their optical properties. NATO Advisory Group for Aerospace Research and Development report AGARD-CP-183 (October 1975)

Shettle EP, Fenn RW (1979) Models for the aerosols of the lower atmosphere and the effects of humidity variations on their optical properties. Environmental research papers no. 676, Air Force Geophysics Laboratory report AFGL-TR-79-0214 (September 1979)

Sundararaman N, St. John DE, Venkateswaren SV (1975) Solar ultraviolet radiation received at the earth's surface under clear and cloudless conditions. Univ of California at Los Angeles, Dept of Meteorology report, June 1975

Zardecki A, Gerstl SAW (1982) Calculations of increased solar UV fluxes and DUV doses due to stratospheric ozone depletions. Los Alamos National Laboratory report LA-9233-MS, Feb 1982

Possible Errors Involved in the Dosimetry of Solar UV-B Radiation

B. L. Diffey

Dryburn Hospital, Durham, United Kingdom

ABSTRACT

Despite the increasing use of artificial sources of optical radiation in industry, medicine, science, and leisure pursuits, the sun still remains the most common source of UV radiation exposure to man. To obtain a quantitative understanding of the relationship between sunlight exposure and photobiological effects caused by exposure demands accurate and reliable dosimetry of solar UV-B radiation.

The most fundamental dosimetric technique is spectroradiometry; that is, determining the irradiance at the point of interest in a series of narrow, contiguous wavelength bands which cover the spectral range of interest. The optical components required to achieve these data are discussed, together with the parameters which affect the accuracy of such spectral measurements. In many cases spectral measurements are not required as ends in themselves, but for application to the calculation of biologically-weighted radiometric quantities. These quantities can be derived from the spectral data by a process of weighted integration over the range of wavelengths present in the spectrum.

An alternative and more common method of measuring solar UV-B radiation is to use a detector whose sensitivity varies with wavelength according to some prescribed weighting function. Probably the most widely-used device has been the Robertson-Berger meter which incorporates an optical filter, a phosphor and a vacuum phototube or photovoltaic cell. This device measures those wavelengths in the global spectrum less than 330 nm with a spectral response which rises sharply with decreasing wavelength, and has been used to monitor natural UV radiation continuously at several sites throughout the world. A different, yet complementary, approach is the use of various photosensitive films as UV dosimeters. The principle is to relate the degree of deterioration of the films, usually in terms of changes in their optical properties, to the incident UV dose. The principal advantages of the film dosimeter are that it provides a simple means of integrating UV exposure continuously and that it allows numerous sites, inaccessible to bulky and expensive instrumentation to be compared simultaneously. The most widely used photosensitive film is the polymer polysulphone. Polymer film UV-B dosimetry has allowed estimates to be made of erythemally-effective solar UV-B doses received by groups of people in different occupations and pursuing different leisure activities.

NATO ASI Series, Vol. G 8
Stratospheric Ozone Reduction, Solar Ultraviolet Radiation
and Plant Life. Edited by R. C. Worrest and M. M. Caldwell
© Springer-Verlag Berlin Heidelberg 1986

INTRODUCTION

The intensity of the solar radiation on a surface normal to the sun's direction, outside the earth's atmosphere and at the earth's mean distance from the sun is called the solar constant. Recent determinations of the solar constant have yielded a value of 1.351 kW m^{-2} (Henderson 1977). About two-thirds of this energy actually reaches the surface of the earth, the remainder being reflected, scattered or absorbed in the atmosphere. Of this energy about 50% lies in the visible spectrum and about 5% is in the ultraviolet. At noon during the summer in the UK, the irradiance on an unshaded horizontal surface is around 40 W m^{-2} in the UV-A waveband and less than 2 W m^{-2} in the UV-B region. The lower wavelength limit of terrestrial sunlight is about 290 nm and so UV-C radiation is not present.

So why is the small UV-B spectral component—less than 0.3% of the terrestrial solar energy—so important, and why is reliable dosimetry of solar UV-B radiation so difficult to achieve? The reason is that many important biological molecules, such as DNA, exhibit significant absorption of radiation with wavelengths shorter than 320 nm. Similarly many macroscopic effects in humans, such as skin erythema and skin cancer, are primarily induced by the same wavelengths. The solar spectrum increases very rapidly with increasing wavelength in the UV-B region (at 305 nm the slope is about 3%/0.1 nm); whereas biological absorption spectra and action spectra generally increase with decreasing wavelength (see Fig. 1). This feature is the principal contributory factor to uncertainties associated with solar UV-B dosimetry.

In most situations in photobiology, the interest lies not so much in the UV-B irradiance, or dose, but rather in the effectiveness of solar UV-B radiation in initiating some photobiological effect. This quantity, termed the biologically-effective UV-B dose and written as UV-B(BE), is obtained by weighting the solar spectral irradiance according to the appropriate function of wavelength and integrating over both the wavelength interval (290-320 nm) and the period of exposure to sunlight. This may be expressed mathematically as:

$$UV\text{-}B(BE) = \int_{T_1}^{T_2} \int_{290}^{320} E_S(\lambda, t)\ \varepsilon(\lambda) d\lambda dt \quad J\ m^{-2} \tag{1}$$

$E_S(\lambda, t)$ is the spectral irradiance (W m^{-2} nm^{-1}) at the point of interest at wavelength λ nm and time of day t seconds. The radiation exposure begins at time T_1 and ends at time T_2. The function $\varepsilon(\lambda)$ is the prescribed weighting function (action spectrum) accorded to the photobiological effect under study, normalized to unity at some reference wavelength λ_0 nm. Although the expression given by Eq. (1) is strictly correct for defining the biologically-effective UV-B dose, it is not appropriate to limit the integration to the UV-B waveband if the photobiological effect occurs outside these limits. In practice, it is only important if $\varepsilon(\lambda)$ is non-zero for wavelengths longer than 320 nm since at wavelengths shorter than 290 nm the solar spectrum [$E_S(\lambda, t)$] will be insignificant.

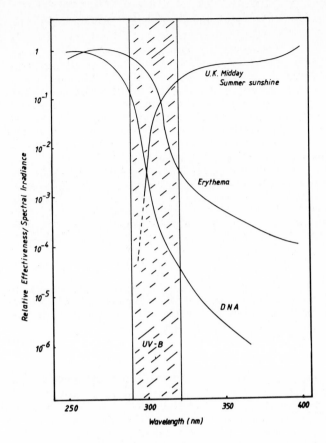

Fig. 1.   Representative examples of the absorption spectrum of DNA, the action spectrum in human skin, and the spectrum of midday summer sunshine in the United Kingdom.

The UV-B(BE) dose can be thought of as equivalent to that hypothetical dose of monochromatic radiation at the reference wavelength λ_0 nm which would produce the same photobiological effect as the exposure to sunlight (assuming no synergistic or protective mechanisms exist between wavelengths).

The UV-B(BE) dose can be determined in two separate ways: by measuring the spectral irradiance of sunlight at one or more times during the period of exposure and then calculating the weighted integral, or by using a device whose relative response to different wavelengths resembles the action spectrum for the particular effect.

A further problem with solar UV-B dosimetry is that, unlike artificial sources of light, the UV-B spectral distribution in sunlight is continually changing with changes in the concentration of the stratospheric ozone mantle, the solar elevation, and usually the weather. These factors considerably complicate solar UV-B dosimetry when the former method is used, i.e., sampling the solar spectral irradiance at infrequent intervals.

SPECTRORADIOMETRY

Spectroradiometry is concerned with the measurement of the spectrum of a source of optical radiation. The three basic requirements of a spectro-radiometer system are (i) the input optics, designed to conduct the radiation from the source into (ii) the monochromator which usually incorporates a diffraction grating as the wavelength dispersion element, and (iii) an optical radiation detector. A block diagram of the components required in a typical spectroradiometer with data processing facilities is shown in Fig. 2.

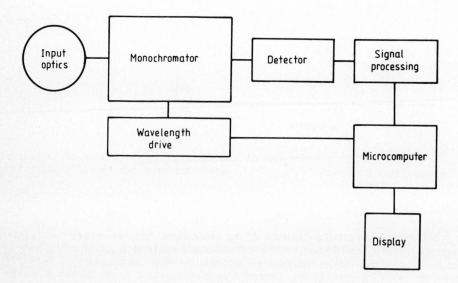

Fig. 2. A schematic diagram of the principal components in a spectro-radiometer.

Input Optics

The spectral transmission characteristics of monochromators depend upon the angular distribution and polarization of the incident radiation as well as the position of the beam on the entrance slit. For measurement of spectral irradiance, particularly from extended sources such as daylight, direct irradiation of the entrance slit should be avoided. There are two types of input optics available to ensure that the radiation from different source configurations is depolarized and follows the same optical path through the system: the integrating sphere or the diffuser (see Fig. 3).

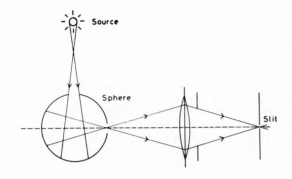

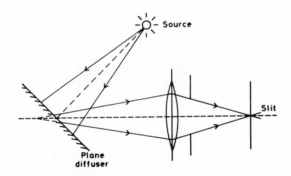

Fig. 3. The input optics to a spectroradiometer using either an integrating sphere (upper figure) or plane reflecting diffuser (lower figure).

The most accurate measurements of spectral irradiance make use of an integrating sphere in which the radiation enters through a small aperture, with the entrance slit of the monochromator located at another aperture on the surface of the sphere. Because of multiple reflections within the sphere it is important that a diffuse coating with a high reflectance in the UV region (e.g., MgO or Eastman 6080 $BaSO_4$ powder paint) be applied to the inside of the sphere to achieve good efficiency. Integrating spheres produce a cosine-weighted response, since the radiance of the source as measured through the entrance aperture varies as the cosine of the angle of incidence.

Alternatively, a plane diffuser of MgO, BaSO$_4$ or ground quartz irradiated normally and viewed at 45° may be used, as may a roughly ground quartz hemispherical diffuser placed at the entrance slit of the monochromator.

Monochromator

A classical monochromator consists of an entrance aperture, a collimating mirror, a dispersion device, a telescope mirror and an exit aperture (see Fig. 4). A blazed ruled diffraction grating is normally preferred to a prism as the dispersion element in the monochromator used in a spectroradiometer, mainly because of better stray radiation characteristics. Any radiation observed between the main diffracted orders in excess of that predicted by Fraunhofer diffraction is regarded as spectral impurity. In general there are four types of unwanted radiation, in addition to Fraunhofer diffraction, that can arise in the spectrum. These are known as ghosts, satellites, grass and diffuse scatter (Palmer et al. 1975) and arise from imperfections in the machinery used to make the gratings.

In more recent years holographic, or interference, gratings have become available. The feature of the interference technique used in the manufacture of these gratings is that there are no short-term errors in groove position and so the gratings generate no ghosts or grass, which results in a stray radiation level of one of two orders of magnitude lower than a contemporary ruled grating.

High performance spectroradiometers, used for determining low-UV spectral irradiances in the presence of high irradiances at longer wavelengths, such as solar UV radiation, demand extremely low stray radiation levels. Such systems may incorporate a double monochromator—that is two single ruled grating monochromators in tandem—or laser-holographically produced concave diffraction gratings can be used in a single monochromator.

Detector

Photomultiplier tubes, incorporating a photocathode with an appropriate spectral response, are normally the detectors of choice in spectroradiometers. However if radiation intensity is not a problem, solid state photodiodes may be used, since they require simpler and cheaper electronic circuitry.

Accuracy of Solar Spectroradiometry

The accuracy of solar spectroradiometry in the UV-B waveband is affected by a number of parameters including wavelength, bandwidth, stray radiation, polarization, angular dependence, linearity and calibration sources (Saunders and Kostkowski 1978).

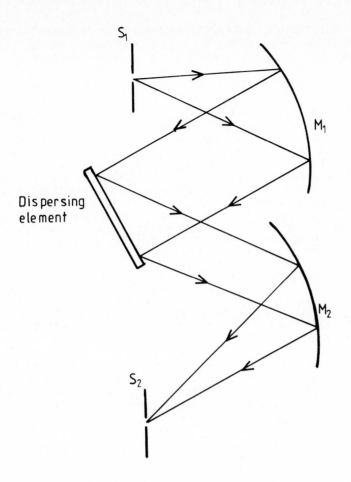

Fig. 4. The basic components of an irradiation monochromator: S_1, entrance slit; M_1, collimating mirror; M_2, telescope mirror; and S_2, exit slit.

High wavelength accuracy is required because the terrestrial solar irradiance is changing rapidly with wavelength in the UV-B region; 3%/0.1 nm at 305 nm and 10%/0.1 nm at 295 nm. Most spectroradiometers have a wavelength uncertainty of no better than 0.1 nm and a temperature coefficient of 0.1 nm/°C or higher.

The bandwidth and stray radiation characteristics of the monochromator are important because these result in output signals at wavelengths for which there is no radiation. The effect of bandwidth can be compensated for by correcting the observed spectrum by the slit function of the monochromator using a mathematical technique known as deconvolution. The effect of stray radiation can be minimized by using good quality optical components, particularly the grating and preferably employing a double monochromator.

Insensitivity to polarization and a cosine-weighted angular response can be realized by using suitable input optics: either an integrating sphere or a properly shaped diffuser.

Since the solar spectral irradiance in the UV-B region covers at least five orders of magnitude, the spectroradiometer must have a responsivity factor whose variation with spectral flux is known, and preferably constant, over this entire range. This response should be linear to not less than a few percent.

The spectral responsivity, or calibration, is determined by measuring the spectroradiometer's response to a source of known spectral irradiance. The secondary standard source which is normally employed is an incandescent tungsten filament operating at a color temperature of about $3000°K$. The spectral irradiance from the lamp is calibrated over the range 250-2500 nm by comparison with the radiance of a blackbody cavity radiator, operated at a known temperature and presumed to have the spectral distribution of power predicted by Planck's law. Alternatively, deuterium lamps can be used as secondary standard sources for calibration in the ultraviolet region. The primary standard for calibrating deuterium lamps is synchrotron radiation.

BIOLOGICALLY-WEIGHTED SOLAR UV-B MEASUREMENTS

Although spectroradiometric measurements are the fundamental approach to solar UV-B dosimetry, there will be circumstances in which this is incon- venient to impossible, and so one must rely on other procedures. One such method is to use a system whose sensitivity varies with wavelength according to the prescribed biological weighting function.

The Robertson-Berger Meter

The measurement system most widely used is probably the so-called Robertson-Berger sunburning ultraviolet meter which has been used to monitor global UV radiation continuously for many years at several sites throughout the world.

The principle of operation (Berger 1976) utilizes a fluorescing phosphor with associated optical filters and detector. A 'black glass' prefilter transmits the UV and near-IR components of the solar spectrum. Those wavelengths less than about 330 nm excite a magnesium tungstate phosphor which fluoresces with green light. A green optical filter then transmits this fluorescent radiation, but absorbs the UV and near-IR radiation trans- mitted by the prefilter. Either a phototube or a photovoltaic cell is used as the detector.

The spectral response of the instrument is governed solely by the phosphor excitation response and the transmission characteristics of the black-glass prefilter. This spectral response is compared in Fig. 5 with an action spectrum for erythema in human skin. Measurements of solar UV-B radiation with the sunburn meter suffers from the fact that its spectral sensitivity only approximates the erythema action spectrum. Because of this difference the meter cannot be calibrated in units of 'MED per hour', but instead a unit called 'sunburn units per hour' is employed. Nevertheless it is still possible to relate 'sunburn units per hour' to 'MED per hour' by employing a correction factor which is calculated by consideration of the solar UV spectrum, the meter's spectral response and the erythema action spectrum. The correction method is applicable to all UV-B meters, regardless of type. It is also possible to relate the response of UV-B meters to other photo-biological effects such as vitamin D production, bacterial killing, and so on.

Polysulphone Film Dosimeter

A different, yet complementary approach to solar UV-B dosimetry is the use of various photosensitive films as UV radiation dosimeters. The principle is to relate the degree of deterioration of the films, usually in terms of changes in their optical properties, to the incident UV dose. The principle advantages of the film dosimeter are that it provides a simple means of continuously integrating UV exposure, and also that it allows numerous sites, inaccessible to bulky and expensive instrumentation to be compared simultaneously. A film which has received wide application in the medical context, because it has a spectral response similar to the erythema action spectrum for human skin, is the polymer polysulphone (Davis et al. 1976).

The polysulphone film approximately 40-μm thick is mounted in cardboard holders, 30-mm square, with a central aperture of 16 x 12 mm (Kodak type 110 mounts) which constitutes the film badge. The badges are worn by subjects, much as photographic film badges are worn as ionizing radiation monitors.

The basis of the method is that when polysulphone film is exposed to UV radiation at wavelengths less than 330 nm, the UV absorption of the film increases. The increase in absorbance measured at a wavelength of 330 nm is proportional to UV dose, and so the film readily lends itself to application as a UV dosimeter. The change is optical absorbance of the film at 330 nm is determined by noting the absorbance of the film badge before and after irradiation. A full description of the properties and uses of poly-sulphone film as a personal dosimeter for solar UV-B exposure is given by Diffey (1984).

Accuracy of UV-B Detectors as 'Biologically-Weighted Detectors'

The best known acute biological effect in man following exposure to ultra-violet radiation is erythema, which is caused principally by wavelengths of less than 320 nm (UV-B and UV-C radiation). Since terrestrial sunlight, which is the most common environmental source of UV radiation, contains no UV-C radiation, UV-B detectors are often synonymous with erythema monitors.

Clearly the only satisfactory UV-B detectors in this context are those whose special response closely matches the erythema action spectrum. By careful attention to optical engineering it is possible to produce a device which satisfies this criterion. However, many UV-B detectors have spectral responses which are significantly different from the erythema action spectrum and can lead to misleading estimates of the erythemal efficacy of a source if they are used injudiciously.

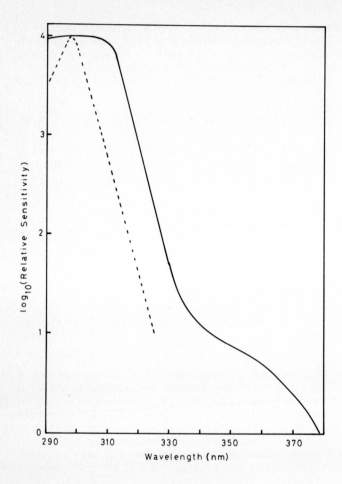

Fig. 5. The spectral sensitivity of the Robertson-Berger meter (solid curve) and the CIE erythema action spectra (dashed curve) (Berger 1976).

One particular commercially-available model of a UV-B detector employs a GaAsP photodiode as the sensor, in conjunction with an optical filter which has a symmetrical wavelength response peaking at 310 nm with a bandwidth of 34 nm (see Fig. 6). The device is calibrated with monochromatic 310-nm radiation and so the meter reading in W m^{-2} only represents true irradiance for this one situation. For a UV source emitting any other spectral distribution, the meter reading will be dependent upon both the spectral irradiance and the spectral response of the detector. Nevertheless it is still possible to relate the meter reading to an erythemally effective irradiance by the relationship

$$\left\{ \begin{matrix} \text{erythemally effective} \\ \text{irradiance (W m}^{-2}) \end{matrix} \right\} = \left\{ \begin{matrix} \text{UV-B detector} \\ \text{reading (W m}^{-2}) \end{matrix} \right\} \times Q \qquad (2)$$

where Q is a correction factor which allows for the difference between the erythema action spectrum and the spectral response of the UV-B detector and is given by

$$Q = \int E_S(\lambda)\epsilon(\lambda)d\lambda / \int E_S(\lambda)v(\lambda)d\lambda \qquad (3)$$

where the integration is over all wavelengths emitted by the source. Hence $E_S(\lambda)$ is the spectral power distribution (i.e., relative spectral intensity) at wavelength λ; $\epsilon(\lambda)$ is the relative effectiveness of radiation of wavelength λ in producing erythema; and $v(\lambda)$ is the spectral response of the UV-B detector at wavelength λ, normalized to unity at 310 nm.

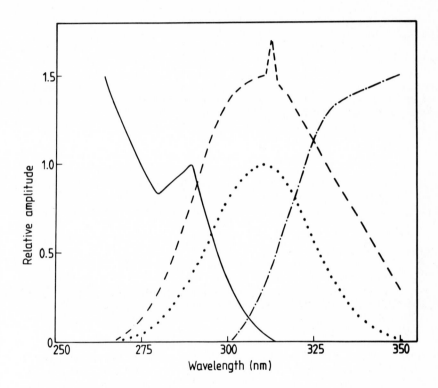

Fig. 6. The spectral power distribution of a Philips TL12 fluorescent sunlamp (dashed curve) and of global UV radiation at a solar altitude of 60° and ozone layer thickness of 0.32 cm at STP (chain curve), together with the spectral response of a UV-B detector (dotted curve) and the erythema action spectrum for human skin (solid curve).

In the following examples the erythema action spectrum which has been used is that published by Mackenzie and Frain-Bell (1973), normalized to unity at a wavelength of 290 nm (Fig. 6). This means that the erythemally effective irradiance given by Eq. (2) is that irradiance of monochromatic 290 nm radiation which would produce the same erythemal effect as the UV source in question. There is nothing especially significant about 290 nm as the reference wavelength, and indeed any wavelength can be chosen within reason. Note, however, that the numerical value of the erythemally effective irradiance will depend upon the reference wavelength.

Consider the result of using this UV-B detector to estimate the erythemally effective irradiance from a fluorescent sunlamp which has the spectral power distribution shown in Fig. 6. The value of the correction factor Q has been calculated to be 0.39. That is, the UV-B meter reading is over-estimating the erythemally effective irradiance of the fluorescent sunlamp by a factor of 2.56. If now the detector is used to estimate the erythemal effectiveness of noontime summer sunshine in the UK (solar elevation 60°, ozone layer thickness equivalent to 0.32 cm at STP), as illustrated in Fig. 6, the correction factor is now only 0.011. The unsuspecting investigator, who naively equates the meter reading of a UV-B detector with erythemally effective irradiance, would overestimate this irradiance by a factor of 100! Clearly it cannot be emphasized too strongly that whenever UV-B detectors (or UV-C or UV-A, for that matter) are used to estimate the effective irradiance of some biological mechanism predominately sensitive to that spectral region, due care and attention must be paid to the action spectrum of the biological effect, the spectral properties of the detector, and the spectral power distribution of the source.

REFERENCES

Berger DS (1976) The sunburning ultraviolet meter: design and performance. Photochem Photobiol 24:587-593

Davis A, Deane GHW, Diffey BL (1976) Possible dosimeter for ultraviolet radiation. Nature 261:169-170

Diffey BL (1984) Personal ultraviolet radiation dosimetry with polysulphone film badges. Photodermatology 1:151-157

Henderson ST (1977) Daylight and its spectrum. Adam Hilger, Bristol

Mackenzie LA, Frain-Bell W (1973) The construction and development of a grating monochromator and its application to the study of the reaction of the skin to light. Br J Derm 89:251-264

Palmer EW, Hutley MC, Franks A, Verrill JR, Gale B (1975) Diffraction gratings. Rep Prog Phys 38:975-1048

Saunders RD, Kostkowski HJ (1978) Accurate solar spectroradiometry in the UV-B. Optical radiation news no 24, US Dept Commerce, NBS, Washington DC

Action Spectra and Their Key Role in Assessing Biological Consequences of Solar UV-B Radiation Change

M. M. Caldwell, L. B. Camp, C. W. Warner and S. D. Flint

Utah State University, Logan, Utah USA

ABSTRACT

Action spectra of UV damage to plants must be used as weighting functions to (1) evaluate the relative increase of solar UV radiation that would result from a decreased atmospheric ozone layer, the radiation amplification factor--RAF, (2) evaluate the existing natural gradients of solar UV irradiance on the earth, and (3) compare UV radiation from lamp systems in experiments with solar UV radiation in nature.   Only if the relevant biological action spectra has certain characteristics is there a potential biological problem that would result from ozone reduction. Similarly the existence of a natural latitudinal solar UV gradient is dependent on action spectrum characteristics.

Several UV action spectra associated with different basic modes of damage to plant tissues all have the common characteristic of decreasing effect with increasing wavelength; however, the rate of decline varies considerably.   Extrapolation from action spectra that have been measured on isolated organelles and microorganisms using monochromatic radiation to effects of polychromatic radiation on intact higher plants is precarious. Development of action spectra using polychromatic radiation and intact higher plant organs can yield spectra that are of more ecological relevance for weighting factors in assessment of the ozone reduction problem.   An example of an action spectrum for photosynthetic inhibition developed with polychromatic radiation is provided in this chapter. This action spectrum has different characteristics, and results in a greater RAF than do action spectra for inhibition of a partial photosynthetic reaction, the Hill reaction, developed with isolated chloroplasts and photosynthetic bacteria. Circumstantial evidence from experiments with plants originating from different latitudes also supports the notion that action spectra with characteristics similar to that of the provisional spectrum, developed with polychromatic radiation, are appropriate. Further work with polychromatic radiation is encouraged.

There are two basic types of error that are associated with the use of action spectra in biological assessments of the ozone reduction problem, the RAF errors and the enhancement errors.   The former are those associated with calculation of the RAF, and the latter are those derived from calculation of the UV radiation enhancement used in experiments with lamp systems. While the RAF errors are recognized, the enhancement errors have not been generally appreciated.   An error analysis is presented showing that the enhancement errors will typically be larger and in the opposite direction

NATO ASI Series, Vol. G8
Stratospheric Ozone Reduction, Solar Ultraviolet Radiation
and Plant Life. Edited by R. C. Worrest and M. M. Caldwell
© Springer-Verlag Berlin Heidelberg 1986

than the RAF errors. The enhancement error should be considerably less in field UV supplementation experiments than in most laboratory experiments which employ fluorescent lamps as the primary UV-B radiation source.

INTRODUCTION

Traditionally, biological action spectra have been employed to elucidate photobiological mechanisms, and specifically to identify potential chromophores. Action spectra are usually assessed by evaluating biological responsiveness to monochromatic irradiation. In order to identify potential chromophores, there has been an emphasis on the fine structure of action spectra and little attention has been directed to the tails of these spectra.

In assessment of the consequences of solar UV-B radiation changes, action spectra serve a very different role. Weighting functions are derived from action spectra to assess the relative biological effectiveness of polychromatic irradiance. Spectral irradiance is weighted and then integrated over a waveband of interest, thus:

$$\text{effective irradiance} = \int I_\lambda \, E_\lambda \, d_\lambda \qquad (1)$$

where I_λ is the spectral irradiance, and E_λ is the relative effectiveness of irradiance at wavelength λ to elicit a particular biological response. The limits of the integration are prescribed by the wavelengths where either I_λ or E_λ approach zero.

Spectral irradiance and the action spectra can be expressed on either a photon or an energy basis. As units of effective irradiance it is most useful to speak of "weighted" or "effective" energy flux or photon flux density (e.g., effective $W \, m^{-2}$ or effective moles photons $m^{-2} \, s^{-1}$) and, of course, to specify the action spectrum used as a weighting function and the wavelength to which the spectrum is normalized. Special units analogous to units of illumination (e.g., lux) have been previously devised for effective UV radiation (such as sunburn units, E-viton and the Finsen for erythemally effective radiation (Robertson 1975; Luckiesh and Holladay 1933) and the G-viton for germicidal UV radiation (Luckiesh and Holladay 1933). However, unlike illumination where a standard spectral luminosity function has been well accepted, UV action spectra for various biological phenomena are continually being refined or abandoned in favor of new action spectra and, thus, either the units need to be redefined or new units described. Therefore, use of effective energy or photon flux with a specified action spectrum is quite preferable to the definition of new units.

The utility of weighting the UV irradiance and, thus, expressing biologically effective irradiance for the ozone reduction problem derives from the highly wavelength-specific absorption characteristics of atmospheric ozone and the wavelength specificity of biological action spectra in the UV-B waveband (National Academy of Sciences 1979; Caldwell 1981; Nachtwey and Rundel 1982). The expression of weighted effective irradiance is useful in addressing three basic issues:

(1) The degree to which biologically damaging solar UV irradiation is increased when atmospheric ozone decreases is very dependent on the action spectrum characteristics. The increment of biologically damaging solar irradiation resulting from a given level of ozone reduction for a specific set of conditions, is known as the radiation amplification factor, RAF (National Academy of Sciences 1979). Without calculation of a RAF, which takes the biological effectiveness of each wavelength into account, the increment of total solar UV-B flux resulting from ozone reduction is trivial, e.g., 1% increase of UV-B radiation for 16% ozone reduction for midday irradiance in the summer at temperate latitudes (Caldwell 1981). The increase of solar UV-B irradiation as a function of ozone reduction only becomes significant when the biological effectiveness of this radiation is calculated and the action spectrum for this calculation has certain characteristics. By the same token, if the action spectrum of a particular biological phenomenon does not exhibit the appropriate characteristics, the RAF will be very small and this phenomenon can be eliminated from concern with respect to the consequences of ozone reduction without the necessity of dose-response studies.

(2) The steepness of the natural latitudinal gradient of solar UV-B irradiance that currently exists on the earth's surface should also be evaluated in terms of potential biological effectiveness. The natural gradient of UV-B radiation serves as a basis for study of organism response to solar UV-B radiation and can provide insight into potential consequences of ozone reduction. This gradient has been used most frequently in the analysis of human skin cancer incidence; however, study of plant adaptation to UV-B radiation at different latitudes can also be instructive (e.g., Caldwell et al. 1982; Robberecht et al. 1980). As is the case with ozone reduction, without taking biological effectiveness of solar UV-B irradiation into account, the natural latitudinal gradient of solar UV-B radiation is virtually nonexistent (Caldwell 1981).

(3) Since spectral irradiance received from commonly-used lamp systems for UV-B studies does not match that of solar irradiance, it is only possible to draw comparisons by calculating "biologically effective" radiation using action spectra as weighting functions. Characteristics of action spectra will thus dictate the amount of radiant flux delivered by lamp systems in experiments designed to evaluate potential consequences of ozone reduction under different conditions.

This chapter will (1) review a few action spectra illustrating different fundamental mechanisms of UV-B damage in plant cells, (2) discuss the use of polychromatic radiation in determining action spectra and illustrate this with a new action spectrum for photosynthetic inhibition, (3) show the nature of radiation amplification factors that result from action spectra of different characteristics, (4) briefly discuss the implications of action spectra for the natural latitudinal gradient of solar UV-B radiation, and (5) develop an error analysis of the types and magnitude of errors that might be encountered in biological assessments by assuming an improper action spectrum.

ANALYSIS

Modes of Damage and Their Spectra

Ultraviolet radiation can damage plant cells by several basic mechanisms
and these pathways of damage likely involve several different chromophores.
To illustrate the diversity of damage pathways, several action spectra are
depicted in Fig. 1. In each case the reaction is identified with damage to
a specific entity or process in the cell. However, this does not mean that
the entity damaged is necessarily the primary chromophore (Peak and Peak
1983). All spectra are normalized to 300 nm. Though these spectra

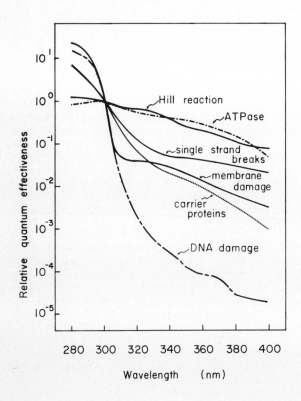

Fig. 1. Ultraviolet action spectra representing basic modes of damage to
plant tissues. The spectra were all developed with monochromatic radiation
and are plotted on a quantum effectiveness basis relative to 300 nm. These
include inhibition of Hill reaction activity in isolated chloroplasts of
spinach (Jones and Kok 1966), inhibition of adenosine triphosphatase
(ATPase) from plasma membranes of Rosa damascena (Imbrie and Murphy 1982),
induction of single strand breaks of Bacillus subtilis DNA (Peak and Peak
1982), lethality of stationary phase cells of E. coli associated with cell
membrane damage (Kelland et al. 1983), inhibition of alanine uptake in E.
coli linked to inactivation of carrier proteins (Sharma and Jagger 1981),
and inactivation of stationary phase cells in E. coli associated with
nucleic acid damage (Webb 1977).

represent different basic pathways of damage, and in many cases presumably very different chromophores, all have the common property of decreasing effect as wavelength increases. These spectra have been developed in the normal manner using monochromatic radiation and in several cases have been pursued through several orders of magnitude. If these were to be plotted with a linear ordinate, the differences in these spectra would be much less apparent. Nevertheless, as will be discussed subsequently, if these spectra are to be used as weighting factors for solar UV radiation, which increases by orders of magnitude with increasing wavelength, the differences in slopes of these action spectra are of key importance. These spectra have been determined with isolated organelles, membranes, or microorganisms, but none involved intact higher plants.

Complications: Intact Higher Plants and Polychromatic Radiation

Since the spectra illustrated in Fig. 1 differ greatly in their decline with increasing wavelength, it is important to know which of these primary damage mechanisms are the most important for intact higher plants exposed to solar radiation. This is not an easily answered question. One would need to know (1) the basic sensitivity of the respective targets (e.g., DNA, membranes, photosystem II of photosynthesis, etc.) for the higher plant species in question, (2) how well shielded these targets are within the plant organ (i.e., the effective spectral fluence reaching the target), (3) to what extent reciprocity would apply, and (4) if radiation at different wavelengths when applied simultaneously and in proportion to their occurrence in sunlight would have strictly additive effects, as predicted by the action spectrum, or whether these effects would be positively or negatively synergistic. Addressing these several issues is hardly a trivial undertaking and satisfactory answers are not likely to be quickly forthcoming. For example, as to the question of synergistic effects of radiation at different wavelengths, much is known about mechanisms of UV radiation effects on nucleic acids and yet it is still difficult to quantitatively predict synergistic effects in higher plants. From in vitro studies and from in vivo experiments with microorganisms, it is clear that there are qualitative differences between UV-B and UV-A damaging photoreactions (Peak and Peak 1983): UV-A radiation can drive photorepair of UV-B-induced damage (Jagger 1981), UV-B radiation can induce other repair systems which can repair at least a fraction of the UV damage (Menezes and Tyrrell 1982), UV-A radiation can cause growth delay in bacteria which allows more time for dark repair systems to operate (Jagger 1981), and, finally, UV radiation at high fluence rates can damage UV repair systems (Webb 1977). The net effect of these phenomena and their applicability to higher plant nucleic acid damage is not easily envisaged.

The primary site of UV damage to photosynthesis is in the reaction center of photosystem II (Renger et al., this volume; Noorudeen and Kulandaivelu 1982) and most likely does not involve nucleic acids. For intact higher plants, interactions between UV-B and longer wavelength radiation have been frequently reported though the mechanisms are not understood. Plant leaves which have developed under low visible radiation conditions are considerably more sensitive to UV-B inhibition of photosynthesis (Teramura et al. 1980; Sisson and Caldwell 1976; Warner and Caldwell 1983). Yet, there is also some evidence that higher visible irradiance during UV-B irradiation increases the inhibition of photosynthesis even though the visible irradiation by itself is not sufficient to inhibit photosynthesis (Warner and Caldwell 1983). At present, one can only speculate about the mechanisms involved in these interactions.

A Photosynthetic Inhibition Spectrum

Predicting the net effect of interactions between radiation of different wavelengths for phenomena such as nucleic acid damage or inhibition of photosynthesis is not possible at present even though the action spectra for some repair processes such as photoreactivation are known. Thus, in order to account for the synergistic effects of radiation at different wavelengths when "damage" action spectra are to be used as weighting functions for polychromatic radiation, one is forced to take a more empirical approach. This can be done by employing polychromatic radiation in the development of action spectra and using intact plant organs. In this process the biological responses to different combinations of polychromatic irradiation are determined and these responses and associated spectral irradiance data can be deconvoluted to yield an action spectrum (Rundel 1983). With respect to the ozone reduction problem, the most logical combinations of polychromatic irradiation involve a constant background of longwave UV-A and visible irradiance with increments of radiation at shorter wavelength intervals. This process can sacrifice some of the fine structure of an action spectrum, but this is of less concern for the purpose of weighting functions.

This series of polychromatic irradiation distributions should be planned to account for the tail of the action spectrum into the UV-A or visible waveband, as the case may be. Characteristics of the action spectrum below 280 nm are of no concern with respect to solar UV radiation changes because of the effectiveness of atmospheric ozone absorption, even in the case of a severely depleted ozone layer (Caldwell 1979). Yet, if a particular lamp system emits shorter wavelengths, the weighting function should include these as well.

To illustrate this polychromatic approach to action spectrum development, reduction of photosynthesis of an intact leaf of Rumex patientia will be depicted. Photosynthetic inhibition of higher plants by UV radiation is of potential concern with respect to ozone reduction. For some species under certain experimental conditions, solar UV radiation at flux rates now received at temperate latitudes has been shown to reduce photosynthesis (Bogenrieder and Klein 1977; Sisson and Caldwell 1977). Yet to be resolved, however, is the extent to which plants experience photosynthetic inhibition under field conditions, or how much they might be inhibited in the event of ozone layer reduction. Nevertheless, there is sufficient impetus to select net photosynthesis for action spectrum analysis because of its potential susceptibility to solar UV radiation and its obvious importance for plant biology. Furthermore, net photosynthesis is an integrated physiological process which requires the integrity of membrane systems and the coordinated action of photochemical and enzymatic processes. Thus, it also serves as a useful indicator of plant response to stress.

Methods: Assessment of photosynthetic depression resulting from UV irradiation involved the determination of net CO_2 uptake by plant leaves, exposure of the leaves to a particular polychromatic irradiation distribution, and then subsequent determination of CO_2 uptake capacity under identical conditions. Ideally, the leaves would be exposed to the inactivating

UV irradiation over a period of several days or weeks as would occur under field conditions. Unfortunately, this involves unreasonable biological and experimental complications. Leaf photosynthetic characteristics change appreciably with leaf age (e.g., Sestak 1977; Sisson and Caldwell 1977). These changes combined with the time and logistic requirements for such experiments render this approach infeasible. Thus, the irradiation periods ranged from one to 16 hours. Seven polychromatic radiation distributions were employed. For each, a dose-response relationship was developed. Even with these irradiation periods, such experiments are very time consuming.

The spectral irradiance for the seven polychromatic irradiation distributions are shown in Fig. 2. These were developed by using a 2.5-kW xenon high pressure lamp and a combination of dichroic and sharp cutoff filters. Much of the visible and infrared radiation was removed by the dichroic filter arrangement in order to prevent excessive leaf temperatures. Yet, there was sufficient visible flux (400 μmoles m^{-2} s^{-1} between 400 and 700 nm) to drive photosynthesis of the plants during the UV irradiation. Leaves still intact with the remainder of the plant were placed on a slowly revolving stage in this radiation field to insure even irradiation of the leaves.

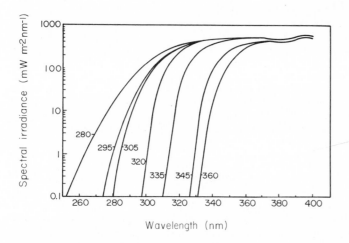

Fig. 2. Spectral irradiance for seven polychromatic irradiation distributions used in the development of an action spectrum for inhibition of photosynthesis. Radiation is supplied by a 2.5-kW xenon high pressure lamp and filtered with a combination of dichroic and absorption filters. The sharp cutoff filters used for each distribution are from the Schott WG series.

Net photosynthetic rates of intact leaves were assessed by measurements of net CO_2 uptake under specific environmental conditions in a gas exchange cuvette. By measurement of simultaneous CO_2 and water vapor flux of the foliage as well as other parameters including leaf temperature, CO_2 concentration in the cuvette air, and water vapor concentration in the cuvette, it is possible to calculate intercellular CO_2 concentrations within the leaf (e.g., Caldwell et al. 1977). Intercellular CO_2 is influenced primarily by photosynthetic rates, CO_2 in the cuvette airstream, and diffusion resistance provided by stomata. Since photosynthesis is normally substrate limited, it is important to manage the cuvette conditions to maintain a constant intercellular CO_2 concentration in the leaf before and after irradiation so that metabolic photosynthetic capacity of a leaf is assessed without the complication of changes in diffusion resistances between the leaf intercellular CO_2 and the air. Direct coupling of the gas exchange system with a computer allows immediate assessment of these parameters so that adjustments in cuvette conditions can be made during the course of these measurements.

Results and discussion: The dose-response relationships for net photosynthetic inhibition of Rumex patientia leaves are shown in Fig. 3 for the different spectral irradiance distributions. Each data point represents a different leaf measured on a different date since each determination required a day to complete. Nevertheless, these dose-response data reveal linear relationships. The coefficients of determination, r^2, range between 0.63 and 0.98 with an average of 0.87. The spectral irradiance combination which included the least short wavelength radiation (WG 360, Fig. 2) did not result in inhibition of photosynthesis under these conditions even after 16 hours of exposure.

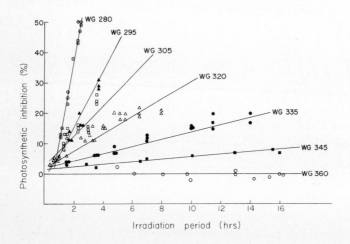

Fig. 3. Relative inhibition of net photosynthesis of intact Rumex patientia leaves exposed to the polychromatic irradiation combinations shown in Fig. 2.

These photosynthetic inhibition data indicate that when there was sensitivity to a particular irradiation distribution, we were dealing with a linear portion of the dose-response relationship. This is important in developing action spectra. If a particular dose-response relationship exhibits diminished slope as with saturation, these data cannot be compared with a dose-response relationship which is in the initial, linear phase. The dose-response relationships presented here indicate that it is suitable

to utilize the slope of these linear regression relationships for biological response in developing these action spectra.

When these slopes are taken as the relative effect for each spectral
irradiance combination, an action spectrum can be deconvoluted from the
slopes and respective spectral irradiance data (Rundel 1983). Several
possible spectra can result depending on the exact procedure followed. The
spectrum shown in Fig. 4 is the monotonic function that provided the best
fit to the data (Rundel 1983).

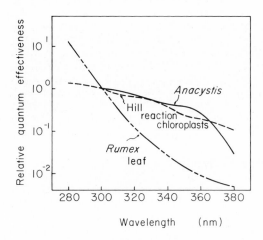

Fig. 4. Action spectrum for inhibition of photosynthesis for an intact
leaf of Rumex patientia deconvoluted from the data of Figs. 2 and 3 by
Rundel (1983) for his monotonic function that provided the best fit.
Spectra for inhibition of Hill reaction in isolated spinach chloroplasts
(Jones and Kok 1966) and of the Cyanobacterium, Anacystis, (Hirosawa and
Miyachi 1983) are also shown. The two Hill reaction spectra were developed
with monochromatic radiation.

The action spectrum deconvoluted from the polychromatic radiation study
with intact leaves is considerably steeper than the spectrum for inhibition
of a partial photosynthetic reaction, the Hill reaction, conducted by
treating isolated spinach chloroplasts with monochromatic radiation (Jones
and Kok 1966). The spectrum for inhibition of fluorescence rise time from
spinach thylakoid suspensions (Björn et al. in this volume) and a spectrum
for Hill reaction inhibition of the Cyanobacterium, Anacystis nidulans,
(Hirosawa and Miyachi 1983), also developed with monochromatic radiation,
correspond closely with the spectrum of Jones and Kok. The two Hill
reaction inhibition spectra are also shown in Fig. 4.

Preliminary results of more recent action spectrum studies for Rumex
patientia and other species suggest that the action spectrum may exhibit
steeper declines with increasing wavelength than the spectrum for Rumex
shown in Fig. 4. This new work involves simultaneous UV irradiation and
leaf gas exchange measurements. With a split-beam arrangement, the leaf is

also simultaneously receiving high visible irradiation (800 to 1200 μmoles m^{-2} s^{-1}, depending on the species being investigated). Though the visible flux employed is still only 40 to 60 percent of that in midday solar radiation, photosynthesis is light saturated. Since subsequent action spectra conducted with higher visible irradiance may reveal steeper declines with increasing wavelength, the action spectrum shown in Fig. 4 for Rumex should be considered as only provisional. The importance of the rate of decline with increasing wavelength when action spectra are used as weighting functions is discussed in a subsequent section.

Although practical limitations do not permit these experiments to be performed with spectral irradiance that exactly matches that of solar radiation, this approach with polychromatic radiation and intact leaves should provide a much more realistic approximation of the ecologically relevant action spectra for photosynthetic inhibition. It also indicates that extrapolation from damage spectra of specific physiological reactions, as shown in Fig. 1, to intact organisms may not provide reliable results. As will be discussed in a subsequent section, there is also indirect evidence that the action spectrum developed for inhibition of photosynthesis by this polychromatic approach may be appropriate for plants in nature.

Action Spectra and Radiation Amplification Factors

One of the functions of action spectra in assessment of the ozone reduction problem is their use in evaluating radiation amplification factors, RAF, i.e., the relative increase in biologically effective UV radiation for a given level of ozone reduction. As noted in a previous section, all UV damage spectra exhibit a general decrease in effectiveness with increasing wavelength, though the rates of decline vary considerably (e.g., Fig. 1). Since solar spectral irradiance increases by orders of magnitude with increasing wavelength, the rate of decline of action spectra has a pronounced effect on the resulting RAF when these spectra are used as weighting functions. Furthermore, even though ozone reduction would only result in increases of solar spectral irradiance in the UV-B part of the spectrum (Caldwell 1981), the weighting function should include all wavelengths of the solar spectrum where the weighted spectral irradiance adds appreciably to the integral of Eq. (1). Although many UV-B biological damage spectra include the UV-A region, few extend far into the visible spectrum. This is not of great consequence if the effectiveness is already so low by 400 nm that the weighted visible flux would not contribute significantly to the total biologically effective flux (such as is the case for DNA damage), or if the action spectrum is already sufficiently flat so as to result in a negligible RAF (as is the case for the Hill reaction spectrum).

To illustrate the behavior of the RAF calculated with different action spectra, a variety of spectra currently in use, as well as the provisional photosynthetic inhibition action spectrum for Rumex, are portrayed in Fig. 5 and the respective RAF values for different total ozone column thickness relative to 0.32 cm in Fig. 6. The model of Green et al. (1980) was used to calculate the spectral irradiance. The action spectra have been chosen because they illustrate the RAF values for spectra with a broad range of slopes.

The DNA damage spectrum is a generalized spectrum compiled by Setlow (1974). The generalized plant damage spectrum is also the result of a compilation of several plant damage spectra that exhibit similar characteristics (Caldwell 1971). The generalized plant damage spectrum terminates at 313.3 nm simply because that was the longest wavelength (an emission line from mercury vapor UV lamps) where information was available for these spectra at the time they were compiled. (If these spectra decline steeply beyond 313 nm as the DNA-damage spectrum does, the contribution of weighted spectral irradiance at wavelengths greater than 313 nm would be negligible for the total integrated biologically effective flux.) The Robertson-Berger, R-B, meter is a widely used integrating dosimeter for solar UV monitoring (Berger et al. 1975). The spectral response of this meter was originally designed to approximate that of human skin erythema. The provisional spectrum for photosynthetic inhibition of Rumex (Fig. 4) is also included; although preliminary data (cited previously) indicate that a steeper spectrum may be more appropriate and that this would result in greater RAF values.

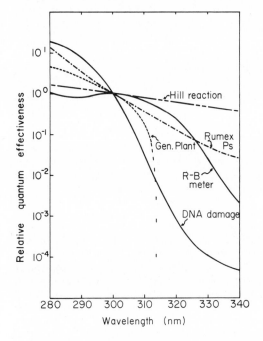

Fig. 5. Action spectra exhibiting different rates of decline with increasing wavelength. These include the spectrum for Hill reaction inhibition of spinach chloroplasts (Jones and Kok 1966), the Rumex leaf photosynthesis inhibition spectrum (Fig. 4), a spectrum for the Robertson-Berger meter (Robertson 1975), a generalized plant damage spectrum (Caldwell 1971) and a generalized DNA damage spectrum (Setlow 1974).

The RAF values in Fig. 6 are calculated over the waveband 290-380 nm. For the DNA and generalized plant damage spectra there would be no significant effect if the integral included longer wavelengths. For the Hill reaction and R-B meter spectra, the RAF values would be even smaller if longer wavelengths were included in the integration. The deconvoluted Rumex spectrum is unreliable at longer wavelengths. As noted in the previous section, polychromatic irradiation which did not include wavelengths shorter than 330 nm had no effect on photosynthesis (Figs. 2 and 3). Also shown in Fig. 6 is a slope indicating a 2% increase of biologically damaging radiation for each 1% decrease of ozone column thickness which has commonly been used as a guideline for the RAF (National Academy of Sciences 1979).

Radiation amplification factors are dependent on the degree of ozone reduction and this dependency is more pronounced for steeper action spectra (Fig. 6). The RAF is also quite dependent on solar angle and, thus, RAF values appropriate for total daily effective radiation would be an integrated function for different times of day. The daily RAF would be dependent on latitude and time of year (National Academy of Sciences 1979). The RAF values in Fig. 6 are plotted for a solar angle from the zenith of 33.6° (air mass of 1.2). For the summer solstice at 40° latitude, these RAF values would also be approximately the same as total integrated daily dose RAF values.

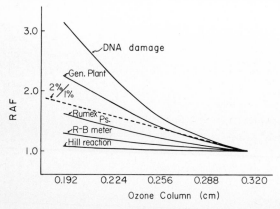

Fig. 6. Radiation amplification factors calculated for different ozone column thickness, relative to 0.32 cm, and a solar angle from the zenith of 33.6° calculated according to the action spectra shown in Fig. 5. The dashed line represents a case of 2% increase in biologically effective radiation for a 1% decrease of ozone. The model of Green et al. (1980) was used to calculate the solar spectral irradiance (direct beam + diffuse) used for those RAF values.

The Natural Latitudinal Gradient of Solar UV-B Radiation

Natural latitudinal gradients of solar UV-B radiation exist on the earth. This is primarily the result of differences in prevailing solar angles and total ozone column thickness at different latitudes (Caldwell 1981). Correlations between the latitudinal UV-B gradient and skin cancer incidence of selected human populations have been used as a tool to predict increases of skin cancer as a function of ozone reduction (e.g., National Academy of Sciences 1979). An analogous correlation between latitude and changes in crop yield or other nonhuman biological phenomena are too confounded with other environmental variables such as soils, temperature and moisture to be of use in predicting the consequences of ozone reduction. Nevertheless, controlled study of plant response to, and tolerance of, UV-B irradiation by species from different latitudes can provide some useful insight as to how plants cope with different levels of UV-B irradiation. However, just as the magnitude of solar UV-B increase resulting from ozone reduction is dependent on the action spectrum used, the steepness of the present latitudinal gradient of solar UV-B irradiance is also quite dependent on the action spectrum used for evaluation.

The effective UV-B radiation for the season of year of maximum solar radiation at different latitudes is shown in Fig. 7. These are plotted relative to the radiation at 40° latitude. (The date of maximum solar radiation at each latitude corresponds to the time when solar zenith angles are minimal. Above 23° latitude this occurs at the summer solstice, but at latitudes less than 23°, this time is progressively closer to the equinoxes, about March 21 and September 23, which are the dates of maximum solar radiation at the equator.) Three different action spectra have been used as weighting functions to illustrate the dependence of this gradient on action spectrum characteristics. Just as the RAF is greater for steeper action spectra (Fig. 6), steeper action spectra also result in a more pronounced latitudinal gradient of solar UV-B irradiance.

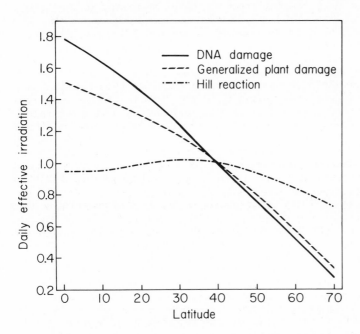

Fig. 7. Daily effective UV-B radiation for the season of year of maximum solar radiation at different latitudes relative to that at 40° latitude calculated according to the generalized DNA damage action spectrum (Setlow 1974), the generalized plant damage action spectrum (Caldwell 1971) and the Hill reaction inhibition spectrum (Jones and Kok, 1966).

If a spectrum has a rather slow rate of decline with increasing wavelength, such as the Hill reaction spectrum, there is no latitudinal solar UV-B gradient. Thus, study of characteristics of plants that have originated from different latitudes can provide some indirect evidence of the type of UV damage action spectrum that is indeed appropriate for plants. For example, if a spectrum such as the Hill reaction is indeed representative of the spectrum for UV damage to plants, one might expect little correlation between the latitude of origin of plants and their sensitivity to UV-B radiation when this radiation is presented at sufficient flux densities to result in damage.

There are several lines of evidence that suggest that a sufficiently steep action spectrum for plant damage exists so that an appreciable latitudinal gradient of solar UV-B radiation has been extant. The first line of evidence is a correlation between the latitude of origin of crop species and their sensitivity to UV-B radiation as measured by a reduction of plant biomass in experiments with UV-B radiation lamps. Over 50% of the crop species which have originated from temperate latitudes were sensitive to UV-B radiation in these tests, whereas for crop species that have originated at low latitudes, only 20% were classified as sensitive (Teramura and Caldwell, unpublished).

The second two lines of evidence are derived from a latitudinal gradient in the arctic-alpine life zone ranging from sea level locations in the Arctic to alpine elevations (3000-4000 m) at mid to low latitudes. In this situation an even steeper gradient should be expected because elevation above sea level is superimposed on the latitudinal gradient (Caldwell et al. 1980). A survey of UV optical properties of leaves from nonagricultural, as well as agricultural, plant species has been conducted by sampling plants in the field at various locations along this gradient and determining the UV epidermal transmittance of the leaves. Plants occurring in areas of high solar UV-B radiation, at low latitudes and high altitudes, exhibited a greater capacity to selectively absorb UV-B radiation in the upper tissue layers of the leaves (Robberecht et al. 1980). If the selective pressure for UV absorption in the epidermis of plant leaves were approximately the same at all locations along this gradient, such as would be the case if an action spectrum like that of the Hill reaction inhibition were appropriate for UV-B damage, one would not expect a correlation between latitude and UV-B filtration capacity of the leaf epidermis.

A third line of evidence involves the inherent differences in sensitivity of the photosynthetic system to UV-B radiation damage that have been demonstrated for species of the same genus or even races of the same species which occur in different locations on the latitudinal gradient of the arctic-alpine life zone (Caldwell et al. 1982). In these experiments, plants were cultured under identical conditions in environmental growth chambers before the sensitivity of their photosynthetic systems to the UV-B radiation was assessed. Arctic races or species consistently showed much greater sensitivity to UV-B radiation and this could not be solely attributed to differences in the optical properties of the leaf epidermis. Such evidence certainly supports the notion that a steeper action spectrum for photosynthetic damage by UV radiation such as that developed using polychromatic radiation (Fig. 4) would be more appropriate than an action spectrum such as that for the Hill reaction. Although all three lines of evidence are circumstantial, they support the notion that an appreciable RAF exists for higher plants.

Biological Assessments and Errors Deriving from Action Spectra

Before radiation amplification factors can be used in assessment of biological and ecological consequences of ozone reduction, it is necessary to demonstrate biological effects of UV-B radiation. Ideally, one should establish a dose-response relationship. In any case, it is at least necessary to show a meaningful response at UV-B flux densities exceeding those currently in sunlight. (Investigations involving the removal of UV-B from solar radiation, such as with filters, have generally not resulted in significant plant responses (Caldwell 1968; Becwar et al. 1982; Caldwell 1981), although there are some notable exceptions (e.g., Bogenreider and Klein 1977).)

Experiments with lamps: One cannot easily simulate the solar spectral irradiance either with or without reduced ozone using UV lamps. Thus, weighted UV irradiance from the lamp system is compared with weighted solar UV irradiance as would occur with a particular solar angle, ozone concentration, etc. The most common lamps used are fluorescent lamps which emit principally in the UV-B region. The spectral irradiance from the most frequently used bulbs is shown in Fig. 8. As with all fluorescent lamps, these are basically low pressure mercury vapor lamps with a phosphor that fluoresces in a continuous spectrum -- in this case, mainly in the UV-B region, but some UV-C and UV-A radiation is included. Though most of the radiant emission from the bulbs comes from the fluorescing phosphor, the emission from the mercury vapor is also present and this appears as distinct lines in the spectrum. The fluorescent lamp manufactured by Philips Co. emits substantially more than the bulb of Westinghouse.

Filters are commonly employed to absorb the UV-C and short-wavelength UV-B radiation emitted by these bulbs. Investigators in Europe have commonly used the Philips bulbs with glass cutoff absorption filters (Schott WG series) for this purpose, while in North America, cellulose acetate plastic film has been used with the bulbs manufactured by Westinghouse (Caldwell et al. 1983). The glass absorption filters have the advantage of being stable in their transmittance, while the plastic film filters are considerably less expensive but exhibit decreased transmittance as they are exposed to UV radiation (Caldwell et al. 1983). The spectral irradiance from filtered and unfiltered lamps shown in Fig. 8 is provided as a comparison of these lamp and filter combinations. Naturally, individual lamps will vary in their emission and, as with all fluorescent lamps, their emission will decrease with age, especially during the first 100 hours of use, and with temperatures that are too warm or too cold for optimal operation (Bickford and Dunn 1972).

If filtered fluorescent lamps are the only source of UV radiation for plant experiments, the discrepancy between spectral irradiance received from the sun and that received from the lamps is considerable (Fig. 9a). When compared with the midday solar irradiance in the summer at temperate latitudes, a bank of filtered Westinghouse fluorescent lamps provide more shortwave UV-B radiation and more than an order of magnitude less longwave UV-B and UV-A radiation. Thus, the action spectrum used to weight the spectral irradiance for comparison of the lamp systems with solar radiation is particularly critical as will be demonstrated subsequently.

When such lamp systems are used to supplement the normal solar radiation, i.e., when plants are exposed both to the solar radiation as well as a

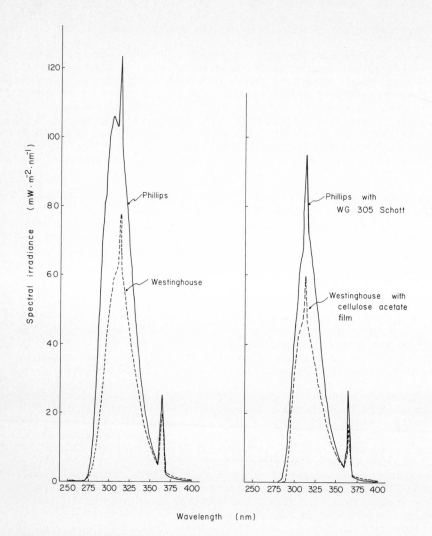

Fig. 8. The spectral irradiance at 30 cm from common fluorescent UV-B lamps with and without filters. The lamps and filters were new and the spectral irradiance was measured with a double monochromator spectro-radiometer (Optronic Laboratories). The Philips TL 40W/12 and the Westinghouse FS40 lamps without filters are shown on the left and with filters on the right. Schott WG 305 sharp cutoff absorption filters (2-mm thickness) were used with the Philips lamp and cellulose acetate plastic film (0.13-mm thickness) with the Westinghouse lamp. These lamp/filter combinations have commonly been used in studies in Europe and North America, respectively.

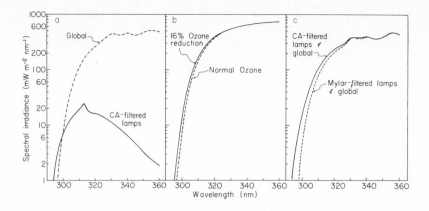

Fig. 9. Global (direct beam + diffuse) UV spectral irradiance and the spectral irradiance from filtered lamp systems used in experiments. In the left panel, a, global spectral irradiance as measured at solar noon on August 20 at 41°45' N latitude and 1460 m elevation and the spectral irradiance from Westinghouse FS40 lamps with cellulose acetate film filters (without background global radiation) are shown. The lamp system is adjusted to provide a UV-B supplement equivalent to a 16% ozone reduction under these conditions calculated according to the generalized plant damage action spectrum. In the middle panel, b, the theoretical global UV spectral irradiance for these conditions, as calculated by the model of Green et al. (1980), is shown for current ozone conditions as well as conditions with a 16% reduction of the atmospheric ozone layer. In the right panel, c, the UV spectral irradiance measured under lamp systems with cellulose acetate and Mylar film filters with the background global radiation is shown. The Mylar-filtered lamp system provides a control with the same UV-A irradiance and obstruction of global radiation as the CA-filtered lamps which constitute the enhancement treatment. The UV-A irradiance in both treatment and control is less than the ambient global shown in the left panel because of the obstruction created by the lamps. The UV-B supplement in the right panel represents the same effective UV-B radiation as that predicted with the Green et al. (1980) model when calculated using the generalized plant damage action spectrum. Adapted from Caldwell et al. (1983).

supplement from the filtered lamp bank, the discrepancy in spectral distribution is much less. Ideally, the supplemented UV spectral irradiance would only be that shown as the difference between solar spectral irradiance under normal conditions and with ozone reduction (Fig. 9b). Since the filtered fluorescent lamps emit longer wavelength radiation than is desired for the supplement, it is necessary to use control lamps which are identical except that they are filtered so as to exclude the UV-B radiation. Thus, the amount of UV-A radiation provided by the control lamps would be the same as that coming from the treatment lamps, and the other microenvironmental influences such as shading by the lamps would also be the same. In Fig. 9c the resulting spectral irradiance under lamp banks with two different plastic film filters, corresponding to treatment and control, is shown with the solar radiation background. The difference between these is still not a perfect simulation of what would be calculated as the difference between solar spectral irradiance with and without ozone reduction; however, the correspondence is considerably better than when the lamps are used by themselves such as would be the case in a glasshouse, growth chamber or laboratory situation.

Error analysis: There are then two basic types of error that are involved in use of action spectra as weighting functions in biological assessments of ozone reduction, namely RAF errors and enhancement errors. The RAF errors are those resulting from an overestimation or underestimation of the radiation amplification factors by assuming an incorrect action spectrum. The RAF errors would be associated with application of dose-response relationships to the solar UV-B irradiance that would result from various scenarios of ozone reduction. For example, if a given amount of UV-B irradiation corresponding to 10% ozone reduction resulted in 5% reduction of photosynthesis, one would employ the RAF to extrapolate to other ozone reduction levels for a given latitude and season of year.

The enhancement error is that associated with relating UV-B radiation supplied by lamp systems in an experiment with solar UV-B radiation for a particular ozone level. For example, if one desires to provide a UV-B radiation supplement under field conditions corresponding to a 15% ozone reduction, and an incorrect action spectrum is used as a weighting factor for comparing UV irradiance from the lamps and the sun, an error in the desired enhancement will result. The enhancement error necessarily includes an RAF error -- namely that associated with calculating the effective UV irradiance corresponding to the 15% ozone reduction.

These two types of error are shown in Fig. 10 for a scenario in which the generalized plant damage action spectrum (Fig. 5) is employed for calculating the RAF and the effective UV radiation enhancement when the DNA-damage spectrum (Fig. 5) is indeed the most appropriate for plant damage. (This is, of course, hypothetical as the most appropriate action spectrum to represent higher plant damage is not known.) The discrepancy between RAF values calculated by the generalized plant damage and DNA damage spectra is illustrated in the upper portion of Fig. 10. With greater ozone reduction, which requires larger RAF values, the divergence from the diagonal 1:1 line increases, i.e., the difference between the RAF values calculated using the two action spectra increases. This is directly from information presented in Fig. 6. The differences in UV radiation enhancement in a field experiment that would result from use of the two action spectra are shown in the lower half of Fig. 10. In field experiments the lamps are used to supplement the solar radiation, and a certain amount of obstruction of the solar global (direct beam + diffuse) radiation inevitably results. The exact spectral distribution of the radiation reaching the plants in the enhancement experiments then depends on the mixture of global and lamp radiation. The relationship in Fig. 10 was determined from actual measurements of spectral irradiance under lamp systems in a field experiment. Each point represents a combination of the UV radiation from the lamps and the background global radiation convoluted using the two action spectra. Relative enhancement is the ratio of the UV-B radiation from the lamps plus the background global radiation to the ambient global radiation without the lamp systems for particular conditions. The range of values shown in

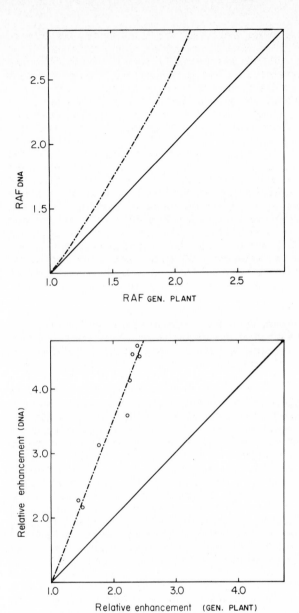

Fig. 10. The relationship between radiation amplification factors calculated with the generalized plant damage action spectrum and the generalized
DNA damage action spectrum for different conditions and levels of ozone
reduction (upper). The relative enhanced UV-B radiation in field experiments calculated according to the generalized plant damage spectrum and the
generalized DNA damage spectrum for different conditions and levels of
ozone reduction based on field spectral irradiance measurements (lower).
The solid diagonal lines in both graphs represent a situation of equiva-
lence for the RAF and relative enhancements calculated from the two
spectra.

Fig. 10 results from different experiments in which enhancements corresponding with different levels of ozone reduction and different solar angles were used. The lamp system is specially modulated so that the amount of UV supplement is adjusted for particular solar conditions such as solar angle, atmospheric turbidity, current ozone concentrations, etc. (Caldwell et al. 1983). A linear regression fits these points adequately ($r^2 = 0.96$). The discrepancy between relative enhancements calculated by the two action spectra is greater than the discrepancy between the RAF values calculated with the two spectra.

For the action spectra represented in this paper (Figs. 1, 4 and 5) and the filtered UV fluorescent lamps commonly used, (Fig. 8), the RAF and enhancement errors are in opposite directions as depicted in Fig. 11. These relative errors are shown as a function of ozone reduction for field enhancement experiments under temperate-latitude conditions in the summer as portrayed in Fig. 10. The scenario is the same, namely, that the generalized plant damage action spectrum was assumed while the DNA-damage spectrum is correct. Although the relative RAF errors are small for situations where the ozone reduction is small, this is not the case for the enhancement errors (Fig. 11).

If a different scenario is pursued such that the generalized plant damage spectrum is assumed, but a spectrum with a slower rate of decline with increasing wavelength, such as the R-B spectrum (Fig. 5), is correct, then the RAF and enhancement errors would be reversed--an overestimation of the RAF and underestimation of the enhancement.

As shown in Fig. 9, field UV enhancement experiments provide a much closer simulation of the spectral irradiance resulting from ozone reduction than do experiments in glasshouses or growth chambers in which UV radiation is provided only by filtered fluorescent lamps, since there is no solar UV background. The enhancement errors that might result from use of an incorrect action spectrum for calculating the enhancement are considerably greater than for field experiments, as shown in Fig. 12. The spectral irradiance from filtered Philips and Westinghouse UV fluorescent tubes following aging of the lamps and filters for about 100 hours was used to calculate the errors shown in Fig. 12. The same spectral distribution is assumed for all enhancement levels in such experiments, which is normally the situation, as doses are typically adjusted by varying the number of lamps, distance between the lamps and the experimental objects, or use of neutral density filters. The errors portrayed here are striking, especially when one considers that the two action spectra used for this scenario are not so very dissimilar (Fig. 5). Other scenarios could easily lead to larger overestimation or underestimation errors for the calculated enhancement. Also, for the errors depicted in Fig. 12 the relative overestimation errors are greater at UV levels corresponding to smaller ozone reductions.

If a source of background UV radiation were present, such as from a xenon arc lamp, these errors could be substantially reduced. However, this would depend on the flux density and spectral distribution received from the background source. An empirical analysis of the errors for particular situations would need to be conducted. However, without knowledge of the most appropriate action spectra for plant damage, the error analysis remains as a heuristic exercise quite dependent on the action spectrum scenarios assumed.

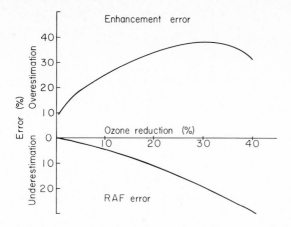

Fig. 11. Relative RAF and enhancement errors as a function of ozone reduction for field UV supplementation experiments under temperate-latitude conditions in the summer as portrayed in Fig. 10. The scenario for these errors is that the RAF values and the enhancements were calculated following the generalized plant action spectrum when the DNA damage spectrum was indeed correct.

CONCLUDING REMARKS

The potential errors resulting from assuming an incorrect action spectrum are compelling reasons for further action spectrum development. This should indeed be pursued. Yet, one should not expect that a single definitive action spectrum for higher plant damage will necessarily be forthcoming. It is unlikely when one considers the many basic modes of damage to plant tissues that UV radiation can effect (Fig. 1), the physiological and morphological diversity of plant species, and the interactions with other environmental factors, such as visible light, that can ensue.

Extrapolation from spectra of component reactions of isolated organelles to spectra of higher plant response in nature is precarious as shown for the case of photosynthetic damage (Fig. 4). Thus, use of intact plant organs and polychromatic radiation should be emphasized in further action spectrum development where these spectra bear on the ozone reduction problem. For damage to higher plant photosynthesis, the provisional action spectrum developed for Rumex patientia (Fig. 4) as well as circumstantial evidence from the natural latitudinal gradient indicate that the appropriate action spectrum is much steeper than the spectra for the Hill reaction. Thus, contrary to what one might conclude from component photosynthetic reactions, an appreciable RAF exists for damage to higher plant photosynthesis. More recent preliminary work with polychromatic radiation suggests that the spectrum may yet be even steeper and, therefore, one should view the Rumex spectrum presented here as only provisional.

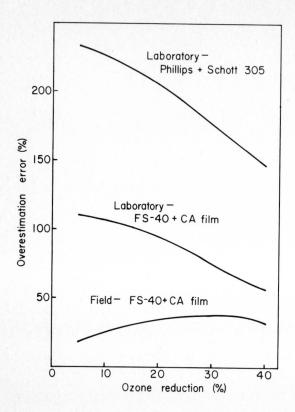

Fig. 12.   Enhancement errors calculated according to the same scenario as for Fig. 11, but with enhancement experiments using different lamp systems. In the field supplementation experiment using Westinghouse FS40 lamps and cellulose acetate film, the errors are the same as portrayed in Fig. 11. Under laboratory conditions (which would correspond to glasshouse or growth chamber conditions with no background UV radiation) the Philips lamps with Schott WG 305 filters and the Westinghouse FS40 lamps with cellulose acetate film render the errors depicted.   The lamps and filters were aged for about 100 hours for these measurements.

In our present state of ignorance concerning the most appropriate action spectra for higher plant damage, we feel it is reasonable to continue to use an action spectrum with an intermediate rate of decline with increasing wavelength, such as the generalized plant damage spectrum.   An intermediate spectrum will, of course, yield intermediate RAF and enhancement values. If the appropriate spectra are later found to be either steeper or flatter, correction from an intermediate spectrum will be easier to conduct.

ACKNOWLEDGMENT

This work has been supported by the U.S. Environmental Protection Agency. Although the research described in this chapter has been funded primarily by the U.S. Environmental Protection Agency through Cooperative Agreement CR808670 to Utah State University, it has not been subjected to the Agency's required peer and policy review and therefore does not necessarily reflect the views of the Agency and no official endorsement should be inferred.

REFERENCES

Becwar MR, Moore FD III, Burke MJ (1982) Effects of deletion and enhancement of ultraviolet-B (280-315nm) radiation on plants grown at 3000 m elevation. J Amer Soc Hort Sci 107:771-774

Berger D, Robertson DF, Davies RF, Urbach F (1975) Field measurements of biologically effective UV radiation. In: Nachtwey DS, Caldwell MM, Biggs RH (eds) Impacts of climatic change on the biosphere, part I, ultraviolet radiation effects, monograph 5. Climatic Impact Assessment Program, US Dept Transportation Report No DOT-TST-75-55, NTIS, Springfield, Virginia, p (2)235

Bickford ED, Dunn S (1972) Lighting for plant growth. Kent State Univ Press

Björn LO, Bornman JF, Olsson E (this volume) Effects of ultraviolet radiation on fluorescence induction kinetics in isolated thylakoids and intact leaves. p 185

Bogenrieder A, Klein R (1977) Die Rolle des UV-Lichtes beim sog. Auspflanzungsschock von Gewächshaussetzlingen. Angew Bot 51:99-107

Caldwell MM (1968) Solar ultraviolet radiation as an ecological factor for alpine plants. Ecol Monogr 38:243-268

Caldwell MM (1971) Solar UV irradiation and the growth and development of higher plants. In: Giese AC (ed) Photophysiology, vol 6. Academic Press, New York, p 131

Caldwell MM (1979) Plant life and ultraviolet radiation: some perspective in the history of the earth's UV climate. BioScience 29:520-525

Caldwell MM (1981) Plant response to solar ultraviolet radiation. In: Lange OL, Nobel PS, Osmond CB, Ziegler H (eds) Encyclopedia of plant physiology, vol. 12A, Physiological plant ecology. I. Responses to the physical environment. Springer, Berlin Heidelberg New York, p 169

Caldwell MM, Osmond CB, Nott DL (1977) C_4 pathway photosynthesis at low temperature in cold-tolerant Atriplex species. Plant Physiol 60:157-164

Caldwell MM, Robberecht R, Billings WD (1980) A steep latitudinal gradient of solar ultraviolet-B radiation in the arctic-alpine life zone. Ecology 61:600-611

Caldwell MM, Robberecht R, Nowak RS, Billings WD (1982) Differential photosynthetic inhibition by ultraviolet radiation in species from the arctic-alpine life zone. Arctic Alpine Res 14:195-202

Caldwell MM, Gold WG, Harris G, Ashurst CW (1983) A modulated lamp system for solar UV-B (280-320 nm) supplementation studies in the field. Photochem Photobiol 37:479-485

Green AES, Cross KR, Smith LA (1980) Improved analytic characterization of ultraviolet skylight. Photochem Photobiol 31:59-65

Hirosawa T, Miyachi S (1983) Inactivation of Hill reaction by long-wavelength ultraviolet radiation (UV-A) and its photoreactivation by visible light in the cyanobacterium, Anacystis nidulans. Arch Microbiol 135:98–102

Imbrie CW, Murphy TM (1982) UV-action spectrum (254–405 nm) for inhibition of K+-stimulated adenosine triphosphatase from the plasma membrane of Rosa damascena. Photochem Photobiol 36:537–542

Jagger J (1981) Near-UV radiation effects on microorganisms. Photochem Photobiol 34:761–768

Jones LW, Kok B (1966) Photoinhibition of chloroplast reactions. I. Kinetics and action spectra. Plant Physiol 41:1037–1043

Kelland LR, Moss SH, Davies DJG (1983) An action spectrum for ultraviolet radiation-induced membrane damage in Escherichia coli K-12. Photochem Photobiol 37:301–306

Luckiesh M, Holladay LL (1933) Fundamental units and terms for biologically-effective radiation. J Opt Soc Amer 23:197–205

Menezes S, Tyrrell RM (1982) Damage by solar radiation at defined wavelengths: involvement of inducible repair systems. Photochem Photobiol 36:313–318

Nachtwey DS, Rundel RD (1982) Ozone change: biological effects. Stratospheric ozone and Man. In: Bower FA, Ward RB (eds) CRC Press, Boca Raton, p 81

National Academy of Sciences (1979) Protection against depletion of stratospheric ozone by chlorofluorocarbons. National Academy Press, Washington, DC

Noorudeen AM, Kulandaivelu G (1982) On the possible site of inhibition of photosynthetic electron transport by ultraviolet-B (UV-B) radiation. Physiol Plant 55:161–166

Peak MJ, Peak JG (1982) Single-strand breaks induced in Bacillus subtilis DNA by ultraviolet light: action spectrum and properties. Photochem Photobiol 35:675–680

Peak MJ, Peak JG (1983) Use of action spectra for identifying molecular targets and mechanisms of action of solar ultraviolet light. Physiol Plant 58:367–372

Renger R, Voss M, Gräber P, Schulze A (this volume) Effect of UV irradiation on different partial reactions of the primary processes of photosynthesis. p 171

Robberecht R, Caldwell MM, Billings WD (1980) Leaf ultraviolet optical properties along a latitudinal gradient in the arctic-alpine life zone. Ecology 61:612–619

Robertson DF (1975) The sunburn unit for comparison of variation of erythemal effectiveness. In: Nachtwey DS, Caldwell MM, Biggs RH (eds) Impacts of climatic change on the biosphere, part I, ultraviolet radiation effects, monograph 5. Climatic Impact Assessment Program, US Dept Transportation Report No DOT-TST-75-55, NTIS, Springfield, Virginia, p (2)203

Rundel RD (1983) Action spectra and estimation of biologically effective UV radiation. Physiol Plant 58:360–366

Sestak Z (1977) Photosynthetic characteristics during ontogenesis of leaves. II. Photosystems, components of electron transport chain, and photophosphorylations. Photosynthetica 11:449–474

Setlow RB (1974) The wavelengths in sunlight effective in producing skin cancer: a theoretical analysis. Proc Natl Acad Sci USA 71:3363–3366

Sharma RC, Jagger J (1981) Ultraviolet (254–405 nm) action spectrum and kinetic studies of alanine uptake in Escherichia coli B/r. Photochem Photobiol 33:173–177

Sisson WB, Caldwell MM (1976) Photosynthesis, dark respiration, and growth of Rumex patientia L. exposed to ultraviolet irradiance (288 to 315 nanometers) simulating a reduced atmospheric ozone column. Plant Physiol 58:563–568

Sisson WB, Caldwell MM (1977) Atmospheric ozone depletion: reduction of photosynthesis and growth of a sensitive higher plant exposed to enhanced UV-B radiation. J Exp Bot 28:691–705

Teramura AH, Biggs RH, Kossuth S (1980) Effects of ultraviolet-B irradiances on soybean. II. Interaction between ultraviolet-B and photosynthetically active radiation on net photosynthesis, dark respiration, and transpiration. Plant Physiol 65:483–488

Warner CW, Caldwell MM (1983) Influence of photon flux density in the 400–700 nm waveband on inhibition of photosynthesis by UV-B (280–320 nm) irradiation in soybean leaves: separation of indirect and immediate effects. Photochem Photobiol 38:341–346

Webb RB (1977) Lethal and mutagenic effects of near-ultraviolet radiation. Photochem Photobiol Rev 2:169–261

Action Spectra for Inactivation and Mutagenesis in Chinese Hamster Cells and Their Use in Predicting the Effects of Polychromatic Radiation

F. Zölzer and J. Kiefer

Strahlenzentrum der Justus-Liebig-Universität Gießen
Federal Republic of Germany

ABSTRACT

If action spectra are to be of any use for risk estimates, they should allow prediction of the effects of polychromatic radiation. Originally we set out to investigate the wavelength dependence of inactivation and muta-tion induction in V79 Chinese hamster fibroblasts. Our results agree quite well with those reported by Rothman and Setlow (1979) for inactivation and dimer formation in the same cells. Both sets of data were then compared with other data obtained by Zelle et al. (1980) by applying two different kinds of polychromatic radiation to Chinese hamster ovary cells. To this end the action spectrum for a given effect had to be multiplied by the emission spectrum of the radiation source in question. Such calculations predicted dimer formation quite well, but considerably underestimated the lethal and mutagenic effects of polychromatic radiation. We suggest, therefore, that the different components of polychromatic radiation act independently with respect to dimer formation, but synergistically with respect to cellular effects.

INTRODUCTION

It is generally assumed that risk estimates for an enhancement of ultra-violet radiation have to be based on action spectra. These were first available for procaryotes and simple eucaryotes (for reviews see Setlow 1974; Webb 1977; Jagger 1981). More recently attention has been focused on mammalian cells (for review see Coohill and Jacobson 1981). The wavelength dependence of inactivation has been studied in V79 Chinese hamster fibro-blasts (Rothman and Setlow 1979), and that of inactivation and mutation induction (5-bromo-2'-deoxyuridine resistance) in L5178 Y mouse lymphoma cells (Jacobson et al. 1981). The latter has so far remained the only investigation of this kind concerning mutagenesis. We present here action spectra for inactivation and mutation induction (6-thioguanine resistance) in V79 Chinese hamster fibroblasts.

MATERIALS AND METHODS

Chinese hamster fibroblasts V79 A received from M. Fox, Manchester, were used for this study.

NATO ASI Series, Vol. G8
Stratospheric Ozone Reduction, Solar Ultraviolet Radiation
and Plant Life. Edited by R. C. Worrest and M. M. Caldwell
© Springer-Verlag Berlin Heidelberg 1986

Cells in an exponential phase of growth were irradiated at 254 nm using a low pressure Hg lamp, and at six different wavelengths between 263 and 313 nm using a Hg arc lamp equipped with a prism monochromator. During the treatment the medium was replaced with buffer. Afterward the cells were trypsinized and replated at appropriate densities to test survival and to allow for expression of mutants. Seven days later resistance to the selective agent 6-thioguanine resistance was used to determine mutagenicity, following the usual selection/viability protocol. Each experiment was carried out at least two times.

RESULTS AND DISCUSSION

The survival curves, although usually not extended to very high doses, are clearly shouldered. The dose, F(0.37), that leaves a surviving fraction of 0.37 was chosen to represent the lethal action of a particular wavelength. The mutation induction curves could be fitted reasonably well to quadratic expressions of the form, mutations per survivor = a F^2 + b F, where F is the dose. The slope of the linear component, b, was chosen to represent the mutagenic action of the radiation. The results are summarized in Tables 1 and 2 (263 nm was used as reference wavelength; the data were accordingly quantum corrected).

Table 1. Inactivation at different wavelengths. For further explanation see text.

Wavelength (nm)	Fluence rate (W m^{-2})	F(0.37) (J m^{-2})	$\dfrac{F(0.37, 263\ nm)}{F(0.37, \lambda)}$	$\dfrac{263\ nm}{\lambda\ nm}$
254	0.41	9.2 ± 0.2	0.75 ± 0.02	
263	0.27	6.7 ± 0.2	1.00 ± 0.03	
273	0.12	6.0 ± 1.0	1.08 ± 0.16	
283	0.13	7.4 ± 0.3	0.84 ± 0.03	
293	0.52	25 ± 2	2.41 ± 0.08 (x 10^{-1})	
303	0.45	280 ± 10	2.08 ± 0.08 (x 10^{-2})	
313	1.66	6500 ± 700	0.87 ± 0.09 (x 10^{-3})	

The action spectra presented here agree quite well with those reported by others for inactivation of V79 Chinese hamster fibroblasts (Rothman and Setlow 1979) and inactivation and mutation induction in L5178Y mouse lymphoma cells (Jacobson et al. 1981). They also essentially resemble the action spectrum for the inactivation of Escherichia coli and phages (Setlow 1974), although they show a broader shoulder around 280 nm and a smaller decline between 290 and 320 nm. It appears that there may be general differences between eucaryotes and procaryotes in this respect.

Table 2. Mutation induction at different wavelengths. For further explanation see text.

Wavelength (nm)	Fluence rate (W m^{-2})	a (J m^{-2})$^{-2}$	$\dfrac{a^{0.5} (\lambda)}{a^{0.5} (263 \text{ nm})} \dfrac{263 \text{ nm}}{\lambda \text{ nm}}$	b (J m^{-2})$^{-1}$	$\dfrac{b(\lambda)}{b(263 \text{ nm})} \dfrac{263 \text{ nm}}{\lambda \text{ nm}}$
254	0.41	2.06 ± 0.07 x 10^{-6}	1.27 ± 0.02	1.17 ± 0.09 x 10^{-5}	1.06 ± 0.08
263	0.27	1.38 ± 0.25 x 10^{-6}	1.00 ± 0.09	1.14 ± 0.05 x 10^{-5}	1.00 ± 0.04
273	0.12	1.97 ± 0.54 x 10^{-6}	1.15 ± 0.17	1.15 ± 0.24 x 10^{-5}	0.97 ± 0.20
283	0.13	1.22 ± 0.03 x 10^{-6}	0.87 ± 0.01	0.97 ± 0.10 x 10^{-5}	0.79 ± 0.08
293	0.52	2.17 ± 0.28 x 10^{-8}	1.13 ± 0.08 x 10^{-1}	2.08 ± 0.31 x 10^{-6}	1.82 ± 0.27 x 10^{-1}
303	0.45	0.50 ± 0.27 x 10^{-9}	1.66 ± 0.33 x 10^{-2}	2.43 ± 0.08 x 10^{-7}	1.85 ± 0.06 x 10^{-2}
313	1.66	0.58 ± 0.03 x 10^{-11}	1.73 ± 0.05 x 10^{-8}	1.01 ± 0.28 x 10^{-8}	0.74 ± 0.21 x 10^{-3}

The action spectra for inactivation and mutagenesis in our hamster cells were not significantly different. This is also evident if one compares mutation rates as a function of surviving fractions. The ratio of mutagenic to lethal action, b F(0.37), is more or less constant over the wavelength range investigated.

If action spectra are to be of any use for risk estimates, they should allow prediction of the effects of polychromatic radiation. To this end, the action spectrum for a given effect has to be multiplied by the emission spectrum of the radiation source in question and integrated over all wavelengths. Such calculations were performed on the basis of our data and those of Rothman and Setlow (1979) for V79 Chinese hamster fibroblasts. We assumed two different spectra of polychromatic radiation, which had actually been applied by Zelle et al. (1980) to Chinese hamster ovary cells, in order to make comparisons between different sets of data possible. The radiation sources used in that study had been a Westinghouse FS20 sunlamp with either a tissue culture dish or a polyethylene terephthalate film as filter.

In the case of the first spectrum, designated '>290 nm', all contributions to the predicted overall effect came from wavelengths less than 313 nm. In the case of the second spectrum, designated '>310 nm', the action spectra were extrapolated to 320 nm in an exponential fashion, which seems justified since deviations from such a course are usually observed only at wavelengths greater than 320 nm (Setlow 1974; Peak and Peak 1982). About half of the predicted overall effect, nevertheless, is accounted for by radiation with wavelengths less than 313 nm.

Results obtained by different groups either experimentally with monochromatic 254-nm radiation or theoretically and experimentally with polychromatic radiation '>290 nm' and '>310 nm' are compared in Table 3. There are only minor differences between corresponding data for monochromatic radiation. However, as they may reflect different sensitivities of the cells used, they were taken into account for the following considerations.

Our action spectra predict the lethal effects of polychromatic radiation fairly well, but they underestimate the mutagenic action by a factor of 2 to 3. The calculated ratio of mutation induction to inactivation is too small by a factor of 1.5 to 2.5. The action spectra of Rothman and Setlow underestimate the lethal effects of polychromatic radiation by a factor of 2 to 5, but they predict dimer formation very well. The calculated ratio of inactivation to dimer formation is too small by the factor of 2 to 6. Whereas a ratio of mutation induction to dimer formation cannot be calculated on the basis of either our data or those of Rothman and Setlow alone, the product of the ratio of mutation induction to inactivation and the ratio of inactivation to dimer formation is given instead. It is too small by a factor of 5 to 9.

It could be argued that our calculations do not take into account possible contributions to inactivation and mutation induction by radiation with wavelengths greater than 320 nm. Lesions other than pyrimidine dimers are thought to play a role in that spectral region (for reviews see Webb 1977; Coohill and Jacobson 1981; Jacobson and Krell 1982) and to be responsible for increased ratios of inactivation and mutation induction as compared to dimer formation (Zelle et al. 1980).

Table 3. Comparison of predicted (Rothman and Setlow 1979; this work) and observed (Zelle et al. 1980) effects of polychromatic radiation.

	254 nm ($5\ J\ m^{-2}$)			>290 nm ($350\ J\ m^{-2}$)			>310 nm ($23000\ J\ m^{-2}$)		
	a	b	c	a	b	c	a	b	c
-ln (surviving fraction)	0.91	0.55	0.50	0.54	0.43	0.50	0.18	0.39	0.50
mutations per survivor $\times 10^5$		11.0	16.5		3.3	14.7		7.1	21.9
% dimers $\times 10^3$	21.2		18.9	11.9		9.5	5.1		3.9[d]
-ln (surviving fraction)	0.043		0.026	0.045		0.053	0.035		0.128
% dimers $\times 10^3$		20.0	33.0		7.7	29.4		18.2	43.8
mutations per survivor $\times 10^5$		0.86[e]	0.87		0.35[e]	1.55		0.64[e]	5.62

[a] From Rothman and Setlow 1979. [b] From this work. [c] From Zelle at al. 1980. [d] Value derived from number of endonuclease sensitive sites. [e] Product of -ln SF/% dimers $\times 10^3$ and (M/S $\times 10^5$)/-ln SF.

Such an explanation also cannot be excluded in the present case. Further studies on the wavelength dependence of cellular effects are clearly needed. It should be noted, however, that the relative action at wavelengths greater than 320 nm is usually more than one order of magnitude smaller than 313 nm (Setlow 1974; Peak and Peak 1982). Thus, it seems improbable that the contribution by that radiation could account for the fact that ratios of cellular effects to dimer formation are larger by a factor of about 5 as expected from our calculation.

In conclusion we suggest that the different components of polychromatic radiation act independently with respect to dimer formation but synergistically with respect to cellular effects. This seems to be less pronounced for inactivation than for mutation induction.

ACKNOWLEDGMENT

This study was supported by the Bundesministerium für Forschung und Technologie of the Federal Republic of Germany through Gesellschaft für Strahlen- und Umweltforschung (KBF 49).

REFERENCES

Coohill TP, Jacobson ED (1981) Action spectra in mammalian cells exposed to ultraviolet radiation. Photochem Photobiol 33:941–945

Jacobson ED, Krell K (1982) Genetic effects of fluorescent lamp radiation on eucaryotic cells in culture. Photochem Photobiol 35:875–879

Jacobson ED, Krell K, Dempsey MJ (1981) The wavelength dependence of ultraviolet light-induced cell killing and mutagenesis in L5178Y mouse lymphoma cells. Photochem Photobiol 33:257–260

Jagger J (1981) Near-UV radiation effects on microorganisms. Photochem Photobiol 34:761–768

Peak MJ, Peak JG (1982) Single-strand breaks induced in Bacillus subtilis DNA by ultraviolet light: Action spectrum and properties. Photochem Photobiol 35:675–680

Rothmann RH, Setlow RB (1979) An action spectrum for cell killing and pyrimidine dimer formation in Chinese hamster V-79 cells. Photochem Photobiol 29:57–61

Setlow RB (1974) The wavelengths in sunlight effective in producing skin cancer: a theoretical analysis. Proc Natl Acad Sci USA 71:3363–3366

Webb RB (1977) Lethal and mutagenic effects of near-ultraviolet radiation. In: Smith KC (ed) Photochemical and photobiological reviews, vol 2. Plenum, New York, p 169

Zelle B, Reynolds RJ, Kottenhagen MJ, Schuite A, Lohmann RHM (1980) The influence of the wavelength of ultraviolet radiation on survival, mutation induction and DNA repair in irradiated Chinese hamster cells. Mutat Res 72:491–509

Dose and Dose-Rate Responses to UV-B Radiation: Implications for Reciprocity

F. R. de Gruijl, H. J. C. M. Sterenborg, H. Slaper and J. C. van der Leun

State University of Utrecht, The Netherlands

ABSTRACT

Solar UV radiation interacts with many organisms and systems within the biosphere. The pathway leading from the entrance of photons into the system to the system's ultimate response is different for various systems and responses. Nevertheless, there are clear-cut parallels between various types of UV responses. This analysis is focused on the effects of various response mechanisms on the quantitative relationship among dose, dose-rate and response, which is the relationship playing a crucial role in risk assessments. In considering these effects, the processes involved are classified into 2 x 2 main categories: deterministic versus stochastic, and reciprocal versus nonreciprocal. The important, but at times vague, distinction between deterministic and stochastic processes is developed in some detail. In connection with this, the often misused concept of a memory effect of an exposure is also considered.

The law of reciprocity (i.e., that the response depends only on the product of dose-rate and exposure time) is important in photobiology. It is what one would expect on the basis of a photochemical reaction. Reciprocity is often found to hold true, to a good approximation, in photobiological experiments. Many responses fail, however, to obey this "law"; this applies especially to long-term responses. Various nonreciprocal processes are treated together with their effects. In connection with reciprocity, the importance of an appropriate definition of "dose" is emphasized.

INTRODUCTION

In this chapter an attempt is made to formulate a general analysis of the response of a biological system to UV radiation, be it plant, animal, human or an ecosystem. As is true for most general analyses, our analysis will inevitably be incomplete, oversimplified and heavily influenced by the field of research of the authors, but it may serve to guide a structured analysis of UV responses and to recognize parallels between various UV responses. At the very least it may evoke criticism, which may yield essential extensions.

Special attention will be given to dose and dose-rate dependencies of responses (reciprocity or nonreciprocity). In Fig. 1 the chain of processes leading from UV irradiance to response is schematically presented.

NATO ASI Series, Vol. G8
Stratospheric Ozone Reduction, Solar Ultraviolet Radiation
and Plant Life. Edited by R. C. Worrest and M. M. Caldwell
© Springer-Verlag Berlin Heidelberg 1986

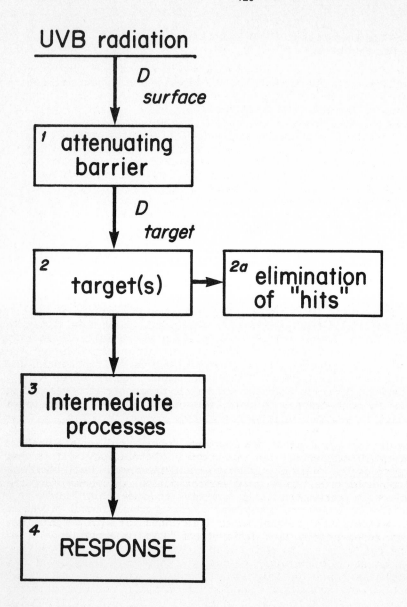

Fig. 1. Schematic chain of processes leading from UV-B irradiance to the ultimate response. Each box may introduce nonreciprocity. D = dose.

RECIPROCITY (Fig. 1, Box 2)

In general, an agent exerts an effect on a biological system by interacting with certain targets (molecules) within the system, and the occurrence of the effect depends on the likelihood of the interaction. For an irreversible interaction the probability of the occurrence of the effect will depend on the concentration of the agent, C, and the exposure time, T. If this primary chemical reaction is a first order reaction, the number of interactions is a function of CxT. The assumption that the magnitude of the ultimate response is a function of the number of interactions (i.e., targets hit) leads us to Harber's rule in toxicology. This rule states that a certain level of response is fully determined by CxT. In photobiology one encounters the equivalent of Harber's rule, viz., the law of reciprocity, which states that a certain level of response is determined by:

$$D' \times T = D \tag{1}$$

where D' stands for dose-rate and D for dose.

In this analysis we limit ourselves to UV responses caused by photochemical reactions, thus, excluding those by thermal effects due to the absorption of radiant energy.

STOCHASTIC OR DETERMINISTIC (Fig. 1, Box 2)

If a single and relatively rare interaction between the agent and a target induces the response, then the response will be subject to chance and the response is said to be of a stochastic nature. If a great many interactions (with a high rate of occurrence) are required, the response will be much less subject to chance. As the number of required interactions increases, the response will take on a deterministic appearance. Compare this with deterministic formulae in chemistry and thermodynamics where the underlying molecular processes are of a stochastic nature.

Examples of deterministic responses to irradiation are photosynthesis, erythema, pigment-darkening and hyperplasia. Examples of stochastic responses are cell killing, mutation and carcinogenesis. Note that the latter responses are a matter of a yes-or-no reaction of a cell; whereas the former responses are gradual and involve the whole tissue. If we consider cell killing in a large population of cells (e.g., tissue) the surviving fraction as a function of dose becomes more and more predictable as the population size increases, i.e., the measured response takes on a deterministic appearance. Thus, the boundaries between stochastic and deterministic responses are not very sharp, and it may be a question of the required number of targets that must be hit.

When measuring responses to UV radiation it is important to be aware of the stochastic or deterministic nature of the response under consideration. It is often said that the dependence of a response on the accumulated dose implies that some kind of cumulative effect (i.e., a memory) should be involved. This may be considered true for a deterministic response--the memory being the number of targets hit. But it is not necessarily true for a stochastic response such as UV-induced carcinogenesis, for which the argument has been used. For a stochastic response it simply becomes less and less likely that an event has not occurred as the dose increases, just as it becomes less and less likely that we will not have thrown a "six" as we increase the number of throws with a die. Whatever has happened before does not affect the next throw. In this sense stochastic processes have no memory.

NONRECIPROCITY

If reciprocity holds, the response is fully determined by D and independent of D' or T. Within certain limits of D' and T reciprocity holds for many responses, for example, for erythema evoked by an irradiation of duration shorter than one hour. Recently, Anderson and Parrish (1980) presented preliminary results which show reciprocity for UV-C-induced erythema over 10 orders of magnitude in D' (with $T \geq 30$ ns and peak value D' = 10^7 W cm^{-2}). Earlier, Claesson et al. (1958) showed that reciprocity holds for capillary leakage (made visible with Evans blue) for a 10^7-fold variation in D' for UV-B radiation.

In spite of these impressive figures on reciprocity, one may not presume that reciprocity is universally valid. If D' becomes very high, deviations from reciprocity may show up, which may be due to interference of thermal effects or two-photon reactions. In the latter case a target molecule simultaneously absorbs two photons and is excited with the sum of the photon energy. The likelihood of such an event is not directly proportional to D' but to D'^{2}. Here, we will limit ourselves to responses caused by one-photon reactions and unaffected by radiation induced thermal effects.

Reciprocity failure is often observed for responses to exposures of long duration as, for instance, in case of tumorigenesis by chronic UV exposure (De Gruijl et al. 1983). Over a sufficiently long period of time the biological system may undergo certain changes which interfere with the reciprocity between D' and T. The sensitivity may change and the induced effects leading to the response may be (partially) reversed.

PROPER DOSE AND DOSE-RATE (Fig. 1, Box 1)

A biologically effective dose and dose-rate with UV-B radiation are in no way uniquely definable. In fact, by simply employing an arbitrary concept of effective UV-B dose, one may be disregarding very fundamental issues. Preferably, one should know where in the system and according to what action spectrum the effective dose should be measured. This knowledge may be crucial for a proper description of the process leading from irradiation to response.

The irradiation may induce changes in the system which in turn may affect the penetration of the radiation to the system's sensitive level (location of the targets). For example, in a marine ecosystem enhanced solar UV radiation could possibly cause shifts of populations of UV sensitive microorganisms to greater depths (most simply by a predominant kill off at lesser depths in the absence of appreciable vertical mixing), thus altering the sensitivity of the system to solar UV radiation entering at the water surface. Another example is the UV-induced epidermal hyperplasia which decreases the penetration of UV radiation to the deeper layers of the epidermis, thus protecting the proliferative basal layer against the tumorigenicity of UV radiation entering the skin surface. A slower administration (low dose-rate) makes the irradiation more efficient (smaller D required). In the case of tumorigenesis by chronic UV exposure, it appears likely that a proper definition of dose at the basal layer of the epidermis may restore reciprocity (De Gruijl and Van der Leun 1982). As the epidermis thickens in a process of chronic UV exposure we have to write:

$$D_b = \int_0^T D'_b(t) \, dt \tag{2}$$

where D_b and D'_b denote D and D' at the base of the epidermis, $D'_b(t)$ represents D'_b as a function of time, t. Note that Eq. (2) is a generalization of reciprocity as presented in Eq. (1), where D' is constant.

In summation, a process which seemingly does not obey the law of reciprocity between T and D', may prove to do so, after all, if D is measured properly. Admittedly, the foregoing argument shifts the problem to a proper measurement of effective dose, which may not be easy or even may be impossible. Nevertheless, if this is the real problem, one should be aware of it.

It should also be mentioned here that it may be crucial to distinguish a measurement of dose as (directed) surface exposure (defined as the total radiant energy per area entering a directed surface) from the quantity "fluence" (defined as the total radiant energy entering an infinitesimal sphere divided by its cross-section [NBS 1976]). In case of diffuse irradiation, the measurement of surface exposure will yield a lower value than that of fluence because of the cosine-dependence in surface exposure. In general, fluence is the proper quantity to describe photochemical reactions; whereas dose (surface exposure) is the more proper quantity to describe photobiological reactions in more-or-less flat objects, such as the human skin. Physical instruments usually have cosine-dependence and are, therefore, suitable to measure surface exposure rather than fluence.

REVERSIBILITY (Fig. 1, Box 2a)

It may well be that an affected target can be converted to its original (pre-hit) state or be replaced by a new unaffected target. Conversion to the original state may be due to enzymatic repair processes: for instance, excision repair of UV-induced dimers in DNA (which can also be considered a replacement of a target). Replacement of targets may be due to cell renewal because of cell turnover in a tissue. The deletion of affected targets (partially) reverses the radiation effects. In general, a characteristic time, τ, can be assigned to this reversal. For $T \ll \tau$ reciprocity holds; whereas for $T \gtrsim \tau$ reciprocity does not hold. Such a reversal will counteract the response. Thus, a larger dose may be required if the exposure is protracted or fractionated than if it is delivered all at once.

To illustrate this we could assume that for dose-fraction d, the number of targets hit is n and that at a time t after exposure $[n \exp(-t/\tau)]$ of these targets have remained affected. If the dose-fractions are separated by a time interval t_s, we find that the number of affected targets after the i-th dose-fraction equals:

$$n \left\{ 1 + \sum_{j=1}^{i-1} f_j \right\} = n \, \frac{1-f^i}{1-f} < n \, \frac{1}{1-f} \qquad (3)$$

where $f = \exp(t_s/\tau) < 1$. If $f = 0.8$ and $3n$ targets need to be affected for the response, then i should at least equal 4. If $f < 2/3$ no response could ever be induced with this mode of exposure, whereas a single dose of $3d$ would induce the response. In this example reciprocity clearly does not hold. Lowering the average dose-rate by increasing t_s decreases the effectiveness of the irradiation. However, we do find reciprocity of dose-rate and exposure time ($T \ll \tau$) within each dose-fraction.

In photobiology there have been reports on UV-A and light enhanced repair, i.e., for higher doses or dose-rates of UV-A radiation and/or light, the repair becomes more effective (Sutherland 1977). Clearly, repair mechanisms can cause significant deviations from reciprocity over a sufficiently large scale.

Cell killing may cause the radiation to become less effective in inducing a response if doses are raised above a certain dose D_m. The killing of cells may counteract the induction of the response and, for higher doses of UV radiation, cell killing may be dominant. If cell killing is a non-reciprocal process (because of repair or replacement of cells), the irradiation may become less effective in inducing the response for higher D' (or dose per dose-fraction).

MULTIPLE HITS (Fig. 1, Box 2)

If n targets in a cell need to be hit by UV radiation for the induction of a response, then the number of cells that responded (a small fraction of a larger population) would be proportional to $(D'T)^n$ and the fraction of plants or animals in which at least one cell responded (i.e., the prevalence, P_n, equals $1 - \exp(-k(D'T)^n)$, where k is a constant. Here reciprocity holds because $D'T = D$ fully and uniquely determines the prevalence. However, if m additional hits, not related to UV radiation, are needed for the induction of the response, then reciprocity does not hold. Assuming that the number of cells with the required m additional hits is directly proportional to T_m, we get:

$$P_{mn}(D',T) = 1 - \exp(-k \, D'^n \, T^{n+m}) \qquad (4)$$

where P_{mn} is the prevalence of plants/animals in which at least one cell responded after n UV-radiation-related hits and m UV-radiation-unrelated hits. Here, P_{mn} is not fully and uniquely determined by $D'T = D$, but by $D'T^p$ where $p = (n+m)/n > 1$. In words, the latter case means that the irradiation becomes more effective for lower D' (i.e., D becomes smaller for reaching a certain value of P_{mn}). For further literature see Emmelot and Scherer (1977).

Note that as n and m become large (say >100), P_{mn} becomes a step function of D' and T. The time at which an animal/plant responds becomes fully determined by D' without any appreciable statistical variance. For such a deterministic response, variance may yet be introduced by biological variation between the animals/plants (for instance, by varying epithelial thickness).

In the literature on carcinogenesis it is often (implicitly) concluded that measuring the relation given in Eq. (4) means that one is dealing with a multiple hit process. This conclusion is not necessarily correct as was pointed out by Pike (1966). A host of stochastic processes can asymptotically approach Eq. (4) (see also De Gruijl 1982).

FROM HIT TO RESPONSE (Fig. 1, Box 3)

This step can also introduce nonreciprocity due to delay or nonlinear processes. One can easily correct for a constant delay, t_d, in the response after the required targets have been hit during a process of chronic exposure. Reciprocity will not exist between T and D', but between $(T-t_d)$ and D'. If t_d is not constant, but subject to statistical variation, one needs to know the distribution of t_d in order to re-establish reciprocity.

We could consider N', hit-rate of targets, and N, number of targets hit, as dose-rate and dose for possible intermediate processes (Fig. 1). N in turn can be interpreted as the concentration of affected targets, which may initiate a toxic reaction (possibly according to Harber's rule). However, Harber's rule is not universally valid. Hence, intermediate processes may introduce nonreciprocity as well as the preceding processes. This may be due to nonlinear physical and/or chemical processes, such as adsorption and enzyme controlled processes, which could introduce a nonlinear relation between N and C, the latter being the concentration of a hypothetical agent, which causes the ultimate response according to Harber's rule.

CONCLUSION

Although the law of reciprocity is a useful concept in photobiology, there are many processes which may introduce nonreciprocity. Locating the non-reciprocal processes, and determining the mechanisms involved, will contribute to a better understanding of the UV response and may serve to formulate a quantitative model of the response. The availability of good quantitative models is essential for a reliable assessment of environmental impacts of enhanced solar UV-B radiation.

REFERENCES

Anderson RR, Parrish JA (1980) A survey of acute effects of UV lasers on human and animal skin. In: Pratesi R, Sacchi CA (eds) Lasers in photomedicine and photobiology. Springer, Berlin Heidelberg New York, p 109

Claesson S, Juhlin L, Wettermark G (1958) The reciprocity law of UV-irradiation effects. Acta Dermato-Venereol 38:123-136

De Gruijl FR (1982) PhD Thesis. Chapter 1, State University Utrecht

De Gruijl FR, Van der Leun JC (1982) Effect of chronic UV exposure on epidermal transmission in mice. Photochem Photobiol 36:433-438

De Gruijl FR, Van der Meer JB, Van der Leun JC (1983) Dose-time dependency of tumor formation by chronic UV exposure. Photochem Photobiol 37:53-62

Emmelot P, Scherer E (1977) Multi-hit kinetics of tumor formation with special reference to experimental liver and human lung carcinogenesis and some general conclusions. Cancer Res 37:1702-1708

NBS (1976) National Bureau of Standards Tech Note 910-1:58

Pike MC (1966) A method of analysis of a certain class of experiments in carcinogenesis. Biometrics 22:142-161

Sutherland BM (1977) Symposium on molecular mechanisms in photoreactivation. Photochem Photobiol 25:413-414

Cellular Repair and Assessment of UV-B Radiation Damage

C. S. Rupert

University of Texas at Dallas, Richardson, Texas USA

ABSTRACT

Repair processes act on the cellular damage produced by UV-B radiation to modify its final effects by a large factor. Consequently any alteration of conditions which changes the extent or efficacy of repairs affects UV-B action on organisms, and the relation of controlled experiments to responses expected in the field. Such repair modulations--well recognized in bacteria--can occur by induction of further repair capacities in the cell, by interactions between different recovery mechanisms, or by changed timing of repairs relative to cellular pathology. They are known, but less well described, in animal cells, and up to now have been little investigated in plant systems. Even without full understanding of these processes, some account can be taken of them in evaluating UV-B effects.

INTRODUCTION

The preceding authors have shown us how changes in atmospheric parameters can change the intensity and wavelength distribution of solar ultraviolet (UV) radiation at the surface of the earth. We now consider what such changes might mean to biological systems. As might be expected, this transition from physics and chemistry to biology introduces considerably larger uncertainties than those involved in predicting atmospheric and radiation changes.

Ultraviolet radiation produces effects in living organisms by inducing photochemical changes in important biomolecules. The primary action is, however, modified by homeostatic mechanisms which tend to diminish the resultant injury, and both aspects weigh heavily in the final consequences. The photochemical process shapes the way in which the amount of primary damage depends on the radiation parameters: the radiant energy, and the wavelength and spatial distributions in the sensitive parts of the organism. The recovery processes determine how such damage finally impinges on metabolic activities: the extent to which the damage is repaired, or compensated for to diminish its consequences. The latter processes can be influenced by various environmental factors, including radiation. Consequently the daylight illumination furnishing the UV radiation can play a dual role in the resulting phenomena: first it causes the primary damage, and then it may influence the way damage is handled. The need to deal with all these factors sets the framework for assessing UV-B radiation injury.

NATO ASI Series, Vol. G 8
Stratospheric Ozone Reduction, Solar Ultraviolet Radiation
and Plant Life. Edited by R. C. Worrest and M. M. Caldwell
© Springer-Verlag Berlin Heidelberg 1986

DOSIMETRY OF DAMAGING RADIATION

The probability that a molecular structure will be raised to an excited state in which it can undergo a photochemical change is directly proportional to both its ability to absorb the radiation and to the radiation fluence in its immediate neighborhood. The fluence F from a single, monochromatic, unidirectional beam is the energy transmitted per unit area normal to the direction of propagation. In a convergent, cross-fire illumination the contributions from all component beam directions simply add directly, making the fluence equal to the energy entering a small sphere around the region of interest, divided by the sphere's cross-sectional area. (Note that this is not the same thing as the energy per unit area on any particular surface.)

Some fraction Φ of the excited molecules (the so-called quantum yield) undergoes photo-reaction to yield an altered structure. If σ is the molecular absorption cross section, λ the wavelength, h Planck's constant and c the velocity of light, it is easy to establish that the number n of molecules reacting in a total population N is given by (see, for example, Rupert 1974)

$$n/N = (\sigma\Phi)\ \lambda F/hc. \tag{1}$$

Evidently one such relation will hold for every kind of photochemical injury[1], and the magnitude of the effect resulting from that injury will be some function of n.

The factor in parentheses depends on properties of the system being irradiated, while the remainder of the expression describes the radiation treatment applied. However, for an intact biological system the fluence referred to is that inside the organism, at the site of the photochemical reaction: $F = F_{int}$. In all but very small, or special structures the radiation has been repeatedly reflected, scattered and filtered before reaching this site, making the fluence considerably different from that at the surface. Although the relation between internal fluence and external radiation depends on many small details, F_{int} at any given point within a strong scattering structure is approximately proportional to the energy per unit area falling on the outer surface--a quantity called the dose D (ICP 1954; CIE 1970). Thus

$$F_{int} \cong KD \tag{2}$$

where K (dimensionless, and usually less than unity) varies greatly with both the particular structure and with the exact internal location. Such a relation is reasonably presumed in UV dosimetry of larger organisms.

While F and D both have the dimensions [energy]/[area], they are rather different quantities. D, the energy per unit area on an actual surface,

---

1 The expression would be simpler if written in photon terms, but energy terminology is so widespread in the literature that it seems appropriate here.

changes as the cosine of the angle of incidence, if the orientation of that surface toward a radiation beam changes. The fluence at a point on the same external surface does not. However, F_{int}--the somewhat lower fluence at a particular spot inside the scattering structure--mirrors any changes in D. Because radiometric devices are usually designed to measure the dose rate dD/dt on some surface related to the instrument, this quantity can often be measured directly for the organism. The same is not generally true of the fluence rate, which specifies the photochemical effectiveness of the radiation field at a point. Only in the case of a roughly unidirectional beam (where the fluence rate is numerically equal to the dose rate on a surface normal to the beam direction) is this quantity equal to an instrumental reading, even for an external fluence. In other cases it must be deduced indirectly.

Eqs. (1) and (2) apply to monochromatic light, but daylight is polychromatic, and σ, Φ and K all vary with wavelength. Consequently the effect of each small wavelength interval must be treated separately, and the total summed. For any one kind of reaction at a particular site in an organism, when we know the spectral distribution of dose $(dD/d\lambda)$ as a function of wavelength, the total amount of primary damage is indicated by

$$n/N = 1/hc \quad K(\lambda)\sigma(\lambda)\Phi(\lambda)\lambda(dD/d\lambda)d\lambda, \tag{3}$$

(the integration to be carried out over all wavelengths which make a significant contribution).

For most biological systems, the quantities in the integrand of this expression are not all known. The usual procedure is therefore to relate them to something experimentally measurable by choosing one particular wavelength λ_0, and rewriting the dose-independent part of the integrand as

$$K(\lambda)\sigma(\lambda)\Phi(\lambda)\lambda = K_0\sigma_0\Phi_0\lambda_0E(\lambda,\lambda_0).$$

K_0, σ_0 and Φ_0 are the values of the corresponding parameters at $\lambda = \lambda_0$; i.e., $E(\lambda,\lambda_0) = 1$ when $\lambda = \lambda_0$. In these terms Eq. (3) then takes the form of Eq. (1), with Eq. (2) substituted into it, i.e.,

$$n/N = (K_0\sigma_0\Phi_0)\lambda_0 \langle D\rangle/hc \tag{4}$$
$$= S_0\langle D\rangle$$

where S_0 measures the sensitivity of the system, and

$$\langle D\rangle = \quad E(\lambda,\lambda_0)(dD/d\lambda)d\lambda \tag{5}$$

serves as an "effective dose".

The function $E(\lambda,\lambda_0)$, known as the action spectrum, is the effectiveness of energy at any wavelength λ compared to that at λ_0.[2] Consequently its numerical value at any wavelength depends on the choice of λ_0. It can be determined from the relative doses required to produce the same magnitude effect, using different, known wavelengths, or wavelength distributions of sufficient variety, when the operation of biological recovery mechanisms can be presumed constant. Since that topic is being discussed in a separate paper in this volume by Caldwell et al., we need not deal with it here.

$\langle \underline{D} \rangle$ is numerically equal to the monochromatic dose which, given at $\lambda = \lambda_O$, would produce the same amount of injury as the polychromatic irradiation it represents. Its value depends on all the features of the photochemical system and location in the organism which affect $\underline{E}$. Therefore the same radiation treatment will correspond to different $\langle \underline{D} \rangle$ when applied to photochemical systems having different action spectra. Furthermore, as noted previously, the value of $\langle \underline{D} \rangle$ changes if the arbitrarily chosen λ_O is changed. ($\underline{S}_O$ also changes in a compensating manner, so that $\underline{n}$ remains independent of the choice of λ_O). These features are rather inconvenient for a quantity used to characterize a radiation treatment.

The foregoing discussion applies to almost any photobiological process. However, we are concerned there with the effects of solar UV on terrestrial ecosystems--particularly in the context of possible depletion of the stratospheric ozone shield. Such depletion would change the solar radiation in only a restricted wavelength range where the absorption of sunlight by stratospheric ozone is partial: that is, between about 290 and 320 nm, in the UV-B spectral region. Since species in terrestrial ecosystems are already adapted to normal daylight, photochemical damage which is caused primarily by radiation at wavelengths outside this spectral region will not change from the amounts currently being tolerated if ozone concentration changes. Only damaging reactions for which $\underline{E} \cdot d\underline{D}/d\lambda$ differs from zero almost exclusively within the UV-B region are relevant.

This focus considerably simplifies the environmental dosimetry problem. As we noted earlier, the dose is affected by the orientation of outer surface with respect to the radiation, and the surfaces of plants and animals are variously oriented. However, since a large part of natural UV-B radiation is skylight, rather than direct sunlight, and comes from many directions, the natural UV-B doses on skin or leaf surfaces do not depend very strongly on the angle these surfaces make with the vertical (Burt and Luther 1979; Caldwell 1981). Consequently, the dose on the horizontal plane provides a useful "engineering estimate" for the average effective dose, even though few surfaces of organisms are actually horizontal. In this limited context, the spectral irradiance (i.e., the energy per unit area, unit time, and unit wavelength interval) on a horizontal plane, specified as a function of wavelength, provides a practical statement of the radiation environment. The uncertainties introduced by such an assumption are smaller than those arising from other causes.[3]

---

2 Many action spectra are reported as $\underline{A}(\lambda,\lambda_O)$, the relative effectiveness of photons at wavelength λ compared with those at λ_O, rather than the relative effectiveness of energy at the two wavelengths. The two forms are related by $\underline{E}(\lambda,\lambda_O) = (\lambda/\lambda_O)\underline{A}(\lambda,\lambda_O)$.

3 On the other hand, it should not be assumed that external surface orientation is never important. As Urbach (1969) has pointed out, skin cancer in humans occurs predominantly on areas whose orientation toward the vertical gives them larger solar UV doses.

THE PHOTOCHEMICAL TARGET

A number of biochemical components of cells are altered by UV exposure. Just which of these lead to significant physiological consequences under the internal fluences (or external doses) commonly encountered in daylight must be determined from action spectra and other analyses of effects. This point is not fully settled, but we cannot go much further with abstract discussion detached from any concrete example. Fortunately an attractive candidate for the target material is available.

There is no question that irradiation with a 250-300 nm UV fluence sufficient to kill most of a population of bacteria or cultured animal cells acts in most cases through its damage to deoxyribonucleic acid (DNA), the repository of cellular genetic information. This view is strongly supported by not only the action spectra, which closely resemble nucleic acid absorption spectra (Rosenstein and Setlow 1982), but also by the physiological effects of the radiation, considered along with studies of the irradiated DNA itself. Damage to DNA inhibits its replication (Setlow et al. 1963) and transcription (Michalke and Bremer 1969), and therefore interferes with cell growth and division as well as with other activities. (Actually several kinds of photochemical damage can be recognized in DNA--see Rahn 1979--but for many purposes any kind of discrete abnormality in the structure has equivalent effects.) Since the DNA absorption cross section declines precipitously with increasing wavelength (by a factor of about 0.7 for each nanometer) over the same spectral region where sunlight intensity is increasing rapidly, the product $E \cdot dD/d\lambda$ peaks in only a narrow region within the UV-B waveband. The amount of photochemical damage in daylight would therefore be affected by ozone concentration. Although we cannot exclude other kinds of cell damage, and are not primarily concerned with effects in which cells are killed (the organisms in which we are interested thrive in daylight), assuming DNA is the UV target provides a useful model for inquiring, "What would be the effect of a modest increase in the average intensity of solar UV-B radiation -- all other radiations and environmental parameters being held constant?" The answer then involves the unique metabolic processes associated with DNA.

DNA REPAIR

We should expect corrective responses in cells following damage to any of their components, simply because organisms tend to maintain themselves in the face of environmental perturbations. The case of DNA is, however, special. Elaborate provisions exist for rectifying errors and assuring accurate transmittal of its encoded information, and among these are very effective mechanisms for repair of chemical and radiation damage (Hanawalt et al. 1979). Such processes reduce the number n of DNA photoproducts from that described by Eq. (4) to some lower number, and the effects of the UV radiation depend only on the residual damage.

A mutant bacterium lacking essentially all normal DNA repair capacities (for example, an Escherichia coli K12 Uvr⁻Rec⁻) requires some three orders of magnitude lower UV fluence for appreciable killing than the wild type cell. On the average a little more than one photochemical damage, (a reaction involving only one or two nucleotides out of the more than 6.5 million in the chromosome) is sufficient to block a cell's proliferation. Repair

deficiencies in higher cells likewise give high sensitization compared with
normal individuals, although not as high for the mutants so far available
as this bacterial example. On the other hand, in the most repair-competent
species known (Micrococcus radiodurans), DNA repair is so effective that it
actually shifts the cell's principal point of vulnerably away from DNA to
different substances. While the primary DNA damage remains similar in
amount and kind to that created in other cells (Setlow and Duggan 1964;
Moseley and Setlow 1968), DNA repair simply accommodates it without loss of
survival up to fluences where the damage to non-DNA components (probably
proteins) gives observable symptoms (Setlow and Boling 1965). These two
examples are extremes illustrating the extent to which DNA repair can
influence radiation sensitivity. In most cells examined, the noxious agent
after fluences which kill an appreciable part of the population is DNA
damage "left over" after repairs, as indicated by "DNA-type" action spectra,
for the effects. Noticeable decreases in their survival require irradiations that initially alter ~ 10^{-4} of the nucleotides present, a considerable number even in small bacteria. Since even a few such altered nucleotides are deadly if not removed, over 99.9% are clearly being repaired.

When most of the primary damage is being repaired, the exact extent of its
removal becomes important. If, for example, 99.9% is usually erased,
physiological changes resulting in repair of 99.95% would reduce the
residual damage to only half of its former value. Conversely, repair of
only 99.8% of the initial damage would double the effective amount "left
over". Changes in survival of microorganisms produced by manipulating the
post-irradiation environment, have long been known, and associated with
variations in repair (see, for example, Harm 1966a; Harm and Haefner 1968;
Patrick and Haynes 1968). The observed survival changes (up to an order of
magnitude) correspond generally to changes in effective fluence of less
than a factor of two, reflecting about this change in the amount of damage
left unrepaired. It is unlikely that any non-lethal stresses on intact
plants or animals perturb the component cells as much as the treatments
given these microorganisms, but such experiments vividly illustrate how the
extent of repair can be changed by physiological conditions after the
irradiation.

A more direct environmental influence comes through daylight radiation
itself, even at wavelengths outside the UV-B region. A very widespread DNA
repair is the photoenzymatic mechanism for removal of pyrimidine dimers
(the most common DNA photoproduct produced by radiation in the 250-300 nm
range). This mechanism--underlying most examples of photoreactivation, in
which a short-wavelength-UV effect is markedly diminished by subsequent
exposure to longer wavelengths (see Sutherland 1981)--requires light to
drive the repair reaction, and fails to occur in the dark. Unaided it can
reduce the amount of primary DNA damage by about an order of magnitude,
leaving as residuals those photoproducts which are not pyrimidine dimers.
However, in the presence of all the other cooperating, light-independent
mechanisms it usually changes the effective fluence by little more than a
factor of two. Since daylight contains a strong complement of the effective radiation, this repair should be strongly stimulated during natural
irradiation. Indeed, a bacterial mutant lacking the mechanism (but
possessing all the other repairs) was killed in strong sunlight (midday,
summer, latitude 33°N) by only about 2/3 the exposure required for the
normal cell (Harm 1966b). Consequently if the normal cell were exposed to
an artificial source of UV-B lacking the longer wavelength radiation
present in daylight, it would appear more sensitive than under an equiv-

alent amount of UV-B radiation plus the "photoreactivating light". This mechanism has been found in cells of plants and animals.

The direct photoreactivation mechanism shares its influence on contingent DNA repair in some bacteria with a different, indirect mechanism, through which exposure to longer wavelengths improves the effectiveness of light-independent repair processes (Jagger and Stafford 1965; Jagger 1983). In both cases exact details of the process are less important to our inquiry than the overall principle that radiation in non-damaging wavelengths can influence the effectiveness with which UV-B damage is repaired. Assessment of UV-B effects must deal with this in comparing controlled experiments with natural irradiation.

Curves showing colony-forming survival of microorganisms or cultured animal cells vs. ultraviolet fluence almost universally have "shoulders" at low fluences, in which the survival diminishes more slowly with irradiation than at higher fluence. These "shoulders" are diminished, or are missing where there are known deficiencies in repair (see Harm 1980), indicating that repair in normal cells is most effective at the lower fluences where killing is low. This is, of course, precisely the range of UV exposure of interest to our considerations.

There is another aspect of DNA repair at low injury levels. We have spoken so far about an internal fluence (or the corresponding dose on an organism's exterior surface) producing photochemical changes as though some definite amount of damage were first generated, and then was either repaired, or left to produce its consequences. However, both the damaging action of solar UV-B radiation on an organism and the subsequent repair extend and overlap in time. During the daylight period solar irradiation takes place for several hours at a rate which, at latitudes of the southern United States, alters $\cong 10^{-6}$ of DNA nucleotides per minute (Harm 1969). The size of a haploid genome in higher cells varies somewhat, but since it is mostly greater than 10^9 nucleotides, between 10^5 and 10^6 discrete damages are created in it per day, day after day, and are simultaneously undergoing repair.

The effectiveness of radiation for killing a considerable fraction of the population is independent of the irradiation rate when the fluence is delivered over a few seconds or minutes. However, when delivered at a lower rate producing DNA damage about as rapidly as sunlight, its effect is much reduced by concurrent repairs which keep the accumulated damage low. Bacterial cells held in non-growing conditions, and irradiated at these "natural" rates, survive fluences many times larger than those which can be tolerated as one massive injury (Harm 1968). An analogous phenomenon has been noted for algae (Nachtwey 1975), although the experimental details in that case were somewhat different. Thus, not only the extent of irradiation, but the irradiation rate is a factor influencing effectiveness of repair.

Whatever the extent of repair, it makes the radiation response quantitatively a function--not of the initial amount of cellular damage $\underline{n}$, described by Eq. (4)--but of a much smaller residual amount

$$\underline{n}_r = \underline{S}\langle\underline{D}\rangle. \tag{6}$$

The sensitivity factor S is correspondingly smaller than S_0 in Eq. (4). If this is true, we should expect similar magnitude responses for similar values of n_r (i.e., for similar values of $S<D>$). This view is strongly supported by studies of photoreactivation, where the amount of repair can be deliberately controlled. Maximum photoreactivation, in which the most abundant kind of ultraviolet damage to DNA has been completely removed, changes the fluence for which any specified level of colony-forming survival occurs by a constant factor over a wide range of survivals (see Harm 1980). This is exactly what is expected if a definite fraction of the DNA damage is of the kind that can be removed by the photorepair process. Detailed analysis shows that, also under partial repair, behavior is consistent with the "dose modification" concept behind Eq. (6) (Harm et al. 1971; Muraoka and Kondo 1969). The absolute value of S is not usually of interest, but changes in its value caused by changes in the amount of repair will change the dose required for any given magnitude effect.

ASSESSMENT OF DAMAGE IN ORGANISMS

Radiation damage and repair mechanisms operate at the cellular level, but the effects that concern us are in a whole organism--an array of many cells, which are differently damaged during irradiation of the individual. To the extent that a particular effect stems from one kind of photochemical damage in all the cells involved, the action spectrum for the composite injury will be that for the damaging reaction itself, and the dosimetry will be essentially as outlined above. Effects of repair should likewise be as described. On the other hand, more complex injuries might require additional analysis.[4]

A small increase in UV-B damage to DNA might be expected to have its greatest effect on immature structures, since delay of cell growth and interference with gene expression can readily disrupt a developmental sequence, even in the absence of cell death. In general, however, the organismal effects of small radiation increases must be determined by testing the organism rather than by deduction. Treatment of plants grown under natural daylight with supplementary sources of UV-B to simulate a hypothetical natural increase, as now being done by several workers, cuts through all the clutter of interacting events directly to the point of concern. Although it is impractical by this means to simulate the actual change in wavelength distribution accompanying a possible ozone depletion, an equivalent dose-rate increment can be approximated with the wavelength distributions experimentally available on the basis of the action spectrum and Eq. (5). Maintaining the organisms otherwise in their normal environment maximizes the likelihood they will display the same sensitivity toward

---

4 Other features can be peculiar to the whole organism, in contrast to single cells. For example, the relative shielding of different cells can change as the wavelength is altered, causing a qualitatively different pattern of injury to the organism at different wavelengths within the effective band. Also, any adaptive response by an organism which decreases epidermal transmission of UV-B [e.g., sun-tanning in humans; see Caldwell (1981) for examples in plants] would change both S and D, since a change of $K(\lambda)$ in Eq. (3) changes both K_0 and $E(\lambda,\lambda_0)$ in Eq. (4). In contrast, varying the amount of repair changes only S.

this increment as they would toward a natural UV-B increase following ozone depletion.

Unfortunately, these tests are tedious, and subject to a number of environmental variables which are difficult to control, making any results come slowly. It would therefore seem desirable to extend and compare a selective set of them with controlled experiments in the laboratory, where variations can presumably be minimized. This procedure, however, requires sufficient knowledge of the factors affecting repair efficiency to assure that the UV-B sensitivity will be comparable to that occurring in nature. Such information is not available. Studies of DNA repair have revealed much about the biochemical mechanisms and regulation of repair processes, but have not so far provided information about their effectiveness in higher plants and animals under the full range of UV-B dose rates, UV-A and visible radiation intensities, and other physiological variables encountered under tolerable growth conditions. In principle, determining the doses required for a given magnitude of some conveniently assayed UV-B effect under these circumstances could delineate the "window" of conditions within which the sensitivities can be compared meaningfully. Lacking this information one is constrained to the field tests alone.

Most of the above discussion has been formulated in terms of DNA damage and its repair, but we cannot be completely certain that DNA is the only radiation target of interest in solar UV-B effects. Although a likely candidate, its extensive protection by repairs might leave some other cellular component equally vulnerable, or more so, at least for some organisms under normal daylight conditions. If a different photochemical injury should turn out to be more significant, the line of thought to be traversed would nevertheless remain much the same as outlined here. One would still need to ask what counteracting biological responses existed, what parameters affected these responses, and how much they altered a sensitivity factor analogous to S in Eq. (6). The experimental determination of conditions under which $\overline{UV}$-B sensitivities of organisms could be compared would thus be largely independent of the particular mechanistic picture one carried in mind while doing it. The general consequences of the model would remain valid, even if its underlying suppositions might eventually have to be discarded.

ACKNOWLEDGMENT

Support by the U.S. Public Health Service through Research Grant GM16547 is gratefully acknowledged.

REFERENCES

Burt JE, Luther FM (1979) Effect of receiver orientation on erythema dose. Photochem Photobiol 29:85-91
Caldwell MM (1981) Plant response to solar ultraviolet radiation. In: Lange OL, Nobel PS, Osmond CB, Ziegler H (eds) Encyclopedia of plant physiology, new series, 12A, physiological plant ecology I. Springer, Berlin Heidelberg New York, p 170

Caldwell MM, Camp LB, Warner CW, Flint SD (this volume) Action spectra and their key role in assessing biological consequences of solar UV-B radiation change. p 87

CIE (1970) International lighting vocabulary. 3rd edn, entries 45-05-160 and 45-05-165. Publication CIE no. 17 (E-1.1). International commission on illumination and International electrotechnical commission Paris

Hanawalt PC, Cooper PK, Ganesan AK, Smith CA (1979) DNA repair in bacteria and mammalian cells. Ann Rev Biochem 48:783-836

Harm W (1966a) The role of host cell repair in liquid-holding recovery of U.V.-irradiated Escherichia coli. Photochem Photobiol 5:747-760

Harm W (1966b) Gene controlled repair processes in phage and bacteria and their role under natural conditions. In: Kohoutová M, Hubáček J (eds) The physiology of gene and mutation expression. Academia, Prague, p 51

Harm W (1968) Effects of dose fractionation on ultraviolet survival of Escherichia coli. Photochem Photobiol 7:73-86

Harm W, Haefner K (1968) Decreased survival resulting from liquid-holding of U.V.-irradiated Escherichia coli C and Schizosaccharomyces pombe. Photochem Photobiol 8:179-192

Harm W (1969) Biological determination of the germicidal activity of sunlight. Radiat Res 40:63-69

Harm W (1980) Biological effects of ultraviolet radiation. Cambridge University Press.

Harm W, Rupert CS, Harm H (1971) The study of photoenzymatic repair of UV lesions in DNA by flash photolysis. Photophysiology 6:279-324

ICP (1954) Proposals concerning definitions and units for the biological and medical use of UV radiation. Proc First Int Photobiol Congr (Amsterdam), Veenman en Zonen, Wageningen, p 440

Jagger J (1983) Physiological effects of near ultraviolet radiation on bacteria. Photochem Photobiol Revs 7:1-75

Jagger J, Stafford RS (1965) Evidence for two mechanisms of photoreactivation in Escherichia coli. Biophys J 5:75-88

Michalke H, Bremer H (1969) RNA synthesis in Escherichia coli after irradiation with ultraviolet light. J Mol Biol 41:1-23

Moseley BEB, Setlow JK (1968) Transformation in Micrococcus radiodurans and the ultraviolet sensitivity of its transforming DNA. Proc Natl Acad Sci USA 61:176-183

Muraoka N, Kondo S (1969) Kinetics of enzymatic photoreactivation studied with phage T1 and Escherichia coli B_{s-1}. Photochem Photobiol 10:295-308

Nachtwey DS (1975) Dose rate effects in the UV-B inactivation of Chlamydomonas and implications for survival in nature. In: Nachtwey DS, Caldwell MM, Biggs RH (eds) Impacts of climatic change on the biosphere, part I, ultraviolet radiation effects, monograph 5. Climatic Impact Assessment Program, US Dept Transportation Report No DOT-TST-75-55, NTIS, Springfield, Virginia, p (3)105

Patrick MH, Haynes RH (1968) Repair-induced changes in yeast radiosensitivity. J Bacteriol 95:1350-1354

Rahn RO (1979) Nondimer damage in deoxyribonucleic acid caused by ultraviolet radiation. Photochem Photobiol Revs 4:267-330

Rosenstein BS, Setlow RB (1982) The critical target in mammalian cells for solar ultraviolet radiation. Comm Mol Cell Biophys 1:223-235

Rupert CS (1974) Dosimetric concepts in photobiology. Photochem Photobiol 20:203-212

Setlow RB, Swenson PA, Carrier WL (1963) Thymine dimers and inhibition of DNA synthesis by ultraviolet irradiation of cells. Science 145:1464-1466

Setlow JK, Duggan DE (1964) The resistance of <u>Micrococcus</u> <u>radiodurans</u> to ultraviolet radiation. I. Ultraviolet-induced lesions in the cell's DNA. Biochim Biophys Acta 87:664-668

Setlow JK, Duggan DE (1965) The resistance of <u>Micrococcus</u> <u>radiodurans</u> to ultraviolet radiation. II. Action spectra for killing, delay in DNA synthesis and thymine dimerization. Biochim Biophys Acta 108:259-265

Sutherland B (1981) Photoreactivation. BioScience 31:439-444

Urbach F (1969) Geographic pathology of skin cancer. In: Urbach F (ed) The biologic effects of ultraviolet radiation. Pergamon, Oxford, p 635

Repair of Genetic Damage Induced by UV-B (290-320 nm) Radiation

R. M. Tyrrell

Swiss Institute for Experimental Cancer Research
Epalinges s./Lausanne, Switzerland

ABSTRACT

Electromagnetic radiation in the UV-B region (290-320 nm) induces genetic damage in cells with both lethal and mutagenic consequences. Furthermore, UV-B radiation can transform eucaryotic cells to an inheritable altered state which can lead to cancer formation in whole organisms. In the following overview, the various types of DNA alteration known to be induced in this region will be summarized together with a description of the various repair systems that can act to ameliorate this damage in procaryotic and eucaryotic cells. We will focus on studies with defined near monochromatic radiations. However, since natural sunlight is composed of a mixture of wavelengths that includes both UV-A (320-380 nm) and visible radiations, interactions that may occur between the longer wavelengths and the shorter UV-B region will also be mentioned.

GENETIC DAMAGE INDUCED BY UV-B RADIATION

Table 1 compares the types of lesion induced by a nonsolar UV-B wavelength (265 nm) with those induced by a monochromatic UV-B wavelength (313 nm). When considering these data it should be borne in mind that if either total lesions or biological parameters are to be expressed in the form of an action spectrum, the values would drop by approximately three orders of magnitude at 313 nm relative to the shorter wavelength. However, since pyrimidine dimers predominate at both wavelengths, the levels of other types of damage are expressed relative to the levels of this lesion to facilitate comparison.

It is clear that types of damage other than pyrimidine dimers are also induced in significant levels by the UV-B wavelength. The production of much of the damage is oxygen-dependent and therefore presumably occurs via an indirect mechanism involving active oxygen species. Included in this class are both single-strand breaks including alkali-labile bonds and ring-saturated thymines (5,6 dihydroxy-dihydrothymines). The latter are apparently induced almost as fast as dimers in human cells but no studies are yet available in bacteria. In vitro assays studying the action of an enzyme that recognizes this group of lesions, endonuclease III, on DNA irradiated at 313 nm, have failed to show this type of damage, a finding consistent with an indirect mechanism of induction via an endogenous sensitization. Pyrimidine adducts (6-4 pyo lesions) are also probably induced in this wavelength region since they are known to have a similar action

NATO ASI Series, Vol. G8
Stratospheric Ozone Reduction, Solar Ultraviolet Radiation
and Plant Life. Edited by R. C. Worrest and M. M. Caldwell
© Springer-Verlag Berlin Heidelberg 1986

Table 1. DNA damage induced by UV-C (265 nm) and UV-B (313 nm) radiations.

	UV-C (265 nm)	UV-B (313 nm)
Pyrimidine dimers (sensitive to denV and M. luteus endonuclease)	1.0	1.0
Single-strand breaks (+ alkali-labile bonds)	low (bacteria) 1.8×10^{-4} (human cells)[a]	2.3×10^{-2} (bacteria)[b] 3.3×10^{-2} (human cells)[c] 1.1×10^{-1} (human cells)[a]
Double-strand breaks	–	4.6×10^{-5} [d]
Ring-saturated thymines (glycols) sensitive to endonuclease III	4.8×10^{-2} (human cells)[a]	7.7×10^{-1} (human cells)[a] 0 (in vitro)[e]
Nonphotoreactivable damage		
Repair defective E. coli	1.0	0.16[f]
Repair proficient E. coli	0.4	all[g]
Haploid frog cells	0.35	0.5[h]

[a]Cerutti and Netrawali (1979); [b]Miguel and Tyrrell (1983); [c]Smith and Paterson (1982); [d]A. G. Miguel (unpublished results); [e]Demple and Linn (1982); [f]Hodges et al. (1980); [g]see this chapter; [h]Rosenstein and Setlow (1980)

spectrum for induction as pyrimidine dimers out to 280 nm (Patrick and Rahn 1976) but appear at a rate approximately 10 times lower. Since this type of damage has been implicated in mutagenesis (Haseltine 1983), this is a prediction that merits validation. Other types of lesions such as DNA-protein crosslinks appear after irradiation with broad spectrum sources (Bradley et al. 1979), but their specific induction by wavelengths in the UV-B region has not yet been established.

Some bacterial, many animal and all plant viruses carry their genetic information in RNA and both pyrimidine dimers and pyrimidine hydrates are induced in the nucleic acids by short-wavelength UV and sunlight (for review see Gordon et al. 1976) but the spectrum of damage induced differs considerably between different phage and viral types. There is also evidence for photoreactivation (pyrimidine dimer specific repair) of embryonic damage to insects which presumably involves damaged messenger RNA (Kaltholf and Jäckle 1982). This interesting finding is mentioned specifically since the action spectrum for inactivation of the embryos peaked at 295 nm (i.e., in the UV-B region) and pyrimidine dimers were induced in the RNA in vivo at 295 nm at a rate 10 times faster than at 265 nm. Related to this finding is the observation that Drosophila eggs and early embryos are apparently not photoreactivated after UV inactivation at 254 nm but are photoreactivated after inactivation at longer UV-B wavelengths (290 nm, 313 nm; Levin and Kozlova 1973).

REPAIR OF UV-B DAMAGE IN BACTERIA

Photoreactivation of UV-B damage to cells has mainly been studied in order to provide information concerning the role of pyrimidine dimers in inactivation in this region. In repair-defective bacterial mutants, photoreactivation of 313-nm-radiation-induced lethal damage is as efficient as photorepair of short wavelength damage (Hodges et al. 1980). However, in fully repair proficient bacteria there is no photoreactivation of 313-nm-induced lethal damage (Fig. 1). This could reflect simultaneous photoreactivation over the long irradiation period required so that cell death could then result from a different lesion. In contrast, 50 percent of the lethal damage induced in haploid frog cells can be photoreactivated (Rosenstein and Setlow 1980). From this type of data and comparisons of action spectra measurements for lesion induction and inactivation of various cell types, it is clear that the pyrimidine dimer is an important lethal lesion in the UV-B region. The picture is far less clear with regard to mutation and transformation.

In addition to photoreactivating enzyme, many of the gene products known to be involved in repair of UV-C damage have also been shown to act on UV-B damage as deduced from survival curve measurements. A summary of this information is shown in Table 2. After irradiation at 302 nm and 313 nm, the lack of genes controlling excision and post-replication repair of pyrimidine dimers (uvrA and recA) sensitizes the bacteria to at least as great an extent as after irradiation at 254 nm (see Table 3). The repair sectors (see Jagger 1967) derived from complete survival curves of populations irradiated at the longer wavelengths are larger than after irradiation at 254 nm. This is also true for natural sunlight. However, this probably reflects simultaneous photoreactivation by the UV-B wavelengths themselves (see above) which will be minimal in the totally repair-defective strain because of the short irradiation times involved.

142

Table 2. E. coli mutants and bacteriophage showing modified resistance to damage induced by radiation at 313 nm.

Gene	Gene product	Repair pathway(s)	Effect	Notes
Bacteria				
uvrA	DNA damage specific endonuclease	excision repair	mutant UV-B sensitive[ab]	similar response at 254 nm, 302 nm
recA	single-strand binding protein	post-replication repair strand-break rejoining	mutant UV-B sensitive[ab]	similar response at 254 nm, 302 nm
polA	polymerase 1	excision repair strand-break rejoining	mutant UV-B sensitive[c]	similar response at 254 nm
phr	photoreactivating enzyme	photoreactivation	photoreactivation of repair deficient strains only[de]	
nur	unknown	unknown	mutant UV-B sensitive[f]	not sensitive to 254 nm, possibly leads to excess endogenous sensitizer
xthA	exonuclease III AP endonuclease	repair of oxidized DNA	mutant sensitive to UV 300–400 nm[g]	not sensitive to 254 nm, sensitivity to UV-B not established
Bacteriophage				
uvrA (host)	damage specific endonuclease	host cell reactivation (excision)	UV-B irradiated phage sensitive in mutant[h]	similar response at 254 nm
recA	UV inducible recA gene product	inducible repair	only slightly enhanced survival of UV-B (302 nm, 313 nm) damaged bacteriophage[i]	large enhancement of survival of 254 nm radiation damaged bacteriophage

[a]Tyrrell and Souza-Neto (1981); [b]Tyrrell (1982a); [c]Miguel and Tyrrell (1983); [d]Hodges et al. (1980); [e]this chapter; [f]Webb and Tuveson (1982); [g]Sammartano and Tuveson (personal communication); [h]Peak and Peak (1978); [i]Menezes and Tyrrell (1982)

Table 3. Relative fluences of UV-C (254 nm), UV-B (302 nm, 313 nm) and solar radiation to inactivate repair-proficient and repair deficient strains of E. coli to the 1% survival level.[a]

	K12 mutants			
Radiation	recA uvrA	recA	uvrA	wild type
254 nm	1	4.0	15	480
302 nm	1	5.6	34	900
313 nm	1	11	59	1100
solar	1	4.9	35	280 (summer) 170 (winter)

[a]data from Tyrrell and Souza-Neto (1981)

Lack of a gene designated nur (product unknown) also leads to a sensitiza-tion of cells to 313 nm radiation but not to short wavelength UV (Webb and Tuveson 1982). Although this was originally believed to reflect a repair defect, it now seems more probable that the mutated gene leads to elevated levels of a sensitizing chromophore (Peak et al. 1983). Lack of the gene coding exonuclease III (xthA) also results in enhanced sensitivity to broad spectrum near-UV (300-400 nm) but not to shorter wavelengths (Sammartano and Tuveson personal communication). However, it is not known whether this sensitization will be observed in the UV-B region. This finding is of particular interest not only because it could reflect repair of damage uniquely induced in the solar-UV region but also because the enzyme is believed to be the major apurinic/apyrimidinic repair endonuclease of E. coli (Lindahl 1982).

Another approach to studying repair of UV-B damage has been to examine the ability of damaged bacteriophage to survive in treated or nontreated host cells. The T4 coded repair genes denV and x act effectively on damage induced by radiation at wavelengths as long as 365 nm (Tyrrell 1979). Constitutive host cell reactivation is also as efficient on radiation damage induced in phage by 313 nm as at shorter wavelengths (Peak and Peak 1978). However important differences are seen when the inducible Weigle reactivation process is considered. Although wavelengths as long as 313 nm (but not longer wavelengths) are able to induce host-cell repair of 254 nm bacteriophage as efficiently as radiation at 254 nm (Menezes and Tyrrell 1982), bacteriophage damaged by UV-B wavelengths show a much reduced (302 nm) or almost absent (313 nm) susceptibility to 254-nm radiation induced host-cell repair. The mechanism underlying this finding is unclear, although presumably it reflects the different lesion spectrum found in this region. However, the difference could be important since this form of inducible repair is believed to be a major pathway for muta-tion induction following UV treatment (Witkin 1976). UV-B radiation at

313 nm induces negligible levels of untargeted mutagenesis (mutation of infecting unirradiated bacteriophage λ) compared with treatment of the host cell at 254 nm (Tyrrell 1984).

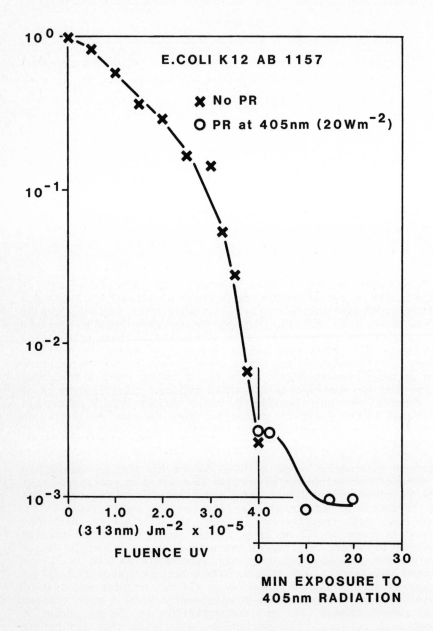

Fig. 1.   Inactivation of repair-proficient E. coli K12 by radiation at 313 nm and photoreactivation (inset) as a function of irradiation with a saturating fluence of 405 nm radiation.

There is not a large literature relating to biochemical measurements of removal of UV-B damage in bacteria. Based on comparative inactivation data with mutants, pyrimidine dimers are presumed to be removed by dark repair systems as efficiently as after irradiation at 254 nm. Recently we have studied strand-break rejoining in various E. coli mutants following irradiation at 313 nm (Miguel and Tyrrell 1983). When immediate induction is compared in a wild-type and a polA strain, breaks are apparently induced at a rate five times faster in the mutant. This presumably represents a fast polymerase 1 dependent repair of this type of damage. Kinetic studies in complete growth medium have shown that there is also a recA gene dependent repair of these breaks. Thus it would appear that the repair of breaks induced by 313 nm radiation is similar to the fast polA dependent and slow recA full-medium dependent repair of x-ray induced strand breaks (Town et al. 1973; Youngs and Smith 1973).

REPAIR OF UV-B DAMAGE IN HUMAN CELLS

Although large numbers of strand breaks and ring-saturated thymine photo-products have been detected in human fibroblasts (see Table 1) the pyrimidine dimer is still considered to be responsible for at least 50 percent of the lethal effects out to a wavelength of 313 nm. Removal of pyrimidine dimers after 313 nm radiation has been measured directly in human fibroblasts (Niggli and Cerutti 1983) and the kinetics are similar to those observed after 254 nm radiation. Recently it has been shown that excision defective cells from patients with xeroderma pigmentosum are as sensitized to UV-B radiation as they are to short wavelength radiation. This is true for both complementation group D (310 nm; Smith and Paterson 1982) and complementation group A (290 nm, 300 nm 313 nm; Keyse et al. 1983). In the latter study, the apparent repair sector diminishes sharply after irradiation at wavelengths of 325 nm and longer. In a different approach, we have looked at excision repair of UV-B damage by employing the α polymerase inhibitor, aphidicolin, to inhibit excision repair in fibroblasts. Conditions have been developed whereby the sector of repair that is inhibited by aphidicolin can be estimated from survival measurements (Tyrrell 1983). We estimate that the sector of repair of damage that can be accounted for by α polymerase dependent excision repair is between 75 and 80 percent after irradiation at either 254 nm, 302 nm or 313 nm (Tyrrell and Amaudruz 1984). These studies demonstrate a major role of excision repair in removal of UV-B damage in human fibroblasts. At longer wavelengths (Keyse et al. 1983; Tyrrell and Amaudruz 1984), a decreasing role for excision with increasing wavelength is evident.

INTERACTION BETWEEN UV-A (320-380 NM) AND UV-B RADIATIONS

UV-A radiation has been shown to modify the action of UV-B or shorter wavelength radiations in several ways, generally via a direct or indirect action on repair.

Photoreactivation

The lethal and mutagenic consequences of damage induced in the UV-B region can be ameliorated via the operation of the photoreactivation process that is mediated via the longer wavelengths in sunlight. In repair-proficient bacteria, the process is so effective that at ambient temperature it would be expected that essentially all the dimers induced by the shorter wavelengths in sunlight would be rapidly split by the longer wavelengths. The precise role of photoreactivation to eucaryotic, particularly human cells is more difficult to estimate. This is partly because the process is generally less efficient but also because in human cells, except for limited experiments with virus (Wagner et al. 1975), the biological consequences of photoreactivation have proved difficult to demonstrate in in vitro systems. Nevertheless experiments using human skin in vivo and a sensitive photoproduct assay suggest that photoreactivation can function extremely rapidly under the appropriate conditions (D'Ambrosio et al. 1981). Further information in this area is essential for predicting the biological consequences of polychromatic irradiation.

Photoprotection (Low Fluence)

In bacteria, exposure to the longer wavelengths in sunlight leads to growth delays (see Jagger 1972, for review) which may reduce the lethal effects of UV-B damage by allowing more time for excision repair processes to operate. Recently it has been shown that the suppression of the appearance of UV-induced suppressor mutations by long wavelengths acts through a similar mechanism (De Moraes and Tyrrell 1983). In human cells, UV-A-induced growth delays are negligible compared with the doubling time, although suppression of DNA synthesis for up to a few hours has been observed (Ortega, this laboratory).

Nevertheless we have recently obtained evidence in human lymphoblastoid cells that exposure to low fluence of UV-A radiations (334 nm and 365 nm) leads to a slight photoprotection from the lethal effects of both UV-C (254 nm) and UV-B (313 nm) radiations (Tyrrell et al. 1984).

Sensitization (High Fluence)

Exposure of bacteria to higher fluences of UV-A radiation sensitizes them to the lethal action of a variety of DNA damaging agents including UV-B radiations (Tyrrell 1982b). There is reasonable evidence that this arises from interference with or damage to repair systems by the longer wavelengths. Recently, we have observed that high fluences of UV-A (365 nm) and a visible wavelength sensitize human lymphoblastoid cells to the lethal action of UV-B (313 nm) radiations (Tyrrell et al. 1984). Studies by Holmberg (1983) with primary human fibroblasts have indicated that radiation in the UV-A region suppresses the levels of unscheduled DNA synthesis (a measure of repair) induced by short wavelength UV radiation, thereby suggesting one possible mechanism for the sensitization effect.

SUMMARY

The spectrum of lesions induced in DNA by UV-B radiation is different from that observed at shorter wavelengths and includes additional damage that is probably largely formed by intermediate oxygen species. Several bacterial gene products have been shown to be involved in repair of UV-induced damage including not only these previously known to be involved in repair of UV-C (254 nm) or x-ray like damage, but also some additional ones. The susceptibility of UV-B damage to UV-inducible repair appears quite different from the susceptibility of UV-C damage. Damage induced by UV-B radiation in human cells is repaired as efficiently by excision repair as after irradiation with UV-C radiation. The pyrimidine dimer appears to play a major role in UV-B inactivation, at least in eucaryotic cells and repair deficient bacteria. However in assessing the biological consequences of UV-B radiation from a polychromatic source such as sunlight, interactions with the longer wavelengths must also be considered. These include not only the direct effect of photoreactivation but also photoprotective effects resulting from growth delays and, at higher UV-A and visible fluences, sensitizations that may result from damage to repair pathways.

ACKNOWLEDGMENT

The research of the author is supported by grants from Swiss National Science Foundation (3'008.81 and 3.397.083) and the Swiss League Against Cancer.

REFERENCES

Bradley MO, Horn IC, Harris CC (1979) Relationships between SCE, mutagenicity, toxicity and DNA damage. Nature 34:313-320

Cerutti PA, Netrawali M (1979) Formation and repair of DNA damage induced by indirect action of ultraviolet light in normal and xeroderma pigmentosum skin fibroblasts. In: Okada S, Imamura M, Terashimo T, Yamaguchi M (eds) Proc 6th int congr rad res. Japanese Association for Radiation Research, Tokyo, p 423

D'Ambrosio SM, Whetstone JW, Slazinski L, Lowney E (1981) Photorepair of pyrimidine dimers in human skin in vivo. Photochem Photobiol 34:461-464

De Moraes EM, Tyrrell RM (1983) Mutational interaction between near-UV radiation and DNA damaging agents in E. coli: the role of near-UV induced modifications in growth and macromolecular synthesis. Photochem Photobiol 38:57-63

Demple D, Linn S (1982) 5,6-saturated thymine lesions in DNA: Production by ultraviolet light or hydrogen peroxide. Nucleic Acids Res 10:3781-3789

Gordon MP, Huang C, Hurter J (1976) Photochemistry and photobiology of ribonucleic acids, ribonucleoproteins and RNA viruses. In: Wang SY (ed) Photochemistry and photobiology of nucleic acids, vol 2. Academic Press, New York, p 265

Haseltine WA (1983) Ultraviolet light repair and mutagenesis revisited. Cell 33:13-17

Hodges NDM, Moss SH, Davies DJG (1980) The role of pyrimidine dimers and non-dimer damage in the inactivation of E. coli by UV radiation. Photochem Photobiol 31:571-577

Holmberg M (1983) Prior exposure of human cells to near-UV radiation gives a decrease in the amount of the unscheduled DNA synthesis induced by far UV radiation. Photochem Photobiol 37:293-296

Jagger J (1967) Introduction to research in ultraviolet photobiology. Prentice Hall, Englewood Cliffs, New Jersey

Jagger J (1972) Growth delay and photoprotection induced by near-ultraviolet light. In: Gallo V, Santamaria L (eds) Research progress in organic, biological and medical chemistry, vol 3. American Elsevier, New York, p 383

Kalthoff K, Jäckle H (1982) Photoreactivation of pyrimidine dimers generated by a photosynthesized reaction in RNA of insect embryos (Smittia spec.). In: Hélène C, Charlier M, Montenay-Garestier Th, Laustriat G (eds) Trends in photobiology. Plenum, New York, p 173

Keyse SM, Moss SH, Davies DJG (1983) Action spectra for inactivation of normal and xeroderma pigmentosum human skin fibroblasts by ultraviolet radiations. Photochem Photobiol 37:307-312

Levin VL, Kozlova MA (1973) Photoreactivation of the Drosophila embryonic cells irradiated by ultraviolet with 254 and 313 nm wavelengths. Tsitologia (Leningrad) XV:1415-1420

Lindahl T (1982) DNA repair enzymes. Ann Rev Biochem 51:61-87

Menezes S, Tyrrell RM (1982) Damage by solar radiation at defined wavelengths: involvement of inducible repair systems. Photochem Photobiol 36:313-318

Miguel AG, Tyrrell RM (1983) Induction of oxygen-dependent lethal damage by monochromatic UV-B (313 nm) radiation: strand breakage, repair and cell death. Carcinogenesis 4:375-380

Niggli HJ, Cerutti PA (1983) Cyclobutane-type pyrimidine photodimer formation and excision in human skin fibroblasts after irradiation with 313 nm ultraviolet light. Biochemistry 22:1390-1395

Patrick MH, Rahn RO (1976) Photochemistry of DNA and polynucleotides: photoproducts. In: Wang SY (ed) Photochemistry and photobiology of nucleic acids, vol 2. Academic Press, New York, p 35

Peak MJ, Peak JG (1978) Action spectra for ultraviolet and visible light inactivation of phage 17: Effect of host cell reactivation. Radiat Res 76:325-330

Peak JG, Peak MJ, Tuveson RW (1983) Ultraviolet action spectra for aerobic and anaerobic inactivation of Escherichia coli strains specifically sensitive and resistant to near-ultraviolet radiations. Photochem Photobiol 38:541-544

Rosenstein BS, Setlow RB (1980) Photoreactivation of ICR 1A frog cells after exposure to monochromatic ultraviolet radiation in the 252-313 nm range. Photochem Photobiol 32:361-366

Smith PJ, Paterson MC (1982) Lethality and the induction and repair of DNA damage in far, mid or near UV-irradiated human fibroblasts: comparison of effects in normal, xeroderma pigmentosum and Bloom's syndrome cells. Photochem Photobiol 36:333-344

Town CD, Smith KC, Kaplan HS (1973) The repair of single-strand breaks in E. coli K12 x-irradiated in the presence and absence of oxygen: The influence of repair on cell survival. Radiat Res 55:334-345

Tyrrell RM (1979) Repair of near-ultraviolet damage induced in bacterio-phage. Photochem Photobiol 29:963-970

Tyrrell RM (1982a) Cell inactivation and mutagenesis by solar ultraviolet radiation. In: Hélène C, Charlier M, Montenay-Garestier Th, Laustriat G (eds) Trends in photobiology. Plenum, New York, p 155

Tyrrell RM (1982b) Biological interactions between wavelengths in the solar-UV range: implications for the predictive value of action spectra measurements. In: Calkins J (ed) The role of solar ultraviolet radiation in marine ecosystems. Plenum, New York, p 629

Tyrrell RM (1983) Specific toxicity of aphidicolin to ultraviolet irradiated excision proficient human skin fibroblasts. Carcinogenesis 4:327-329

Tyrrell RM (1984) Lack of induction of non-targeted mutations in intact bacteriophage by UVB (313 nm), UVA (313 nm, 365 nm) and visible (405 nm) irradiation of host cells. Mutation Res 139:45-49

Tyrrell RM, Amaudruz FA (1984) Alpha polymerase involvement in excision repair of damage induced by solar radiation of defined wavelengths in human fibroblasts. Photochem Photobiol 40:449-452

Tyrrell RM, Souza-Neto A (1981) Lethal effects of natural solar-ultraviolet radiation in repair proficient and repair deficient strains of Escherichia coli: Actions and interactions. Photochem Photobiol 34:331-338

Tyrrell RM, Werfelli P, Moraes EC (1984) Lethal action of ultraviolet and visible (blue-violet) radiations at defined wavelengths on human lympho-blastoid cells: action spectra and interaction studies. Photochem Photobiol 39:183-189

Wagner EK, Rice M, Sutherland BM (1975) Photoreactivation of herpes simplex virus in human fibroblasts. Nature 254:627-628

Webb RB, Tuveson RW (1982) Differential sensitivity to inactivation of NUR and NUR$^+$ strains of Escherichia coli at six selected wavelengths in the UV-A , UV-B and UV-C ranges. Photochem Photobiol 36:525-530

Witkin EM (1976) Ultraviolet mutagenesis and inducible DNA repair in Escherichia coli. Bacteriol Rev 40:869-907

Youngs DA, Smith KC (1973) X-ray sensitivity and repair capacity of a polA exrA strain of Escherichia coli K12. J Bacteriol 114:121-127

Physiological Responses of Yeast Cells to UV of Different Wavelengths

J. Kiefer, M. Schall, A. Al-Talibi

Strahlenzentrum der Justus-Liebig-Universität, Giessen
Federal Republic of Germany

ABSTRACT

Liquid-holding recovery was investigated in diploid yeast after exposure to UV radiation of different wavelengths in the 254-313 nm range. It was found that the kinetics were wavelength dependent. Further experiments using a different technique showed that recovery was absent with 303- and 313-nm radiation, but was mimicked in liquid medium by divisions of surviving cells. This division may be facilitated by materials released from irradiated cells. This process is more pronounced with longer wavelengths. Cell progression was also studied by measuring budding and division delay. While 254-nm radiation induces division delays which are considerably greater than those for budding, suggesting an additional G2 block, this difference was not found with 313-nm radiation. With 303-nm radiation division was even less delayed than budding, suggesting a general distortion in the sequence of cellular events.

INTRODUCTION

It has been shown (Zölzer and Kiefer 1983) that the action spectrum for killing and mutation induction in yeast coincides essentially with DNA absorption. This must not be taken to mean, however, that other targets are unimportant in the wavelength range of 254-313 nm. Additional effects may remain undetected by the techniques used. Recovery from potentially lethal damage, as usually measured by recovery following holding cells in suboptimal medium before plating, "liquid holding recovery", may be a more sensitive parameter. It was recently demonstrated that recovery does not occur in E. coli after exposure to near-UV radiation (Tuveson and Violante 1982) although it was found after far-UV exposure. The situation in yeast was not then known. We report experiments which essentially confirm the quoted results but show some additional features. When UV-irradiated cells are kept in liquid suspension they release UV-absorbing materials. These could facilitate division of surviving cells, thus mimicking recovery. We therefore measured the amount of released substances after exposure to UV of different wavelengths. Recovery ability may also be influenced by alterations in cell cycle progression, which can be recorded at high survival levels, thus constituting a sensitive means to monitor disturbances of physiological processes.

NATO ASI Series, Vol. G8
Stratospheric Ozone Reduction, Solar Ultraviolet Radiation
and Plant Life. Edited by R. C. Worrest and M. M. Caldwell
© Springer-Verlag Berlin Heidelberg 1986

MATERIALS AND METHODS

A wild type diploid strain of Saccharomyces cerevisiae (211, originally obtained from W. Laskowski, Berlin) was used for all experiments. Cells were grown on YEDP-agar plates (20 g l^{-1} glucose, 20 g l^{-1} bactopeptone, 10 g l^{-1} yeast extract, 20 g l^{-1} agar) for three days, to reach stationary phase. For UV exposure they were washed off and resuspended in distilled water at a density of 2 x 10^8 ml^{-1}. Survival was assessed in the usual way by plating on YEDP agar. Macrocolonies were scored after 3-4 days incubation at 30°C.

"Liquid holding recovery" was investigated by incubating irradiated cells in 9 g l^{-1} NaCl, 9.1 g l^{-1} K_2HPO_4 and 10 g l^{-1} glucose (pH 5.5) for various lengths of time at 30°C in a shaking waterbath. For "agar-holding," cells were plated on agar containing the same ingredients plus 20 g l^{-1} agar. After the recovery period they were overlayed with soft (0.7%). YEDP-nutrient agar and incubated for 3-4 days.

Irradiated cells were grown in liquid YEDP medium at 30°C under vigorous shaking. Cell counts and percentages of budding cells were determined microscopically. Delay times were estimated as shown in Fig. 1.

Cells were incubated at a density of 2 x 10^7 cells ml^{-1} in liquid holding medium. After centrifugation the absorption of the supernatant was determined spectrophotometrically. The maximum was found around 265 nm, suggesting release of nucleotides or similar material. The optical density at 265 nm was used as the parameter for quantitation of the material released.

Irradiation was performed by means of a high-pressure mercury lamp (6293/1000, Oriel) equipped with a prism monochromator. The samples were exposed in cylindrical 1-cm quartz cuvettes agitated by an air stream. Fluence rates were determined by chemical actinometry as described by Hatchard and Parker (1956). To produce a wavelength of 254 nm, a low-pressure mercury lamp was also used.

RESULTS

Figure 2 shows the apparent increase in cell survival by liquid holding recovery after exposure to UV-radiation of different wavelengths. To facilitate comparison the data were normalized to the initial surviving fractions, which were typically on the order of 10%. It can be seen that there is a remarkable difference in recovery kinetics. With 254-nm radiation one finds the usual prompt increase as already reported by Patrick et al. (1964). This changes to a biphasic pattern with increasing wavelength. It is particularly striking with 293 nm radiation where the initial rise is followed by a drastic decline--even below the initial survival level. Although this feature was consistently found in parallel experiments, it certainly requires further confirmation. With radiation of $\lambda > 296$ nm there is no immediate rise although the apparent cell survival increased after longer incubation times.

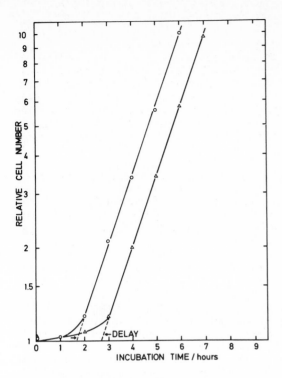

Fig. 1(a). Relative cell number as a function of incubation time. Circles = controls, triangles = after 1247 J m^{-2} at 303 nm.

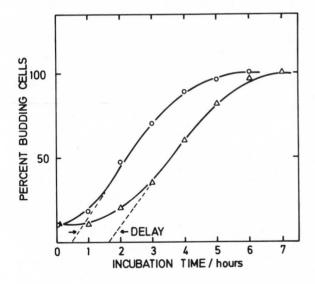

Fig 1(b). Percentages of budding cells as a function of incubation time. Symbols as in Fig. 1(a). Delay times as used for quantification are indicated.

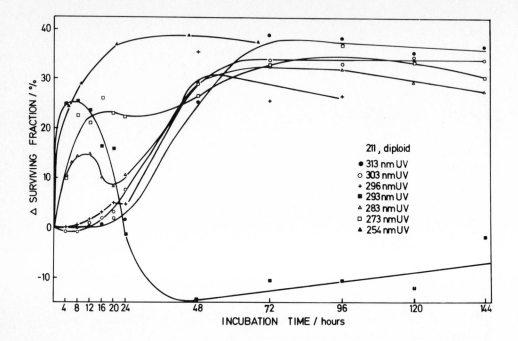

Fig. 2. "Liquid holding recovery" after exposure to UV-radiation of different wavelengths.

Since it may be possible that recovery is mimicked by cell multiplication as reported with near UV for E. coli (Tuveson and Violante 1982), the technique was changed to an "agar-holding" procedure. As the cells are fixed in their positions any division should not increase the number of colonies counted. An increase of colony formers should, therefore, reflect real recovery. Figure 3 shows the results for 254-nm and 313-nm radiation. It is seen that at the former wavelength recovery is clearly demonstrable, while it is absent at the latter. Liquid-held cells, which were investigated in the same experiment, however, increased in number, suggesting the involvement of multiplication. This is not likely to occur in the buffer medium used, because of the lack of a nitrogen source. It was therefore determined whether irradiated cells release material that could facilitate a few divisions. It was found that substances with an absorption maximum around 265 nm, which may consist of nucleotides or similar compounds, could be detected after exposure. Figure 4 demonstrates that the amounts and kinetics differ with wavelength. At 254 nm, the release was slow and the total amount less than with 303-nm radiation (which had to be used instead of 313-nm radiation because of technical reasons). It appears possible that the apparent recovery seen after 313- and 303-nm exposure and extended incubation in liquid suspension is due to multiplication of surviving cells.

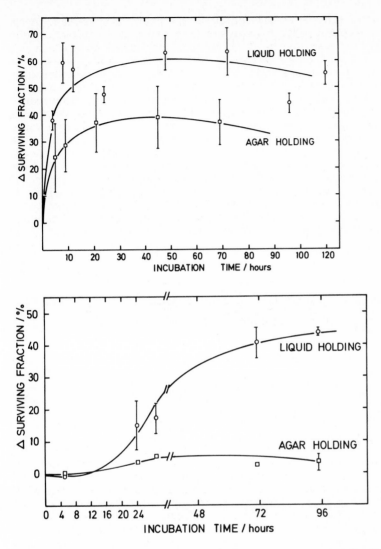

Fig. 3. Comparison of "liquid-holding" and "agar-holding". Upper figure = 254 nm; lower figure = 313 nm.

If this assumption is correct it can nevertheless not explain the lack of recovery with longer wavelengths as evidenced by the "agar-holding" experiments. Since a distortion in cell cycle progression might play a role we investigated budding and division delay. It was found that both increase with UV fluence at all wavelengths studied, but there are interesting differences, as shown in Fig. 5. With 254 nm, division was considerably more delayed than budding. As the emergence of buds coincided roughly with the start of DNA synthesis in the undisturbed yeast cell cycle, it may be concluded that the entry into S phase was delayed and the subsequent phases were also retarded. With 313-nm radiation, budding and division were equally delayed, indicating that only entry into S phase is affected. This was confirmed by cytofluorometric analysis (Fingerhut and Kiefer, unpublished). If equal survival levels are compared, it can be seen (data not shown) that both types of progression delay are considerably more pronounced with 313-nm radiation. With 303-nm radiation, a surprising result was found: budding was significantly less delayed than division. This can be explained only by assuming that the synchronization of cellular events is disturbed, or in other words, budding no longer indicates the start of S phase. If the entry into S were more delayed than cell division, one would postulate that progression through S and G2 is accelerated, which seems rather unlikely. But apart from this it seems clear that the G2 block, which is seen after 254-nm exposure, is absent with 303- and 313-nm radiation. Cellular repair may occur during this period. The pattern of UV-induced changes in cell-cycle progression is thus consistent with the established lack of recovery. Even if this speculation turns out not to be tenable, it is obvious from our results that exposure to longer wavelengths induces qualitatively different effects compared to those at 254 nm.

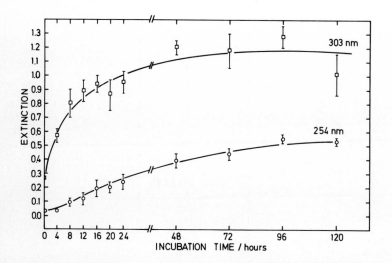

Fig. 4. Release of UV-absorbing materials after exposure to 254- and 303-nm UV-radiation.

DISCUSSION

The main results of our study are:

1. There is no recovery from potentially lethal damage from 303- and 313-nm radiation.
2. UV-irradiated yeast cells release UV-absorbing substances into the medium; the quantities of which are considerably greater with longer wavelengths, if approximately equal survival levels are compared.

3. The budding process is more influenced by 303- and 313-nm radiation as compared to 254-nm radiation.
4. There appears to be no G2 arrest with longer wavelengths.

All of these findings taken together suggest that there may be an additional chromophore for UV-B action. Without further knowledge only speculations are possible. Membrane damage would be consistent with our results but direct evidence is lacking. It is noteworthy, however, that membrane damage has been suggested as the reason for the lethal action of broad-band far-UV radiation (Ito and Ito 1983). Imbrie and Murphy (1982) found that membrane-bound ATPase is inactivated by UV-B radiation with an efficiency maximum around 293 nm. Whether this is also the case in our system remains to be seen.

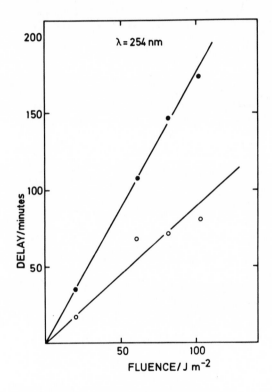

Fig. 5(a). Budding (open circles) and division (closed circles) delay as a function of UV dose. λ = 254 nm.

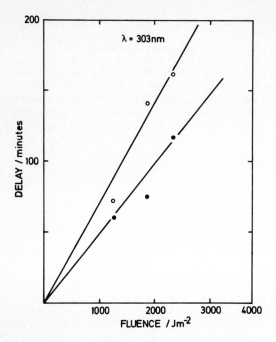

Fig. 5(b). Budding (open circles) and division (closed circles) delay as a function of UV dose. λ = 303 nm.

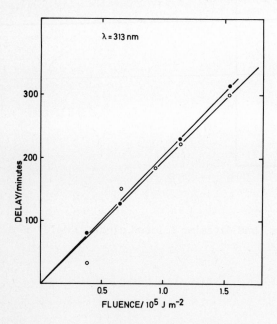

Fig. 5(c). Budding (open circles) and division (closed circles) delay as a function of UV dose. λ = 313 nm.

ACKNOWLEDGMENTS

We thank P. Bepler for skillful technical assistance and K. Brieden for actinometric measurements. This study was supported by the Bundesministerium für Forschung und Technologie of the Federal Republic of Germany through the Gesellschaft für Strahlen- und Umweltforschung, Bereich Projekkträgerschaften (KBF 49).

REFERENCES

Hatchard CG, Parker CA (1956) A new sensitive chemical actinometer. Roy Soc (London) Proc A 235:518-536

Imbrie CW, Murphy TM (1982) UV-action spectrum (254-405 nm) for inhibition of K^+-stimulated adenosine triphosphatase from the plasma membrane of Rosa damascena. Photochem Photobiol 36:537-542

Ito A, Ito T (1983) Possible involvement of membrane damage in the inactivation by broad-band near-UV radiation in Saccharomyces cerevisiae cells. Photochem Photobiol 37:395-402

Patrick MH, Haynes RH, Uretz RB (1964) Dark recovery phenomena in yeast. I. Comparative effects with various inactivating agents. Radiat Res 21:144-163

Tuveson RW, Violante EV (1982) The effects of liquid holding recovery on near UV (300-400 nm) irradiated Escherichia coli strains differing in near and far-UV sensitivity. Photochem Photobiol 35:845-848

Zölzer F, Kiefer J (1983) Wavelength dependence of inactivation and mutagenesis in haploid yeast cells of different sensitivities. Photochem Photobiol 37:39-48

Effects of UV-B Radiation on Photosynthesis

William B. Sisson

New Mexico State University, Las Cruces, New Mexico USA

ABSTRACT

Several component reactions of photosynthesis, and net CO_2 assimilation rates have been shown to be detrimentally affected by UV-B radiation levels equal to or above those presently received at various locations on the earth's surface. Many of these studies, however, were completed under environmental conditions that deviated from ambient field conditions to such an extent that extrapolation of results for predictive purposes would probably overestimate the harmful effects of reduced atmospheric ozone levels and the concomitant increases in UV-B radiation. Nevertheless, the photosynthetic process has been shown to be sensitive to UV-B radiation and, therefore, warrants further study. The ability of plants to tolerate increased levels of UV-B radiation, relative to photosynthetic capacity, may be dictated by acclimation processes that would intensify UV-B attenuation prior to its absorbance by sensitive components or reactants of photosynthesis.

INTRODUCTION

Experimentally increasing the quality and quantity of polychromatic ultra-violet-B (UV-B, 280-320 nm) radiation equivalent to atmospheric ozone reductions of 50% or less adversely affects various component reactions of photosynthesis and reduces carbon assimilation rates in several native and agronomically important plants. Although the underlying objective of this research has been to develop and refine predictive capabilities relative to the impact of specific increases in terrestrial UV-B radiation on plant growth, productivity, and physiological processes such as photosynthesis, this objective has yet to be achieved. This is particularly true regarding a relatively subtle increase in UV-B radiation corresponding to the projected 3 to 5% reduction in atmospheric ozone (National Academy of Sciences 1984).

Although relatively few plants have been evaluated under UV-B radiation levels corresponding to specific decreases in atmospheric ozone concentrations, a wide range in sensitivity to UV-B radiation is apparent between plant species (Caldwell et al. 1975; Biggs and Kossuth 1978), as well as within cultivars of single species (Krizek 1978; Biggs et al. 1981). Similarly, sensitivity of the component reactions of photosynthesis varies among plant species (Vu et al. 1982a) and is probably associated with differential UV-B radiation attenuation properties expressed within various plants (Robberecht and Caldwell 1978; Robberecht et al. 1980). Unfortunately many studies addressing the effects of UV-B radiation on carbon assimilation rates by intact plant tissues, or the component reactions of the photosynthetic apparatus, utilized very high UV-B radiation levels and/or other environmental parameters deviated considerably from those present under ambient field conditions. Thus, many studies need to be repeated with low levels of supplemented UV-B radiation corresponding to the reduction in atmospheric ozone projected to occur over the next century. Recent development of a modulated lamp system (Caldwell et al. 1983) permits UV-B radiation enhancement corresponding to specific reductions in atmospheric ozone under field conditions. Although this lamp system may not be feasible in studies such as evaluating crop productivity within large plots, every effort should be made to employ this type of lamp system in studies addressing the effects of UV-B radiation on photosynthesis.

The purpose of this chapter is to present a summary of the research that has been conducted over the past decade and contributed to the present knowledge relative to UV-B radiation effects on carbon assimilation of native and agronomically important plant species.

CHLOROPHYLL

Low visible light intensities concomitant with UV-B radiation is a rather well documented interaction that tends to reduce plant growth and enhance damage to physiological processes (Sisson et al. 1974; Sisson and Caldwell 1976; Teramura et al. 1980; Biggs et al. 1981). Most reports of chlorophyll reduction were determined under experimental conditions where very low visible light regimes (e.g., Vu et al. 1981, 1982a) were used or with plants exposed to very high UV-B irradiance levels (e.g., Tevini et al. 1981). Vu et al. (1982a), for example, observed a 40% reduction in chlorophyll content and a 35% reduction in carotenoids when soybean plants were exposed to high doses of UV-B radiation in a greenhouse where visible light was approximately 200 μmol m^{-2} s^{-1} (PAR, 400-700 nm). Teramura et al. (1980), however, did not observe reduced chlorophyll concentrations within soybean plants exposed to various UV-B radiation and visible light levels (530 to 1600 μmol m^{-2} s^{-1} PAR). Even though chlorophyll concentrations remained similar between the UV-B radiation treated and control plants, photosynthetic rates were shown to be clearly inhibited at low visible light levels. Similarly, photosynthetic rates were shown to be depressed in pea (Pisum sativum) (Brandle et al. 1977) and Rumex patientia L. (Sisson and Caldwell 1976) exposed to moderate UV-B radiation levels. No chlorophyll reductions were noted in these studies when PAR levels were greater than 600 μmol m^{-2} s^{-1}. An increase in chlorophyll concentration has been reported in soybean plants exposed to moderate UV-B radiation levels and 750 μmol m^{-2} s^{-1} PAR, and in field-grown spinach (Spinacia oleracea L.) exposed to high UV-B radiation levels (Esser 1980).

Photostability of chlorophyll appears to be somewhat species-dependent (Monfort 1950) and is readily destroyed by visible light (Shirley 1945), intense UV-C radiation at 254 nm (Cline and Salisbury 1966), and reportedly by UV-B radiation (Basiouny et al. 1978; Vu et al. 1981, 1982a). Tanada and Hendricks (1953) suggested that chlorophyll destruction by UV-C radiation is the result of a disturbance of some basic metabolic process rather than an immediate and direct destruction of chlorophyll by 254 nm radiation. This was later supported by the temperature dependent destruction of chlorophyll noted by El-Mansy and Salisbury (1971). Whether a similar effect on the biosynthesis of chlorophyll or its precursors leads to reduced chlorophyll concentrations in plant tissues exposed to UV-B radiation is not presently known. However, perhaps the initial question that warrants addressing relative to photosynthetic pigments is whether chlorophyll concentrations are altered in plants exposed to UV-B radiation levels equivalent to a reduction in atmospheric ozone of less than 20% under field conditions.

COMPONENT REACTIONS OF PHOTOSYNTHESIS

Most research to date on the effects of UV radiation on photosynthesis has dealt with measurements of photophosphorylation, electron transport, and carbon assimilation. Early studies evaluating the effects of UV-C radiation (primarily 254 nm radiation) showed that electron transport and photophosphorylation were sensitive to this radiation. Jones and Kok (1966a,b) reported that UV-C radiation (254 nm) inhibited photoreduction of DPIP and NADP with H_2O as the electron donor and that both cyclic and noncyclic photophosphorylation were inhibited. The loss of PS II activity and its associated reactions was later suggested to result from the loss of structural integrity of the lamellar membranes of the chloroplast (Mantai et al. 1970).

The disruption of chloroplast membranes by UV-B radiation has been shown to occur in soybean (Campbell 1975), Rumex patientia (Campbell et al. 1975), and pea (Brandle et al. 1977). In all these studies, damage appeared to accumulate with duration of dose. In pea, ultrastructural damage was detected after 0.25 h of UV-B radiation treatment, and 26% of the cells examined exhibited damage after 16 days treatment. The structural damage to the chloroplast lamellar membrane in this study was found to coincide with the decrease in PS II activity. However, the physical disruption that occurred may not have been the only factor causing a decrease in PS II activity since O_2 evolution was restored to within 12% of that of the control plants by an artificial donor (ascorbate-reduced DCPIP). Thus, UV-B radiation was suggested as an inhibitor of PS II, or some specific step in the electron transport associated with PS II apart from the resultant effects of chloroplast disruption. The latter is supported by the results of Okada et al. (1976). In this study, 254 nm radiation directly inhibited the primary photochemistry at the reaction center chlorophyll of PS II.

More recently, Noorudeen and Kulandaivelu (1982) and Iwanzik et al. (1983) showed that the impairment of photosystem II activity by UV-B radiation is due to blockage of photosystem II reaction centers rather than an inhibition of the water-splitting enzyme system. Photosystem I, on the other hand, was found to be unaffected by moderate levels of UV-B radiation in these studies. Renger et al. (in this volume) presents considerable evidence suggesting that UV radiation disrupts the plastoquinones of PS II leading to a loss of activity in the primary acceptor. Results of this study also suggest that, except for quantum efficiency, UV-B and UV-C radiation act similarly at this site.

Comparing the effects of UV-B radiation on C_3 and C_4 plants, Basiouny et al. (1978) found that of the four C_3 plants tested, Hill reaction activity was significantly reduced in collards (<u>Brassica oleracea</u>), oats (<u>Avena sativa</u>), and soybeans. They reported no reductions of Hill activity in peanuts (<u>Arachis hypogaea</u>) (C_3), sorghum (<u>Sorghum bicolor</u>) (C_4), and maize (<u>Zea mays</u>) (C_4). This lower sensitivity displayed by C_4 plants may have been due to sclerification of tissue, nearly vertical orientation of leaves with protective based sheaths, and/or to their narrow leaves. Ultra-violet-B radiation exposure resulted in significant reductions in plant height and fresh or dry weight of the C_3 plants tested; no significant biomass reductions occurred in the C_4 plants.

Monochromatic radiation at 296 nm and 298 nm was found to inhibit RuBP carboxylase activity in tomato (<u>Lycopersicum esculentum</u>) and pea (Thai 1975). Recently, Vu et al. (1982a) reported significant reductions in the activity of this enzyme in soybean (C_3), pea (C_3), and tomato (C_3) exposed to polychromatic UV-B radiation: the reductions in RuBP carboxylase activity correlated well ($r^2 > 0.79$) with reductions in photosynthetic rates in both soybean and tomato. Activity of PEP carboxylase was signifi-cantly decreased in maize (C_4) exposed to the higher UV-B radiation levels while low UV-B dose rates enhanced the activity of this enzyme. Soybean and pea soluble protein levels were significantly decreased at the higher levels of UV-B radiation. These reductions were suggested to be a reflec-tion of the RuBP carboxylase reductions noted, since this enzyme makes up a large fraction of the total leaf protein. Thus, it was suggested that UV-B radiation inhibits protein synthesis leading to the observed reductions in RuBP carboxylase activity. These results were not, however, consistent with the decrease in RuBP carboxylase and increase in soluble leaf proteins found in tomato plants exposed to UV-B radiation. In a similar study, Vu et al. (1982b) suggested that the inhibition of RuBP carboxylase activity by UV-B radiation in soybean was due mainly to protein destruction rather than inactivation.

Tevini et al. (1981) found large increases in soluble proteins in the leaves of maize, bean (<u>Phaseolus vulgaris</u> L.), barley (<u>Hordeum vulgare</u> L.), and radish (<u>Raphanus sativus</u> L.) exposed to very high UV-B radiation levels. Similarly, leaf protein increases were found in potato (<u>Solanum tuberosum</u> L.), spinach, radish, and bean exposed to moderate levels of supplemented UV-B radiation under ambient field conditions (Esser 1980). As suggested by Vu et al. (1982a), more definitive studies are needed to more clearly define the effects of UV-B radiation on soluble protein, and RuBP carboxylase and PEP carboxylase activity.

CARBON ASSIMILATION

The effect of UV-B radiation on carbon assimilation rates has been deter-mined for a relatively large number of agronomically important, and native plant species. With few exceptions, published research suggests that nearly all plants evaluated will be detrimentally affected, to some degree, by an increase in terrestrial UV-B radiation. As previously stated, however, many of these studies need to be repeated under ambient conditions and more subtle levels of UV-B radiation (e.g., < 20% atmospheric ozone reduction simulated). Nevertheless, the noted reductions in photosynthetic capacity of native and agronomically important crop plants may indicate a detrimental effect by an increase in terrestrial UV-B radiation levels. As such, careful experimentation is needed to explore further the potential deleterious effect, if any, and to quantify any observed reductions in the photosynthetic capacity of plants. Of particular importance is the need to carry out any future experiments under high visible fluxes similar to or approaching ambient levels. It is well known that low visible light enhances the deleterious effects of UV-B radiation on the photosynthetic apparatus (Sisson and Caldwell 1977; Teramura et al. 1980; Teramura 1982). Low visible flux has been shown to result in thin leaves, lower chlorophyll a/b ratios, and less UV-B-absorbing pigments (Warner and Caldwell 1983). These factors, and perhaps others mediated by low visible flux levels would tend to decrease the attenuation of UV radiation prior to its absorption by the chloroplast. Thus, the net effect of low visible irradiation levels would be to increase the sensitivity of plant photosynthesizing surfaces to UV-B radiation.

Reported effects on guard cell activity, as measured by stomatal diffusion resistances, has been variable. Brandle et al. (1977) reported that moderate levels of UV-B radiation had no effect on the stomatal resistance of pea. Teramura et al. (1983), on the other hand, reported that the stomatal resistances of radish and cucumber (Cucumis sativus L.) increased; a 3-fold increase developed in cucumber for approximately 8 days, and thereafter, stomatal activity ceased. Bennett (1981) also reported a small increase in stomatal resistance in cucumber, bean, and soybean; the increases noted coincided with photosynthetic rate reductions. Because carbon assimilation is dependent upon functional guard cells of the stomata, and the diffusion of CO_2 to the stomatal cavity, further research seems warranted to determine the impact, if any, of UV-B radiation on guard cell activity.

Perhaps the most intriguing question being addressed relative to carbon assimilation is whether the repression of photosynthesis is dose-dependent and cumulative (i.e., whether reciprocity is maintained). A reciprocal relationship was demonstrated in component reactions of photosynthesis for isolated spinach chloroplasts exposed to UV-C radiation at 254 nm (Jones and Kok, 1966a). They suggested that UV photoinhibition was independent of oxygen and thus would not be similar to the oxygen-dependent photobleaching of chlorophyll by intense visible radiation.

Teramura et al. (1980) demonstrated reciprocity in soybeans exposed to four levels of UV-B radiation. The effect of UV-B radiation was more efficient in reducing photosynthetic rates when plants were exposed to concomitant low visible radiation. Even UV-B irradiance corresponding to ambient levels reduced photosynthesis when PAR intensities were low (528 $\mu E\ m^{-2}\ s^{-1}$ PAR). Trocine et al. (1981) also demonstrated that reciprocity occurred in the photosynthetic inhibition of two of three seagrasses. The differential degree of UV-B sensitivity within these species was suggested to be a function of epidermal cell wall thickness and the associated transmittance properties. Stimulated synthesis of anthocyanin and other flavonoid UV-B radiation-absorbing pigments was also evident in the plants exposed to UV-B radiation.

Reciprocity has also been demonstrated in R. patientia exposed to four levels of UV-B radiation (Sisson and Caldwell 1977). The reciprocal relationship shown in this plant, however, may have partially resulted through a gradual deterioration of the epidermis (Robberecht and Caldwell 1978), thereby increasing the transmittance of UV-B radiation to the sites of the component reactions of photosynthesis during leaf ontogeny. Reductions in photosynthesis were especially evident during the early stages of leaf ontogenesis, and leaf longevity was reduced by increased UV-B radiation levels. However, leaf growth did not display a reciprocal relationship to UV-B radiation, but was found to be dependent upon dose rate and was affected primarily during the early state of leaf ontogeny (Sisson and Caldwell 1976, 1977). In a similar experiment using Cucurbita pepo, photosynthesis was repressed in a cumulative manner in fully expanded leaves but not through leaf ontogenesis (Sisson 1981). Although absorbance of UV-B radiation by extracted pigments (flavonoids and other UV-B radiation-absorbing pigments) increased substantially, UV-B radiation attenuation was apparently insufficient to protect completely the photosynthetic apparatus or leaf growth process.

ACCLIMATION

The ability of plants to acclimate to increased levels of terrestrial UV-B radiation provides one possible avenue for plants to tolerate this radiation without a decrease in productivity or displacement by a more resistant plant. As pointed out by Caldwell (1981), if acclimation results from a phenotypic response already available to plants, the rate of UV-B radiation increases over time may be of little consequence. If, on the other hand, this acclimation process involves a change in genetic composition that may involve considerable time to evolve, then the rate of UV-B radiation increase may be of considerable importance. That plants do possess the ability to acclimate to the UV-B environment was amply supported by a study conducted by Bogenrieder and Klein (1977). They exposed Rumex alpinus seedlings to ambient irradiance after being grown in an environment free of UV-B radiation. The seedlings displayed severely depressed photosynthetic rates and some plants were killed after a 3-day exposure period. Thus, plants are apparently tolerating ambient levels of UV-B irradiance by acclimation processes induced by environmental stimuli that include UV radiation. Although plants appear to possess mechanisms by which they can respond to environmental stimuli and acclimate to changes in radiation, neither the rates nor limits of acclimation are known for any plant. The result of such acclimation will probably determine whether or not any one plant species will be deleteriously affected. Since plants are known to possess a wide range of sensitivities to UV-B radiation, morphological and biochemical responses to increased levels of ambient UV-B radiation might be species dependent. Alternatively, if even sensitive plants possess the capacity to acclimate to small increases in UV-B radiation, the resultant effects may be nonexistent.

REFERENCES

Basiouny FM, Van TK, Biggs RH (1978) Some morphological and biochemical characteristics of C_3 and C_4 plants irradiated with UV-B. Physiol Plant 42:29-32

Bennett JH (1981) Photosynthesis and gas diffusion in leaves of selected crop plants exposed to ultraviolet-B radiation. J Environ Qual 10:271-275

Biggs RH, Kossuth SV (1978) Effects of ultraviolet-B radiation enhancements under field conditions. In: UV-B Biological and climatic effects research (BACER), final report

Biggs RH, Kossuth SV, Teramura AH (1981) Response of 19 cultivars of soybeans to ultraviolet-B irradiance. Physiol Plant 53:19-26

Bogenrieder A, Klein R (1977) Die Rolle des UV-Lichtes beim sog. Auspflanzungsschock von Gewächshaussetzlingen. Angew Bot 51:99–107

Brandle JR, Campbell WF, Sisson WB, Caldwell MM (1977) Net photosynthesis, electron transport capacity, and ultrastructure of Pisum sativum L. exposed to ultraviolet-B radiation. Plant Physiol 60:165–169

Caldwell MM (1981) Plant response to solar ultraviolet radiation. In: Lange OL, Nobel PS, Osmond CB, Ziegler H (eds) Encyclopedia plant physiology, vol 12A. Physiological plant ecology 1 Springer-Verlag, Berlin Heidelberg New York Tokyo, p 169, ISBN 0-387-10763-0

Caldwell MM, Sisson WB, Fox FM, Brandle JR (1975) Plant growth response to elevated UV irradiance under field and greenhouse conditions. In: Nachtwey DS, Caldwell MM, Biggs RH (eds) Impacts of climatic change on the biosphere, part I, ultraviolet radiation effects, monograph 5. Climatic Impact Assessment Program, US Dept Transportation Report No DOT-TST-75-55, NTIS, Springfield, Virginia, p (4)253

Caldwell MM, Gold WG, Harris G, Ashurst CW (1983) A modulated lamp system for solar UV-B (280–320 nm) supplementation studies in the field. Photochem Photobiol 37:479–485

Campbell WF (1975) Ultraviolet-radiation-induced ultrastructural changes in mesophyll cells of soybeans (Glycine max (L.) Merr.). In: Nachtwey DS, Caldwell MM, Biggs RH (eds) Impacts of climatic change on the biosphere, part I, ultraviolet radiation effects, monograph 5. Climatic Impact Assessment Program, US Dept Transportation Report No DOT-TST-75-55, NTIS, Springfield, Virginia, p (4)166

Campbell WF, Chang DCN, Caldwell MM, Sisson WB (1975) Ultrastructural effects of elevated UV-B radiation on Rumex patientia. In: Nachtwey DS, Caldwell MM, Biggs RH (eds) Impacts of climatic change on the biosphere, part I, ultraviolet radiation effects, monograph 5. Climatic Impact Assessment Program, US Dept Transportation Report No DOT-TST-75-55, NTIS, Springfield, Virginia, p (4)179

Cline MG, Salisbury FB (1966) Effect of ultraviolet radiation on the leaves of higher plants. Radiat Bot 6:151–163

El-Mansey HI, Salisbury FB (1971) Biochemical responses of Xanthium leaves to ultraviolet radiation. Radiat Bot 11:325–328

Esser G (1980) Einfluss einer nach Schadstoffimission vermehrten Einstrahlung von UV-B licht auf Kulturpflanzen. 2. Versuchjahr. Bericht Battelle Institut E. V. Frankfurt, BF-4-63, 984-1

Iwanzik W, Tevini M, Dohnt G, Voss M, Weiss W, Gräber P, Renger G (1983) Action of UV-B radiation on photosynthetic primary reactions in spinach chloroplasts. Physiol Plant 58:401–407

Jones LW, Kok B (1966a) Photoinhibition of chloroplast reactions. 1. Kinetics and action spectra. Plant Physiol 41:1037–1043

Jones LW, Kok B (1966b) Photoinhibition of chloroplast reactions. 2. Multiple effects. Plant Physiol 41:1044–1049

Krizek DT (1978) Differential sensitivity of two cultivars of cucumber (Cucumis sativus L.) to increased UV-B irradiance. I. Dose-response studies. In: UV-B biological and climatic effects research (BACER), final report

Mantai KE, Wong J, Bishop NI (1970) Comparison studies of the effects of ultraviolet irradiation on photosynthesis. Biochim Biophys Acta 197:257–266

Monfort C (1950) Photochemische Wirkung des Hohenklimas auf die Chloroplasten photolabiler Pflanzen im Mittel- und Hochgebirge. Z Naturforsch 56:221–226

National Academy of Sciences (1984) Causes and effects of stratospheric ozone reduction: an update 1983. National Academy Press, Washington, DC, p 105, ISBN 0-309-03443-4

Noorudeen AM, Kulandaivelu G (1982) On the possible site of inhibition of photosynthetic electron transport by ultraviolet-B (UV-B) radiation. Physiol Plant 5:161-166

Okada M, Kitajima M, Butler WL (1976) Inhibition of photosystem I and photosystem II in chloroplasts by UV radiation. Plant Cell Physiol 17:35-43

Renger G, Voss M, Gräber P, Schulze A (this volume) Effect of UV irradiation on different partial reactions of the primary processes of photosynthesis. p 171

Robberecht R, Caldwell MM (1978) Leaf epidermal transmittance of ultraviolet radiation and its implications for plant sensitivity to ultraviolet-radiation induced injury. Oecologia 32:277-287

Robberecht R, Caldwell MM, Billings WD (1980) Leaf ultraviolet optical properties along a latitudinal gradient in the arctic-alpine life zone. Ecology 61:612-619

Shirley HL (1945) Light as an ecological factor. Bot Rev 11:497-532

Sisson WB (1981) Photosynthesis, growth, and ultraviolet irradiance absorbance of Cucurbia pepo L. leaves exposed to ultraviolet-B radiation (280-315 nm). Plant Physiol 67:120-124

Sisson WB, Caldwell MM (1976) Photosynthesis, dark respiration, and growth of Rumex patientia L. exposed to ultraviolet irradiance (288-315 nm) simulating a reduced atmospheric ozone column. Plant Physiol 58:563-568

Sisson WB, Caldwell MM (1977) Atmospheric ozone depletion: reduction of photosynthesis and growth of a sensitive higher plant exposed to enhanced UV-B radiation. J Exp Bot 28:691-705

Sisson WB, Campbell WF, Caldwell MM (1974) Photosynthetic and ultrastructural responses of selected plant species to an enhanced UV (280-320 nm) irradiation regime. Am Soc Agron Abs 66:76 (abstract)

Tanada T, Hendricks SB (1953) Photoreversal of ultraviolet effects in soybean leaves. Amer J Bot 40:634-637

Teramura AH (1982) The amelioration of UV-B effects on productivity by visible radiation. In: Calkins J (ed) The role of solar ultraviolet radiation in marine ecosystems. Plenum, New York, p 367, ISBN 0-306-40909-7.

Teramura AH, Biggs RH, Kossuth SV (1980) Effects of ultraviolet-B irradiance on soybean. II. Interaction between ultraviolet-B and photosynthetically active radiation on net photosynthesis, dark respiration, and transpiration. Plant Physiol 65:483-488

Teramura AH, Tevini M, Iwanzik W (1983) Effects of ultraviolet-B irradiance on soybean. I. Effects on diurnal stomatal resistance. Plant Physiol 65:483-488

Tevini M, Iwanzik W, Thoma U (1981) Some effects of enhanced UV-B irradiation on the growth and composition of plants. Planta 153:388-394

Thai VK (1975) Effects of solar ultraviolet radiation on photosynthesis of higher plants. PhD diss, Univ Florida, Gainesville

Trocine RP, Rice JD, Wells GN (1981) Inhibition of seagrass photosynthesis by ultraviolet-B radiation. Plant Physiol 68:74-81

Vu CV, Allen LH Jr, Garrard LA (1981) Effect of supplemental UV-B radiation on growth and leaf photosynthetic reactions of soybean (Glycine max). Physiol Plant 52:353-362

Vu CV, Allen LH, Garrard LA (1982a) Effects of supplemental UV-B radiation on primary photosynthetic carboxylation enzymes and soluble proteins in leaves of C_3 and C_4 crop plants. Physiol Plant 55:11-16

Vu CV, Allen LH Jr, Garrard LA (1982b) Effects of UV-B radiation (280-320 nm) on photosynthetic constituents and processes in expanding leaves of soybean (Glycine max (L.) Merr.). Environ Exp Bot 22:465-473

Warner CW, Caldwell MM (1983) Influence of photon flux density in the 400-700 nm waveband on inhibition of photosynthesis by UV-B (280-320 nm) irradiation in soybean leaves: separation of indirect and immediate effects. Photochem Photobiol 38:341-346

Effect of UV Irradiation on Different Partial Reactions of the Primary Processes of Photosynthesis

G. Renger, M. Voss, P. Gräber and A. Schulze

Technische Universität Berlin, Federal Republic of Germany

ABSTRACT

Spinach (Spinacia oleracea L. cv. Matador) chloroplasts were irradiated with several levels of UV-B and UV-C radiation. The site of impairment of the electron transport in photosystem II was localized. Electrochromic absorption changes at 520 nm, absorption changes of the primary photosystem II acceptor at 320 nm, dichlorophenolindophenyl reduction by photosystem II, oxygen yield per flash, and variable fluorescence and herbicide binding (atrazine, 3-(3,4-dichlorodiphenol)-1,1 dimethylurea [DCMU]) were measured in normal and UV-treated chloroplasts. A decrease of electron transport and variable fluorescence, and also a decreased binding constant for DCMU-type inhibitors was observed after UV treatment. It is suggested that UV irradiation primarily modifies the binding protein of the primary and secondary plastoquinones of PS II acceptor side (Q_A-Q_B apoprotein), simultaneously leading to a functional blocking of the primary acceptor. Except for the quantum efficiency, UV-B and UV-C act at this site in a similar way.

A striking difference between UV-C and UV-B action was observed in the correlation between variable fluorescence and 2,6-dichlorophenol-indophenol (DCIP) reduction as a function of irradiation time: both activities are linearly related in UV-B-treated chloroplasts but nonlinearly in the case of treatment with UV-C radiation. If one accepts that UV-B and UV-C radiation transform PS II reaction centers into dissipative exciton sinks, the differences can be understood with the additional assumption that UV-B radiation interrupts exciton migration between different photosynthetic units, while this migration remains unaffected in UV-C-treated chloroplasts.

UV-C-treated leaves reveal no recovery of the variable fluorescence reflecting electron transport activity, even 5 days after irradiation (5 or 30 min duration).

INTRODUCTION

The exploitation of solar radiation as unique free energy source of the biosphere is realized by the primary processes of photosynthesis. Therefore, any detrimental effect, e.g., caused by UV irradiation, on this reaction sequence implies serious consequences for all other metabolic activities.

NATO ASI Series, Vol. G8
Stratospheric Ozone Reduction, Solar Ultraviolet Radiation and Plant Life. Edited by R. C. Worrest and M. M. Caldwell
© Springer-Verlag Berlin Heidelberg 1986

Previous investigations led to the conclusion that irradiation with UV light with a significant UV-C content predominately affects photosystem (PS) II (Shavit and Avron 1963; Trebst and Pistorius 1965; Jones and Kok 1966b; Mantai and Bishop 1967; Yamashita and Butler 1968; Lichtenthaler and Tevini 1969; Katoh and Kimimura 1974; Okada et al. 1976; Renger et al. 1982), the system which carries out the essential steps of water cleavage (for recent review see Renger 1983). Under conditions excluding UV-C radiation, UV-B radiation also was found to attack primarily the reaction sequence of PS II (Brandle et al. 1977; Van et al. 1977) and its site of action was found to be close to the reaction center (Noorudeen and Kulandaivelu 1982; Iwanzik et al. 1983). However, the detailed mechanism of inhibition still remains to be clarified. Therefore, this chapter attempts to specify the mechanism of UV-radiation-induced deterioration of PS II in more detail. Figure 1 presents a preliminary and simplified scheme of the functional and organizational set-up of PS II (for details see Renger et al. 1984). It shows that two different targets exist for UV radiation: (a) the functionally active pigment components, and (b) the protein matrices acting as apoenzymes. Different methods were applied to analyze effects on both of these possible targets.

In addition to these mechanistic studies, the effect of UV irradiation on whole leaves was investigated.

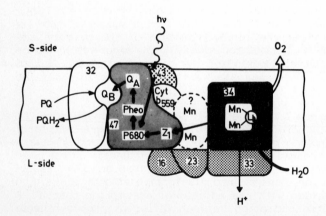

Fig. 1. Simplified scheme of PS II electron transport in photosynthesis. P 680 = photoactive chlorophyll complex; Pheo = pheophytin; Q_A, Q_B = primary and secondary plastoquinone acceptor; D_1 = redox carrier which connects the water oxidizing enzyme system Y with P 680; Mn-Mn = functional manganese complex of system Y. L- and S-side refer to lumenal and stromal side of the thylakoid membrane. The mole masses of the protein components are indicated.

MATERIALS AND METHODS

Chloroplasts were isolated from market spinach according to the method of Winget et al. (1965). Additionally, 10 mM ascorbate was present in the grinding medium. For storage in liquid nitrogen, dimethylsulfoxide (DMSO) was added as cryoprotective agent. Chloroplasts were exposed in quartz cuvettes to the radiation of a germicidal lamp, equipped either with a WG 230 Schott filter (transparent to the 253.7 nm lines of the lamp) or with a WG 305 Schott filter to exclude UV-C radiation. Irradiance at the surface of the petri dishes surrounded by an ice bath (measurements with WG 305) was 1.2 W m^{-2} (UV-B), or 4 W m^{-2} (UV-C + UV-B) when using WG 230.

The standard reaction mixture contained chloroplasts at a concentration of 10 µM, 100 µM or 1 mM chlorophyll (as indicated in the legends of the figures), 2 mM MgCl$_2$, 10 mM KCl and 20 mM 2-(N-morpholino) ethanesulfonic acid (MES-NaOH) buffer, pH = 6.5.

The irradiated samples were transferred into the cuvettes used for measurement of fluorescence induction, 2,6-dichlorophenol-indophenol (DCIP) reduction, polarographic oxygen detection, or absorption changes (the assay conditions are described in the figure legends).

Chlorophyll fluorescence of dark adapted (5 min) chloroplasts was excited with an He-Ne laser (633 nm, actinic light intensity at the sample 4.4 W m^{-2}). Fluorescence was detected at 680 nm by a photomultiplier and recorded by a storage oscilloscope. Fluorescence measurements in whole plants were performed as described by Voss et al. (1984). DCIP reduction under flashing light (200 flashes, duration of flash 15 µs, dark time between flashes 250 ms) was monitored at 580 nm. Absorption changes were measured as described previously (Iwanzik et al. 1983).

Oxygen measurements with an unmodulated Joliot-type O$_2$ electrode and numerical evaluation of the data were performed as described by Dohnt (1984). The herbicide binding was measured by using ^{14}C-labelled atrazine as described previously (Hagemann et al. 1984). Samples incubated with the inhibitor were irradiated with UV light for the period indicated in the figure legends.

RESULTS

The effect of UV irradiation on overall electron transport capacity, including excitation energy transfer, has been analyzed in isolated spinach chloroplasts by comparative measurements of the fluorescence induction and absorption changes reflecting the turnover of photosystem I and II, respectively. The data depicted in Fig. 2 indicate that the photosystem II activity monitored by flash-induced DCIP reduction appears to be less sensitive to UV irradiation than the variable fluorescence. These results are in line with previous findings (Renger et al. 1982). On the other hand, photosystem I activity measured via electrochromic absorption changes at 520 nm in the presence of the PS II inhibitor 3-(3,4-dichlorophenyl)-1,1-dimethylurea (DCMU) and the PS I donor couple DCIPH$_2$/ascorbate appears to be much more resistant to UV-C-induced damage. The 520 nm absorption changes measured when PS I and PS II are operating reveal a dependence on UV-C irradiation similar to the DCIP reduction by PS II (the small differences will not be discussed here). The decrease of variable fluorescence is almost exclusively caused by a decrease of maximum fluorescence, while the initial fluorescence is not significantly affected. This result corresponds to previous findings (Katoh and Kimimura 1974; Renger et al. 1982). The nonlinear relationship between variable fluorescence and PS II activity

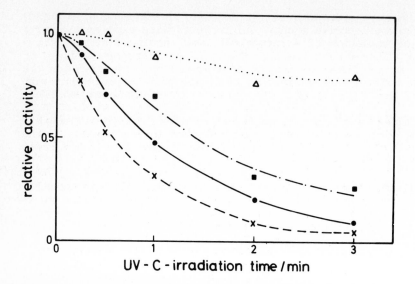

Fig. 2. Relative activity of different partial reactions of photosynthetic electron transport as a function of the UV-C irradiation time: initial amplitude of the electrochromic absorption change at 520 nm reflecting photosystem I activity (triangles), initial amplitude of electrochromic absorption changes at 520 nm reflecting photosystem I and photosystem II activity (squares), extent of flash-induced DCIP reduction (circles), and extent of variable fluorescence (crosses).

(measured as DCIP reduction) can be simply explained using the assumption that UV-C irradiation transforms the reaction centers into dissipative exciton sinks, so that no linear electron flow takes place through PS II, and the fluorescence yield remains constant. If one takes into account the possibility of exciton transfer between the PS II antenna systems (described by the energy transfer probability p) the fluorescence emission of the total ensemble can be calculated by (Joliot and Joliot 1964):

$$F_{max}^{UV}(\underline{q}) = \frac{F_O \, \underline{q} + F_{max}^C \, (1 - \underline{q}) \, (1 - \underline{p})}{1 - \underline{p}(1 - \underline{q})}$$

(1)

where $\underline{q}$ represents the fraction of the PS II centers modified by UV-C irradiation; $F_{max}^{UV}(\underline{q})$, F_{max}^C, maximum fluorescence in UV irradiated and control samples, respectively, and F_O, initial fluorescence. If $[DCIP]_{red}^{UV}$ and $[DCIP]_{red}^C$ symbolize the activity of flash-induced DCIP reduction with water as electron donor in UV-treated and control chloroplasts, respectively, one obtains, with $\underline{q} = 1 - [DCIP]_{red}^{UV}/[DCIP]_{red}^C$ and $F_v = F_{max} - F_O$, by rearrangement of Eq. (1):

$$1 - \frac{[\text{DCIP}]^{\text{UV}}_{\text{red}}}{[\text{DCIP}]^{\text{C}}_{\text{red}}} = \frac{(1 - \underline{p})(1 - F^{\text{UV}}_{\text{V}}(\underline{q})/F^{\text{C}}_{\text{V}})}{1 - \underline{p}(1 - F^{\text{UV}}_{\text{V}}(\underline{q})/F^{\text{C}}_{\text{V}})} \qquad (2)$$

Figure 3 shows that the results in Fig. 2 can be satisfactorily described within the framework of the above model if the exciton transfer probability $\underline{p}$ is on the order of 0.5. This value is in agreement with data obtained in normal chloroplasts (Joliot and Joliot 1964). Thus, it can be inferred that UV-C irradiation does not seriously affect exciton transfer between the antenna systems of different PS II reaction centers. A markedly different pattern, however, is observed if chloroplasts are UV-B irradiated (triangles in Fig. 3). The inhibition of PS II activity and the concomitant decrease of variable fluorescence are much slower under these conditions (about 70% inhibition after 1 h irradiation, data not shown).

A remarkable difference to UV-C treatment, however, lies in the occurrence of an almost linear relationship between $[\text{DCIP}]^{\text{UV}}_{\text{red}}/[\text{DCIP}]^{\text{C}}_{\text{red}}$ and $F^{\text{UV}}_{\text{V}}/F^{\text{C}}_{\text{V}}$. If one accepts that UV-B radiation also transforms the reaction centers into inactive dissipative sinks for excitons (Iwanzik et al. 1983), the results in Fig. 3 (triangles) can be interpreted by the assumption that UV-B radiation, in contrast to UV-C radiation, also distorts the exciton flux between different PS II. According to the tripartite model (Butler and Strasser 1977), the energy transfer between PS II antennae systems could be realized via the light harvesting chlorophyll complexes. Within

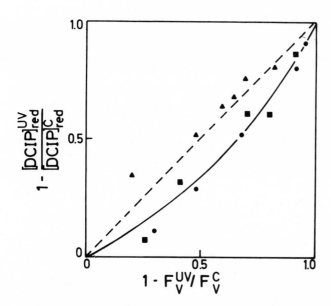

Fig. 3. Relative extent of DCIP reduction as a function of variable fluorescence in UV-B-treated (triangles) and UV-C-treated (circles, squares) samples. Data from Fig. 2 and additional sets of experiments. Squares = 1 mM Chl in suspension medium during irradiation, circles = 100 µM chlorophyll, other experimental conditions as described in Materials and Methods.

the framework of this hypothesis, our data might suggest that UV-B radiation directly attacks the light harvesting complex (LHC) or indirectly modifies the functional connection between LHC and the PS II-antennae system via changes of the overall membrane organization. Further experiments are required to clarify this point.

It has been reported previously that UV-C radiation primarily attacks the water-oxidizing enzyme complex (or Y system) (Yamashita and Butler 1968; Renger et al. 1982). However, a selective inhibition of the Y system, as achieved by other procedures (e.g., tris-(hydroxymethyl) aminomethane (Tris)-washing, was not observed. This was demonstrated by the comparatively small reactivation of PS II electron transport after the addition of typical PS II electron donors (Renger et al. 1982).

If one assumes that the destruction by radiation is a statistical process which noncooperatively destroys the activity of the water oxidizing enzyme system Y with a probability $p_I(Y)$, and the activity of the reaction center with a probability $p_I(RC)$, the sensitivity ratio $p_I(Y)/p_I(RC)$ is given by:

$$\frac{p_I(Y)}{p_I(RC)} = \frac{M(+D) - M(-D)}{M(+D)(1 - M(+D)/M^c)} \tag{3}$$

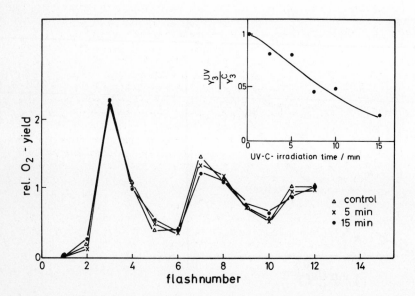

Fig. 4. Normalized pattern of oxygen yield induced by a train of single turnover flashes in normal and UV-C-irradiated chloroplasts. Inset: Relative oxygen yield of the third flash as a function of UV-C radiation. Irradiation time as indicated in the figure, other experimental conditions as described in Materials and Methods.

where M(-D) and M(+D) represent the activities measured in the absence and presence of PS II donors and M^C is the control value. Using the data of Yamashita and Butler (1968) one obtains a ratio close to 1, while our data (not shown) give a ratio of less than 1, i.e., an even higher sensitivity of the reaction center to UV-C radiation than the Y system. Based upon the classic experiments of Joliot et al. (1969, 1971) and Kok et al. (1970) which revealed a characteristic oscillation pattern of oxygen evolution induced by a train of single turnover flashes, each Y system was inferred to be functionally coupled with only one reaction center. Accordingly, the complete blockage of either the reaction center or Y system, or both, should not affect the oscillation pattern, as was experimentally confirmed (Joliot et al. 1971; Kok et al. 1970). The results in Fig. 4 show that UV-C radiation also does not modify the oscillation pattern normalized to the oxygen yield caused by the third flash, Y_3, despite the absolute Y_3-values. Absolute Y_3-values, as well as the steady state yields, are inhibited by more than 75% after 15 min UV-C irradiation (see Fig. 4 insert). The numerical evaluation of the oscillation pattern within the framework of Kok's model confirms that neither the probability of misses nor that of double hits is affected by UV-C irradiation.

Another sensitive target for destruction by UV radiation could be the acceptor side of system II, which contains two functional plastoquinone molecules with absorption bands in the UV-C range, referred to as Q_A and Q_B (see Fig. 1). Electron transfer between these molecules was inferred to be regulated by a protein matrix (Renger 1976) which could also be modified by UV radiation. To test the effects on Q_A, measurements of absorption changes at 320 nm were performed, which mainly reflect the turnover of Q_A and, to a minor extent, reactions in system Y (Renger and Weiss 1983). Absorption changes at 320 nm in UV-B-treated samples are depicted in Fig. 5. Obviously, the total amplitude decreases with prolonged UV-B irradiation. The inhibition of the fast phase exhibits the same dependence on irradiation time as the DCIP reduction activity and the variable fluorescence (data not shown). This effect could be caused either by a blockage of the reaction center itself or by a disconnection of Q_A from Q_B due to a modification of the Q_A-Q_B apoprotein. The latter effect can be achieved by trypsin treatment, which also causes a decrease in the variable fluorescence (Renger et al. 1976). If UV-B treatment primarily attacks the Q_A-Q_B apoprotein leading to a blockage of the electron transfer from Q_A to Q_B, then Q_A^- reoxidation should occur via a slow back reaction with the donor side of system II, provided that the latter remains unaffected by UV-B treatment. However, measurements at a flash frequency of 0.2 Hz and using pulsed measuring light reveal that in contrast to trypsinized chloroplasts (Renger and Weiss 1982) the amplitude of the 320 nm absorption change is not enhanced in UV-B irradiated chloroplasts under conditions which kinetically favor a high degree of back reaction (data not shown). Accordingly, the UV-B effect cannot be ascribed exclusively to a modification of the Q_A-Q_B apoprotein. However, the markedly lower sensitivity of the slower relaxation kinetics could still reflect some changes in the Q_A-Q_B-apoprotein leading to a retardation of the Q_A^- reoxidation in a certain fraction of the PS II units. In UV-C-treated chloroplasts the sensitivity difference between linear electron transport (measured as DCIP reduction) and 320 nm absorption changes appears to be more pronounced.

A structural modification of the Q_A-Q_B apoprotein should also give rise to a change in the action of PS II herbicides (Pfister and Arntzen 1979; Renger 1979; Trebst 1979). Figure 6 shows the DCMU effect on DCIP reduc-

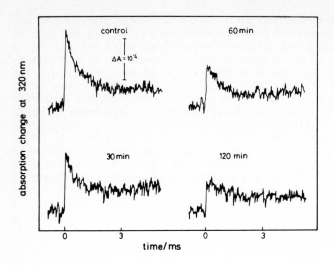

Fig. 5. Flash-induced absorption changes at 320 nm in normal and UV-B-irradiated chloroplasts. Irradiation time as indicated in the figure; 80 signals were averaged, other conditions as described in Materials and Methods.

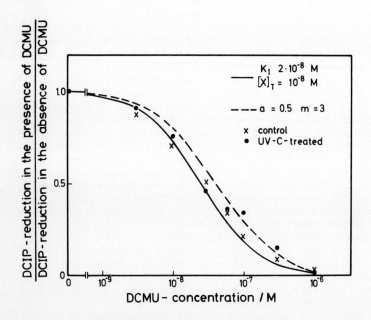

Fig. 6. Dependence of DCIP-reduction activity on DCMU concentration in normal and UV-C-irradiated chloroplasts. In the absence of DCMU the activity of the irradiated sample was 40%–50% of that of the normal chloroplasts (irradiation time 1 min, see Fig. 2). Theoretical curves are calculated according to Eq. (4) and Eq. (5).

tion in normal and UV-C-treated chloroplasts. Only slight differences are observed. For comparison, theoretical curves are presented which are calculated on a simple model including only one herbicide binding site per system II, characterized by a dissociation constant $K_I = [X][I]/[XI]$, where [X] and [XI] are the concentrations of free and occupied binding sites and [I] is the concentration of free inhibitor molecules. Furthermore, it is assumed that each occupied site completely blocks the electron transfer to DCIP. [DCMU changes the binding site very slowly, so that effects due to the dynamics of herbicide binding can be neglected, (Vermaas et al. 1984).] If, in a fraction 1-a of system II, the K_I value is changed by a factor m, and if the number of inhibitor molecules unspecifically bound to the thylakoid membrane is negligibly small, one obtains, with c_T = total inhibitor concentration and $[X]_T$ = total concentration of binding sites:

$$\frac{[DCIP]_{red}^{DCMU}}{[DCIP]_{red}^{c}} = \frac{a \, K_I}{K_I + [I]} + \frac{(1-a) \, m \, K_I}{m \, K_I + [I]} \tag{4}$$

where [I] can be calculated from the equation of mass balancy:

$$[I] = c_T - \frac{a \, [X]_T[I]}{K_I + [I]} + \frac{(1-a)[X]_T[I]}{m \, K_I + [I]} \tag{5}$$

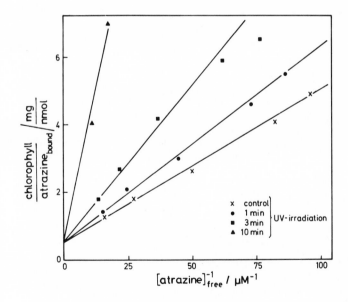

Fig. 7. Reciprocal concentration of bound atrazine as function of the reciprocal concentration of free atrazine in normal and UV-C irradiated chloroplasts. UV-C-irradiation time as indicated in the figure, other conditions as described in Materials and Methods.

A comparison of the experimental data with the theoretical curves (Renger and Schulze unpublished) suggests that UV-C irradiation affects DCMU action slightly, but an unambiguous conclusion cannot be drawn from these results. Therefore, the binding of DCMU-type inhibitors was measured directly using ^{14}C-labeled atrazine. The results obtained are depicted in Fig. 7, which shows that with increasing duration of UV-C irradiation the slope of the curves becomes steeper, while the intersection with the ordinate remains unaffected. This indicates (Tischer and Strotmann 1977) that UV-C irradiation causes a decrease in inhibitor binding strength without changing the number of binding sites. Accordingly, the Q_A-Q_B apoprotein appears to be a likely target of the damaging effects induced by UV-C radiation. A modification of the binding capacity has also been observed in UV-B irradiated chloroplasts (Hagemann and Renger unpublished).

The data presented so far were obtained with isolated chloroplasts. It was found that UV irradiation (either UV-B or UV-C) generally leads to a decrease of variable fluorescence. Accordingly, it appears worthwhile to study the effect of UV radiation on fluorescence induction in intact leaves. We studied the question of whether a plant leaf exposed to a temporary UV-C irradiation possesses a repair mechanism. Plants were irradiated with a germicidal lamp for different times, and fluorescence induction curves were monitored each day after the UV exposure. The data in Fig. 8 show that variable fluorescence drastically decreases with increasing irradiation time. These findings correspond with recently reported data on other UV-sensitive plants (Smillie 1983). After heavy damage no recovery could be observed.

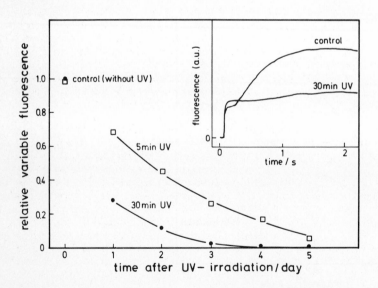

Fig. 8. Relative variable fluorescence of UV-C-irradiated spinach leaves as a function of time after irradiation. Inset: Typical traces of normal and UV-C-treated leaves.

DISCUSSION

The results presented in this study confirm previous conclusions that UV irradiation primarily attacks the reaction center of PS II (for recent discussion of this point see Iwanzik et al. 1983). The underlying mechanism is still an open question. Phenomenologically, the UV effect resembles in many respects the effect of trypsin treatment (Renger 1976; Renger et al. 1976), i.e., decrease of linear electron flow and of variable fluorescence as well as decreased binding capacity for DCMU-type inhibitors (Tischer and Strotmann 1979; Trebst 1979). This suggests that UV effects on the reaction center of PS II is due to a destruction of the Q_A-Q_B apoprotein. However, there exists an essential difference between the actions of trypsin and UV irradiation, that is, the reactivity of Q_A. In trypsinized chloroplasts, Q_A remains functional and the reoxidation of Q_A^- can occur either via internal back reactions with the donor side (Renger and Weiss 1982) or by use of $K_3Fe(CN)_6$ as exogenous electron acceptor (Renger 1976). In contrast, in UV-irradiated chloroplasts the turnover of Q_A appears to be totally blocked. A total blockage has also been observed in part of the system II reaction centers after prolonged trypsin treatment of Tris-washed chloroplasts. This effect was explained by the assumption that extensive trypsinization functionally disconnects Q_A (Renger and Eckert 1981). Accordingly, on the basis of the present results it is tentatively assumed that UV irradiation primarily destroys the Q_A-Q_B apoprotein, leading simultaneously to a blockage of Q_A function. Apparently no significant differences exist in the effects of UV-B and UV-C radiation, except in the quantum efficiency of the damaging processes (Jones and Kok 1966a).

However, mechanistic differences between UV-B and UV-C radiation were found in relation to their inhibitory effects on exciton migration and electron transport activity, respectively. In UV-C-treated samples the nonlinear relationship could be simply explained by the assumption that the modified PS II units act as dissipative exciton sinks, whereas the application of the same model for the explanation of UV-B effects requires an additional assumption, namely the interruption of exciton migration between different PS II units. This effect will be analyzed in a forthcoming paper.

Another difference in the action of UV-B and UV-C radiation is the effect on the water oxidizing enzyme system Y. Under UV-C treatment both units, system Y and the reaction centers, are destroyed, with the reaction center being the more sensitive target. In contrast to that, system Y appears to be rather insensitive to UV-B radiation (Iwanzik et al. 1983), unless the reaction center remains unaffected. It remains to be clarified whether a destruction of the reaction center function also modifies the stability of system Y to UV-B irradiation. The differences in the action of UV-C and UV-B irradiation might be due to the fact that polypeptides which are assumed to constitute system Y or are essential for the functional coupling with the reaction center (Renger and Akerlund 1983) exhibit absorption maxima at 277 nm, while the absorption at $\geq$ 300 nm appears to be rather small (Jansson 1984).

The present analysis favors the idea that UV radiation attacks the Q_A-Q_B apoprotein. However, if this were the only effect, a rapid recovery of photosynthetic activity could be anticipated when the 32 kDa herbicide-binding polypeptide is mainly involved in the damage process, because this protein turns over rather rapidly (Mattoo et al. 1981). The results obtained in the UV-C treated leaves show that no recovery takes place. Therefore, irreversible effects occur in addition to the effects on the 32 kDa polypeptide.

ACKNOWLEDGMENTS

The authors would like to thank Dipl.-Phys. G. Dohnt, Dipl.-Chem. R. Hagemann, Dipl.-Chem. M. Völker and Dipl.-Phys. W. Weiss for their contribution to this paper and T. Wesolowski for very skillful assistance. The financial support of the Bundesministerium für Forschung und Technologie (KBF 46) is gratefully acknowledged.

REFERENCES

Brandle JR, Campbell WF, Sisson WB, Caldwell MM (1977) Net photosynthesis, electron transport capacity, and ultrastructure of Pisum sativum L. exposed to ultraviolet-B radiation. Plant Physiol 60:165-168

Butler WL, Strasser RJ (1977) Tripartite model for the photochemical apparatus of green plant photosynthesis. Proc Natl Acad Sci USA 74:3382-3385

Dohnt G (1984) Polarographische und fluorometrische Untersuchungen am Photosystem II. Thesis TU Berlin

Hagemann R, Vermaas W, Renger G (1984) Participation of a lysine-containing polypeptide in herbicide binding at the acceptor site of photosystem II. In: Sybesma C (ed) Advances in photosynthesis research, vol 4. Martinus Nijhoff/Dr. W. Junk Publ, The Hague, p 17

Iwanzik W, Tevini M, Dohnt G, Voss M, Weiss W, Gräber P, Renger G (1983) Action of UV-B radiation on photosynthetic primary reactions in spinach chloroplasts. Physiol Plant 58:401-407

Jansson C (1984) Proteins involved in photosynthetic oxygen evolution isolation and characterization. In: Sybesma C (ed) Advances in photosynthesis research, vol 4. Martinus Nijhoff/Dr. W. Junk, The Hague, p 375

Joliot A, Joliot P (1964) Etude cinétique de la réaction photochimique libérant l'oxygène au cours de la photosynthèse. C R Acad Sci Paris 258:4622-4625

Joliot P, Barbieri G, Chabaud R (1969) Un nouveau modele des centres photochimique du système II. Photochem Photobiol 10:309-324

Joliot P, Joliot A, Bouges B, Barbieri G (1971) Studies of system II photocenters by comparative measurements of luminescence, fluorescence and oxygen emission. Photochem Photobiol 11:287-305

Jones LW, Kok B (1966a) Photoinhibition of chloroplast reactions. I. Kinetics and action spectra. Plant Physiol 41:1037-1043

Jones LW, Kok B (1966b) Photoinhibition of chloroplast reactions. II. Multiple effects. Plant Physiol 41:1044-1049

Katoh S, Kimimura M (1974) Light-induced changes of C-550 and fluorescence yield in ultraviolet-irradiated chloroplasts at room temperature. Biochim Biophys Acta 333:71-84

Kok B, Forbush B, McGloin M (1970) Cooperation of charges in photosynthetic O_2 evolution - I. A linear four step mechanism. Photochem Photobiol 11:457-475

Lichtenthaler HK, Tevini M (1969) Die Wirkung von UV-Strahlen auf die Lipochinon-Pigmentzusammensetzung isolierter Spinatchloroplasten. Z Naturforsch 24:764-769

Mattoo AK, Hoffmann-Falk H, Pick U, Edelman M (1981) The rapidly metabolized 32000-Dalton polypeptide is the "proteinaceous shield" regulating photosystem II electron transport and mediating diuron herbicide sensitivity. Proc Natl Acad Sci USA 78:1572-1576

Mantai KE, Bishop NI (1967) Studies on the effects of ultraviolet irradiation on photosynthesis and on the 520 nm light-dark difference spectra in green algae and isolated chloroplasts. Biochim Biophys Acta 131:350-356

Noorudeen AM, Kulandaivelu G (1982) On the possible site of inhibition of photosynthetic electron transport by ultraviolet-B (UV-B) radiation. Physiol Plant 55:161-166

Okada M, Kitajima M, Butler WL (1976) Inhibition of photosystem I and photosystem II by UV radiation. Plant Cell Physiol 17:35-43

Pfister K, Arntzen CJ (1979) The mode of action of photosystem II specific inhibitors in herbicide-resistant weed biotypes. Z Naturforsch 34c:996-1009

Renger G (1976) Studies on the structural and functional organization of system II of photosynthesis. The use of trypsin as a structurally selective inhibitor at the outer surface of the thylakoid membrane. Biochim Biophys Acta 440:287-300

Renger G (1979) Studies about the mechanism of herbicidal interaction with photosystem II in isolated chloroplasts. Z Naturforsch 34c:1010-1014

Renger G (1983) Photosynthesis. In: Hoppe W, Lohmann W, Markl H, Ziegler H (eds) Biophysics. Springer, Berlin Heidelberg New York, p 515

Renger G, Akerlund HE (1983) On the functional role of the 32 kDa polypeptide for photosynthetic water oxidation. In: Inoue Y, Crofts AR, Govindjee, Murata N, Renger G, Satoh K (eds) Oxygen-evolving system of plant photosynthesis. Academic Press, Japan, p 209

Renger G, Eckert HJ (1981) Studies on the functional organization of photosystem II. The effect of protein-modifying procedures and of ADRY-agents on the reaction pattern of photosystem II in Tris-washed chloroplasts. Biochim Biophys Acta 638:161-171

Renger G, Weiss W (1982) The detection of intrinsic 320 nm absorption changes reflecting the turnover of the water-splitting enzyme system Y which leads to oxygen formation in trypsinized chloroplasts. FEBS Lett 137:217-221

Renger G, Weiss W (1983) Spectral characterization in the ultraviolet region of the precursor of photosynthetically evolved oxygen in isolated trypsinized chloroplasts. Biochim Biophys Acta 722:1-11

Renger G, Erixon K, Döring G, Wolff C (1976) Studies on the nature of the inhibitory effect of trypsin on the photosynthetic electron transport of system II in spinach chloroplasts. Biochim Biophys Acta 440:278-286

Renger G, Gräber P, Dohnt G, Hagemann R, Weiss W, Voss M (1982) The effect of UV irradiation on primary processes of photosynthesis. In: Biological effects of UV-B radiation. Gesellschaft für Strahlen- und Umweltforschung mbH, München, p 110, ISSN 0721-1694

Renger G, Völker M, Weiss W (1984) Studies on the nature of the water-oxidizing enzyme: 1. The effect of trypsin on the system-II reaction pattern in inside-out thylakoids. Biochim Biophys Acta 766:582-591

Shavit N, Avron M (1963) The effect of ultraviolet light on photophosphorylation and the Hill reaction. Biochim Biophys Acta 66:187-195

Smillie RN (1983) Chlorophyll fluorescence in vivo as a probe for rapid measurements of tolerance to ultraviolet radiation. Plant Sci Letters 28:283–289

Tischer W, Strotmann H (1977) Relationship between inhibitor binding by chloroplasts and inhibition of photosynthetic electron transport. Biochim Biophys Acta 460:113–125

Tischer W, Strotmann H (1979) Some properties of the DCMU-binding site in chloroplasts. Z Naturforsch 34c:992–995

Trebst A (1979) Inhibition of photosynthetic electron flow by phenol and diphenylether herbicides in control and trypsin-treated chloroplasts. Z Naturforsch 34c:986–991

Trebst A, Pistorius E (1965) Photosynthetische Reaktionen in UV-bestrahlten Chloroplasten. Z Naturforsch 20b:885–889

Van TK, Garrard LA, West SH (1977) Effects of 298 nm radiation on photosynthetic reactions of leaf discs and chloroplast preparations of some crop species. Environ Exp Bot 17:107–112

Vermaas WFJ, Dohnt G, Renger G (1984) Binding and release of inhibitors of Q_A^- oxidation in thylakoid membranes. Biochim Biophys Acta 765:74–83

Voss M, Renger G, Gräber P (1984) Fluorometric detection of reversible and irreversible defects of electron transport induced by herbicides in intact leaves. In: Sybesma C (ed) Advances in photosynthesis research, vol 4. Martinus Nijhoff/Dr. W. Junk Publ, The Hague, p 53

Winget G, Izawa S, Good NE (1965) Stoichiometry of photophosphorylation. Biochem Biophys Res Commun 21:438–443

Yamashita T, Butler WL (1968) Inhibition of chloroplasts by UV irradiation and heat treatment. Plant Physiol 43:2037–2040

Effects of Ultraviolet Radiation on Fluorescence Induction Kinetics in Isolated Thylakoids and Intact Leaves

L. O. Björn, J. F. Bornman and E. Olsson

University of Lund, Sweden

ABSTRACT

Action spectra were determined for the UV-induced delay of fluorescence induction (expressed as increase of half-rise time) in leaves of _Elodea densa_ Casp. and _Oxalis deppei_ Lodd. _Elodea_ leaves showed increasing sensitivity with wavelength decreasing from 310 to 280 nm, while the _Oxalis_ leaves, when irradiated from the abaxial side, gave a rather flat spectrum with lower sensitivity than _Elodea_ throughout the range. When _Oxalis_ leaves were irradiated from the adaxial side, UV sensitivity was even lower. The difference is partly explainable by a higher content of water-soluble, UV-absorbing substances in the adaxial epidermis. The action spectra are compared to that determined earlier for isolated spinach thylakoids. We also analyzed in more detail the effect of ultraviolet radiation on fluorescence kinetics at selected wavelengths. The fluorescence induction in isolated spinach thylakoids is the sum of three first order processes with different time constants. The main effect of ultraviolet radiation (280 nm) is to decrease the amplitude of the intermediate component. It also slightly retards the slow component. Results with 320-nm radiation were similar.

Fluorescence induction in _Elodea_ leaves can be described as the sum of two first order components. The main effect of 320-nm radiation was to decrease the rate constant of the slow component.

The fluorescence rise in _Oxalis_ leaves (and also in spinach leaves) has a biphasic pattern with a plateau in the middle. It is interpreted using a model with two kinds of photosystem II units: one set of unconnected units and one set with an energy transfer probability of about 0.65. The main effect of ultraviolet radiation is to diminish the fluorescence from the unconnected units, which have the largest rate constant.

INTRODUCTION

In another paper (Bornman et al. 1984) we described an action spectrum for the UV-induced decrease of the rate of fluorescence induction in isolated spinach thylakoids. In this chapter we analyze the kinetics of the induction more carefully. Spectral and kinetic analyses of the UV effects on the fluorescence induction in intact leaves of one aquatic (_Elodea densa_) and one terrestrial (_Oxalis deppei_) plant were carried out and internal screening effects were also evaluated.

NATO ASI Series, Vol. G8
Stratospheric Ozone Reduction, Solar Ultraviolet Radiation and Plant Life. Edited by R. C. Worrest and M. M. Caldwell
© Springer-Verlag Berlin Heidelberg 1986

MATERIALS AND METHODS

Spinach (Spinacia oleracea L.) was grown and thylakoids isolated as described elsewhere (Bornman et al. 1984). Elodea densa Casp. was purchased from a commercial supplier and kept in an aquarium at about 22.5°C. Illumination was about 40 µmol m^{-2} s^{-1} photosynthetically active radiation from a Philips TL F20W/33 fluorescent tube for 8 h per day, and the plants were kept under these conditions for at least a week before use. Oxalis deppei Lodd was grown in pots outside (unusually sunny and dry summer conditions in Lund, southern Sweden).

The apparatus and procedure for UV irradiation and fluorescence measurements are described in Bornman et al. (1984). Whole Elodea leaves were inserted into the water-filled quartz cuvette. From Oxalis leaves 3-mm wide strips were cut with razors and inserted into the air-filled cuvette. Before UV irradiation two control measurements were done on the unirradiated sample to ensure stable conditions.

The temperature of the cuvette holder was 10 $\pm$ 0.2°C for spinach thylakoids, 20.0 $\pm$ 0.2°C for Elodea and 18.0 $\pm$ 0.2°C for Oxalis leaves.

The influence of wavelength on the effect of UV irradiation was investigated for the leaves as previously for isolated thylakoids. Due to the amount of numerical work involved we have analyzed the spectral data in terms of amplitudes and rate constants of different kinetic components only at selected wavelengths, and for other wavelengths only recorded the half-rise time (the time taken for the fluorescence to rise from the initial (F_O) level to $F_O + (F_{max} - F_O)/2$ where F_{max} is the maximum fluorescence. It should be noted that the action spectrum for the effect of UV on spinach thylakoids described in Bornman et al. (1984) is based on the time to reach $F_O + (F_{max} - F_O)$ x 0.75; so allowance has to be made for this (by multiplication by the factor 0.659, computed from data in Table 1) when absolute sensitivities are compared.

In all cases the time was expressed as percentage increase over the control value for the same sample before irradiation, and this percentage increase was divided by the UV fluence to obtain ordinates for the action spectra. For Elodea leaves results for various irradiation times (i.e., for various fluences) were simply averaged. In the case of Oxalis the values tended to decrease with increasing fluence, and therefore a linear regression analysis (least squares fit) with extrapolation to zero fluence was carried out. In the case of Elodea we did not distinguish between irradiations from the adaxial and abaxial sides of the leaf, but in the case of Oxalis such a distinction was necessary as the leaves were much less sensitive to irradiation from the adaxial side. Water-soluble UV-absorbing substances were assayed in the following way: epidermal strips were peeled from the abaxial and adaxial sides separately, air-dried to constant weight and extracted with quartz distilled water for about 1 h with occasional gentle agitation (no homogenization). The absorbance of the supernatant was recorded from 250 to 400 nm, after dilution when necessary and with quartz distilled water as reference.

RESULTS AND DISCUSSION

Fluorescence from the spinach thylakoids rose from the F_O level to the F_{max} level along a curve without any maximum or inflection point (Figs. 1A and 1B). The induction curve (F_{max} − F vs. time) was the sum of three exponential (first order) components with time constants 840 ± 160 s^{-1}, 95 ± 1 s^{-1} and 23 ± 2 s^{-1}.

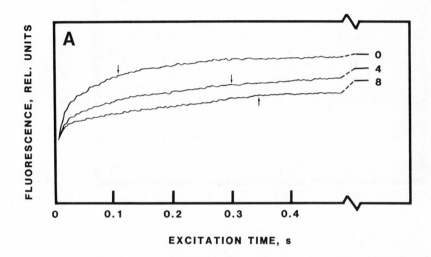

Fig. 1A. Typical recordings of fluorescence induction in spinach thylakoids. Numbers on the curves indicate minutes of exposure to 22.9 μmol m^{-2} s^{-1} of 320-nm radiation. The maximum fluorescence (F_{max}) is indicated to the right of the curves which start at the F_O level. The gap between the left frame and the start of the curves represents the opening time of the shutter. The arrows indicate the times required for 75% of the fluorescence rise.

The main effect of 280-nm radiation was a decrease of the amplitude of the intermediate component (Table 1). This occurred in such a way that the increase in time taken for the combined variable fluorescence components to rise to 75% of the maximum value was linear with UV fluence (linear with UV-irradiation time). The increase in time required to rise to 50% of the maximum value was, on the contrary not linear with UV fluence, but leveled off with increasing fluence. The effect of 320-nm radiation was qualitatively similar.

In the case of Elodea leaves the fluorescence induction (Figs. 2A and 2B) could adequately be described as the sum of two exponential functions. UV radiation (320 nm) decreased the rate constant, and to a lesser extent the amplitude of the slow component (Table 2). With 280-nm radiation the amplitude of the fast, but not that of the slow component decreased (results not shown).

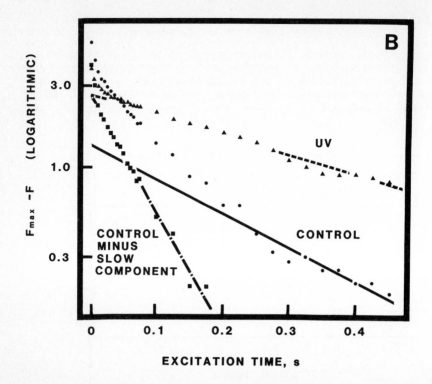

EXCITATION TIME, s

Fig. 1B. Semilogarithmic plot of data from Fig. 1A. Dots show data for nonirradiated (control) sample, and triangles data from sample treated for 8 min with 320-nm irradiation. The corresponding slow first order components are indicated by solid and dashed lines. Squares indicate the difference between measured control data and the solid line, and the dash-dot line is the intermediate first order component of the control (note that the first squares to the left lie above this line, indicating the presence of an even faster component). The intermediate component is practically absent from the UV-irradiated sample.

The fluorescence induction in <u>Oxalis</u> leaves exhibited more complicated kinetics. To show the changes over a wide time range we replotted the original recordings using a logarithmic time scale. In this way we could combine the results from 3 to 4 experiments in different time domains into a single diagram.

The fluorescence of <u>Oxalis</u> leaves (Fig. 3) rose rapidly from the initial (F_O) level to the intermediate level (I; cf. Govindjee and Papageorgiou 1971) where it leveled out. Somewhat later there was a second, slower rise to the F_{max} level, and then a biphasic decline. It was found that the blue light used for exciting fluorescence was strong enough to cause damage to the leaf when kept on for about the 100 s required for a curve like that in Fig. 3. Thus the final decline in fluorescence was partly an irreversible process. The "control" curve in Fig. 3 was recorded as second in a series of measurements; the variable fluorescence for the first control curve (not shown) was about 40% larger. To avoid this damage during extended irradiation with blue light, it was decided to study more closely only the rising part of the curve and the effect of UV on this rise. This required only 2-s exposures to the blue light, which was not long enough to cause irreversible change. Nevertheless, Fig. 3 shows that the UV exposure employed (in this case 22.5 mmol m^{-2} of 340 nm radiation) did not eliminate the first phase of fluorescence decline, rather it only (together with the blue light exposure) decreased it in approximate proportion to the decrease in fluorescence rise.

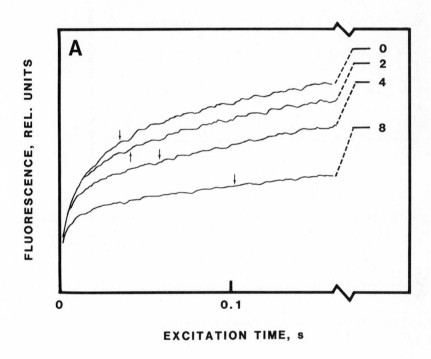

Fig. 2A. Typical recordings of fluorescence induction in an Elodea leaf. Numbers on the curves indicate minutes of exposure to 22.9 μmol m^{-2} s^{-1} of 320 nm, and arrows show half-rise times. Otherwise as for Fig. 2B.

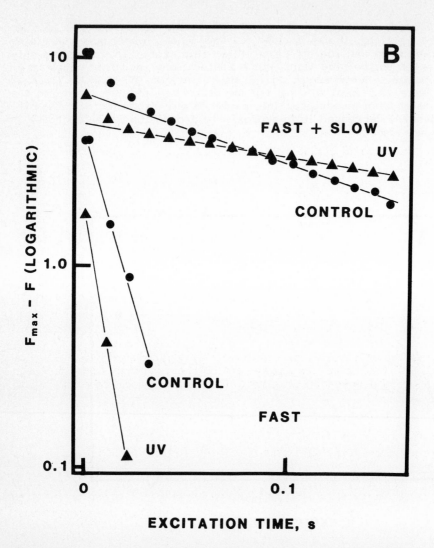

Fig. 2B. Semilogarithmic plot of data in Fig. 2A for 0 (control, dots) and 8 (triangles) minutes 320-nm irradiation (upper part of figure). The lines indicate slow first order component. Symbols in the lower part of the figure show difference between measured data and slow component.

Table 1. Effect of ultraviolet radiation (280 nm, 12.6 µmol m^{-2} s^{-1}) on amplitudes (A$_i$) and rate constants (R$_i$) of the three components of fluorescence induction in spinach thylakoids. The amplitudes are expressed as percentage of the average variable fluorescence of nonirradiated thylakoids, the rate constants in s^{-1}. Means and standard errors from seven experiments.

Irradiation time (min)	Fast component		Intermediate component		Slow component	
	A$_1$(%)	R$_1$(s^{-1})	A$_2$(%)	R$_2$(s^{-1})	A$_3$(%)	R$_3$(s^{-1})
0	27.2$\pm$3.3	840$\pm$160	45.6$\pm$5.3	95$\pm$13	27.2$\pm$2.9	22.9$\pm$1.5
3	23.1$\pm$0.7	793$\pm$55	16.2$\pm$2.7	75$\pm$4	38.1$\pm$1.7	17.4$\pm$0.7
6	21.4$\pm$1.4	853$\pm$50	5.0$\pm$2.4	84$\pm$14	34.9$\pm$4.3	13.2$\pm$1.1

Table 2. Effect of ultraviolet radiation (320 nm, 22.9 µmol m^{-2} s^{-1}) on amplitudes (A$_i$) and rate constants (R$_i$) of the two components of fluorescence induction in leaves of _Elodea_. Five experiments, otherwise as in Table 1.

Irradiation time (min)	Fast component		Slow component	
	A$_1$(%)	R$_1$(s^{-1})	A$_2$(%)	R$_2$(s^{-1})
0	33.5$\pm$5.7	129$\pm$21	59.8$\pm$6.8	13.2$\pm$2.9
2	36.0$\pm$3.9	97$\pm$21	46.0$\pm$7.3	6.8$\pm$1.3
4	33.0$\pm$4.8	108$\pm$28	42.8$\pm$6.0	5.2$\pm$0.8
6	35.5$\pm$9.4	93$\pm$21	40.0$\pm$6.2	3.9$\pm$0.7

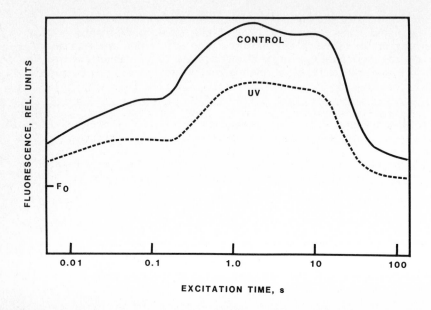

FLUORESCENCE, REL. UNITS

CONTROL

UV

F_0

0.01   0.1   1.0   10   100

EXCITATION TIME, s

Fig. 3. Time course of fluorescence from an Oxalis leaf. The "control" was recorded as second in a series of measurements when the leaf had already been exposed to the strong fluorescence-exciting blue light (90 μmol m^{-2} s^{-1} centered at 440 nm) for 100 s. The first run with the dark-adapted leaf gave about 40% higher amplitude of variable fluorescence. The UV treatment consisted of 22.5 mmol m^{-2} of 340 nm radiation. Note the logarithmic time-scale. The magnitude of initial fluorescence (F_0) is indicated.

Figure 4 shows the fluorescence rise in an Oxalis leaf that had not been damaged by overexposure to blue light (dots). This rise can also be described as the sum of two components. In Oxalis however, contrary to spinach thylakoids and Elodea leaves, only the first component followed a simple exponential (first order) law. The slow component showed a lag. A similar pattern was found by Joliot and Joliot (1964) for the fluorescence induction in Chlorella. They explained this (also explained more fully by Lavorel and Joliot 1972) as being due to energy transfer between photo-synthetic units. It seems that this explanation fits our experiments, and we therefore use their mathematical treatment, although other explanations have been advanced for such a lag. The model is strictly valid only for cells poisoned with DCMU to prevent reoxidation of reduced Q (electron acceptor of photosystem II). However, as Joliot and Joliot (1964) point out, if very strong exciting light is used (as is the case here), this makes very little difference in principle since reduction of Q is much faster than its reoxidation. Therefore we used the equation system of Lavorel and Joliot (1972) to simulate our experimental curves:

$$\Phi_F = [(1 - p)(1 - Q)]/[1 - p(1 - Q)] \tag{1}$$

$$-p + (1 - p) \ln Q + pQ = R_2t \tag{2}$$

In these equations Φ_F is fluorescence yield, Q the proportion of photosystem II traps that are open, p the probability of interunit energy transfer, R_2 a rate constant and t time after switching on the exciting light. Φ_F cannot be expressed as an explicit function of t, but since values of both Φ_F and t can be computed for various values of Q, theoretical $\Phi_F = f(t)$ relations could be plotted together with the experimental values in Fig. 4. For curve fitting the following five parameters could be varied: R_2, p and an amplitude factor A_2 for the slow component, and a rate constant R_1 and an amplitude factor A_1 for the fast component ($A_1 + A_2 = F_{max} - F_o$).

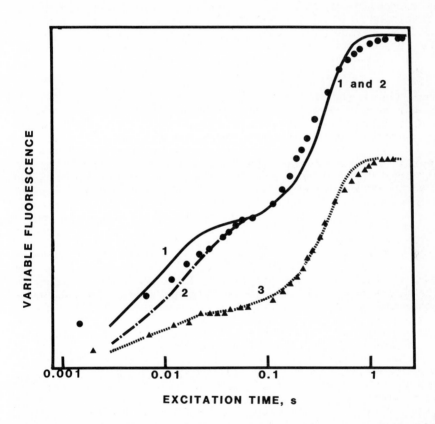

Fig. 4. Effect of UV radiation (11.7 mmol m^{-2} of 310 nm radiation) on the rising part of the fluorescence induction curve of an Oxalis leaf. In this case only brief exposures to blue light were used to avoid photoinhibition by blue light. Experimental values are denoted by dots (control) and triangles (UV irradiated); theoretical models 1, 2 and 3 (cf. Table 3) by solid, dash-dot and dotted lines, respectively. For times exceeding 0.1 s, models 1 and 2 coincide.

Our computing capacity did not permit, within reasonable time, an exact optimization of all parameters. We show two approximations for a non-irradiated leaf (Models 1 and 2 in Table 3) and one for the same leaf after exposure to a fluence of 11.7 mmol m^{-2} from 310-nm radiation (Model 3 in Table 1, cf. Fig. 4). Somewhat surprisingly, it seems that the apparent slowing down in this case is not due to a decrease of the rate constants of the slow component, but mostly to the fact that the amplitude of the fast component is decreased more than the amplitude of the slow one. The effect is distinctly different from the irreversible inhibition of the slow component brought about by extended irradiation with high-intensity visible light (Critchley and Smillie 1981). The latter authors have a different interpretation of the biphasic fluorescence rise than we do. According to them, the rapid component is due to electron transfer from photosystem II (and primary electron donors on its oxidizing side) to Q, while the slow component would represent electron transfer from the water splitting system, or intermediate electron carriers near to it. It is difficult to understand our results in the light of this model, since in our experiments the fast component is eliminated before the slow one. The fast reaction thus cannot be a prerequisite for the slow one as in the model of Critchley and Smillie. Besides, two consecutive reactions cannot result in a product curve with an inflection point, at least not if both reactions are first order.

As seen from Fig. 4, the final approach to the F_{max} level took place too rapidly with all models, as compared to the experimental data. This may be due to the fact that the samples were not DCMU poisoned and that Q reoxidation cannot be neglected when the reduction level is close to 100%.

We consider it likely that the two kinetic components found in Oxalis leaves are due to the existence of two kinds of photosynthetic units, or at least two kinds of photosystem II. The different rate constants could reflect a different antenna size. A possibility would be that one kind of photosynthetic unit is characteristic of grana thylakoids and one of stroma thylakoids. Another possibility is that the components correspond to the two kinds of photosystem II postulated by Arnon et al. (1981).

Table 3. Kinetic models for fluorescence induction in Oxalis leaves. All models follow the equation

$$F = F_o + A_1[1 - \exp(-R_1t)] + A_2 \Phi_F$$

where t is time and Φ_F is defined by Eqs. (1) and (2), and with the following constants (A_1 and A_2 expressed as percentage of $A_1 + A_2$ for model 1).

Model	A_1(%)	$R_1(s^{-1})$	A_2(%)	$R_2(s^{-1})$	p
1	41.0	70	59.0	3.2	0.65
2	41.0	140	59.0	3.2	0.65
3	15.4	140	47.4	3.2	0.65

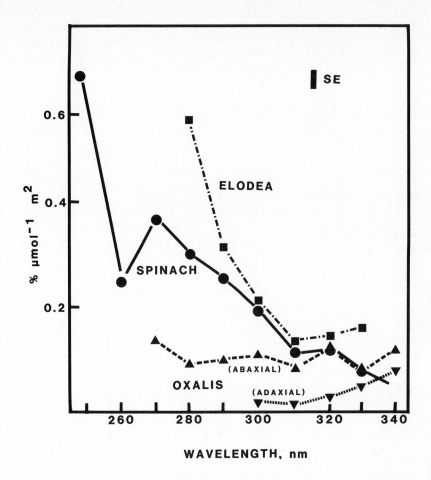

Fig. 5. Action spectrum for UV retardation of fluorescence induction (expressed as half-rise time) in leaves of Elodea and Oxalis. For comparison we also show a spectrum for UV retardation of fluorescence induction in isolated spinach thylakoids, recalculated from data in Bornman et al. (1984) and Table 1 to the same scale and averaged in the same way. The ordinate shows (percentage increase in half-rise time)/(μmol photons m^{-2}).

The half-rise time for irradiated leaves of Elodea and Oxalis was expressed as percentage increase over the control value for nonirradiated leaves. The results of the comparison of sensitivities to radiation of different wavelengths and radiation impinging on different sides of the leaves are summarized in Fig. 5. It is seen that the epidermis-free Elodea has the highest sensitivities throughout the spectral range investigated. Oxalis is more sensitive to radiation impinging on the abaxial side than to radiation impinging on the adaxial side. We suspected that larger amounts of water-soluble, UV-absorbing compounds were the cause of the difference. Extraction of excised epidermis and absorbance measurements on the extracts (Fig. 6) showed, on a weight basis, about 3 times as much absorbing material in the adaxial epidermis. However, the absorbance maximum of the extracts is close to 330 nm, while the difference in UV sensitivity of the leaf is more pronounced at shorter wavelengths. Therefore, other substances than those easily extractable with water are likely to exhibit a more important protective role.

In conclusion it can be said that our spectral measurements have not yet led to an identification of the radiation sensitive target in photosystem II, but some compounds can now be excluded. The UV sensitivity of intact Elodea leaves is surprisingly high at 280 nm, and one can infer that even at the thylakoid level there are sensitivity differences between different plants. In the whole leaf the sensitivity is further modified by absorbing substances in the epidermis, and there may be large differences in sensitivity depending on the direction of the UV radiation in relation to a plant organ. The UV effects on fluorescence kinetics obtained with different plant materials cannot, at the present time, easily be interpreted under a unified model.

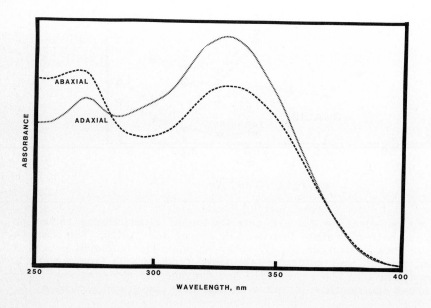

Fig. 6. Absorption spectra of aqueous extracts of the adaxial and the abaxial epidermis of Oxalis leaves. Different amounts of material and different instrument ranges were used; the maximum absorbance at 329 nm was recalculated to a concentration of 1 mg air-dried epidermis per ml extract, and corresponds to the following absorbances: adaxial epidermis 1.1 absorbance units, abaxial epidermis 0.39 absorbance units.

ACKNOWLEDGMENT

The authors gratefully acknowledge financial support from Nordiska Ministerrådet and Hilleshög AB.

REFERENCES

Arnon DI, Tsujimoto HY, Tang GM-S (1981) The oxygenic and anoxygenic photosystems of plant photosynthesis: An updated concept of light-induced electron and proton transport and photophosphorylation. In: Akoyunoglou G (ed) Photosynthesis. Proc Fifth Int Photosyn Congr, Vol II, p 7, ISBN 0-86689-012-2

Bornman JF, Björn LO, Åkerlund H-E (1984) Action spectrum for inhibition by UV radiation of photosystem II activity in spinach [Spinacia oleracea] thylakoids. Photobiochem Photobiophys 8:305-314

Critchley C, Smillie RM (1981) Leaf chlorophyll fluorescence as an indicator of high light stress (photoinhibition) in Cucumis sativus L. Aust J Plant Physiol 8:133-141

Govindjee, Papageorgiou G (1971) Chlorophyll fluorescence and photosynthesis transients. In: Giese AC (ed) Photophysiology, vol 6. Academic Press, New York, p 1

Joliot A, Joliot P (1964) Etude cinétique de la réaction photochimique libérant l'oxygène au cours de la photosynthèse. C R Acad Sci Ser D 258:4622-4625

Lavorel J, Joliot P (1972) A connected model of the photosynthetic unit. Biophys J 12:815-831

Fine structural effects of UV radiation on leaf tissue of <u>Beta</u> <u>vulgaris</u>

J. F. Bornman, R. F. Evert*, R. J. Mierzwa* and C. H. Bornman**

University of Lund, Sweden
*University of Wisconsin-Madison, Madison, Wisconsin USA
**Hilleshög Research AB, Landskrona, Sweden

ABSTRACT

The effect of UV radiation was studied at the ultrastructural level using treated and untreated leaf material from <u>Beta</u> <u>vulgaris</u> L. cv. Primahill. Although interest was centered mainly on changes resulting from treatment with UV-B radiation (280-320 nm), UV-C radiation (<280 nm) was included as an indicator for severe damage. Among the more pronounced effects caused by UV-C irradiation were extreme damage to epidermal and subepidermal cells and extensive chloroplast disorganization. In addition, membrane-bound crystalloid inclusions were found where the cells were relatively undamaged. The more severe changes resulting from UV-B radiation were exemplified by disruptions in the chloroplast envelope, and while no crystalloid inclusions were found as in the UV-C treated tissue, a pronounced crystalloid formation of the stroma was seen in several sections from different leaves. Dilation of thylakoid membranes was discernible in tissue treated with UV-B or UV-C radiation.

When comparing the findings from studies using UV radiation to those of other environmental stresses, the cell membrane seems to be the primary target. The present investigation also lends further support to the idea of the similarity between the effects of UV radiation and those of advanced cellular aging.

INTRODUCTION

Although it is apparent that there are different UV radiation targets, and that there are qualitative and quantitative differences in response depending on the wavelength range, damage to particular targets probably dominates overall damage to the cell or organism. Apart from extensive investigations on DNA and general protein damage, attention has fairly recently focused particularly on the cellular membrane. Reports of loss of membrane integrity as a result of UV radiation come from both ultrastructural (Campbell et al. 1974, Brandle et al. 1977, Skokut et al. 1977, Bornman et al. 1983) and physiological studies (Doughty and Hope 1973, Murphy 1983).

Alterations of membrane properties occur also under stress conditions other than UV radiation, for example chilling, freezing and heat injury (Palta and Li 1978, Levitt 1980), ozone treatment (Pauls and Thompson 1982) as well as during the normal process of senescence (e.g., Dhindsa et

NATO ASI Series, Vol. G8
Stratospheric Ozone Reduction, Solar Ultraviolet Radiation and Plant Life. Edited by R. C. Worrest and M. M. Caldwell
© Springer-Verlag Berlin Heidelberg 1986

al. 1982). More specifically L.A. Wright (Thesis, Univ of California, Davis, USA, 1978) suggested that other stress symptoms, for example ozone and chilling injury had characteristics in common with those caused by UV radiation. For instance, for all three cases, photosynthesis inhibition, ultrastructural disorganization and alteration of membrane permeability have been found.

In this report some of the findings from UV-B (280-320 nm) as well as UV-C (<280 nm) treated sugar beet leaf material are discussed. The possible similarity of the changes induced by UV radiation to those of natural senescence is also considered.

MATERIALS AND METHODS

Sugar beet (Beta vulgaris L. cv. Primahill, derivative 9164) was grown under white light in a climate chamber as previously described (Bornman et al. 1983) for two weeks and thereafter for a further 7 days under the same conditions with or without UV-B radiation (FS20 and FS40 Westinghouse sunlamps, 9.46×10^{-2} W m^{-2} filtered through aged 0.1-mm thick cellulose acetate) for 6 h within a 24-h cycle. Treatment with UV-C radiation (1.63 W m^{-2} for 1.5 h within a 24-h cycle) was included for comparative purposes. At the termination of the dark period fully expanded leaves of both control and treated plants were randomly selected and leaf segments prepared for electron microscopy. Five hundred to 1,000 sections were examined for control and UV treated material, respectively.

RESULTS AND DISCUSSION

Terms such as damage, repair, degradation, integrity, dilation etc. constantly require points of reference if fine structural changes are to be meaningfully analyzed. For this reason it was found necessary to compare UV-B-irradiated tissue not only to untreated material, but also to UV-C-treated tissue, in order to establish criteria of deviation from the normal. It was not intended, however, that a direct correlation be made between effects observed in UV-B and UV-C-treated material, since it is generally accepted that the two wavelength ranges elicit or cause different physiological effects. Rather, UV-C-damaged tissue could be useful in showing the nature and extent of damage which might occur, since except where very large doses of UV-B radiation are given, UV-B radiation appears to be much less destructive as determined at ultrastructural, morphological and physiological levels.

No evidence of structural damage, due either to growth conditions or subsequent preparative procedures was found in leaf sections from control plants (Figs. 1 and 2). Membrane systems were well-defined and appeared intact. Many subcellular changes involving membranes, organelles and macromolecules are beyond the resolution of the electron microscope and although some of these changes admittedly can be detected cytochemically at the level of the electron microscope, many have to be inferred. That UV-C radiation can cause extreme damage to adaxial epidermal and subepidermal cells is illustrated clearly by Figs. 3 and 4. Although its effects may reach deeper into the spongy mesophyll (Figs. 5-7) and vascular bundle (Fig. 8), the structural damage decreases towards the abaxial leaf surface (Bornman et

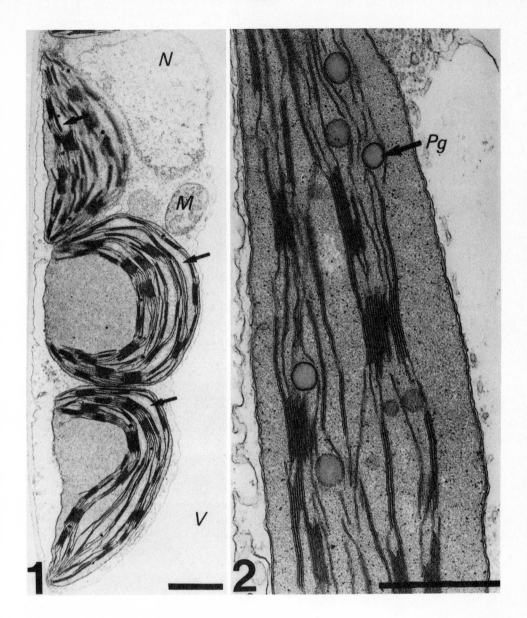

Figs. 1-2. Parts of Beta vulgaris mesophyll cells from control leaf tissue. In Fig. 1 cells were highly vacuolated, containing numerous chloroplasts with well-developed grana and stroma thylakoids. Other organelles were well-defined. Arrows show electron lucent areas of apparently recently mobilized starch. In Fig. 2 part of a chloroplast with finely granular stroma containing plastoglobuli. N, nucleus; M, mitochondrion; Pg, plastoglobuli; V, vacuole. Bar = 1 µm.

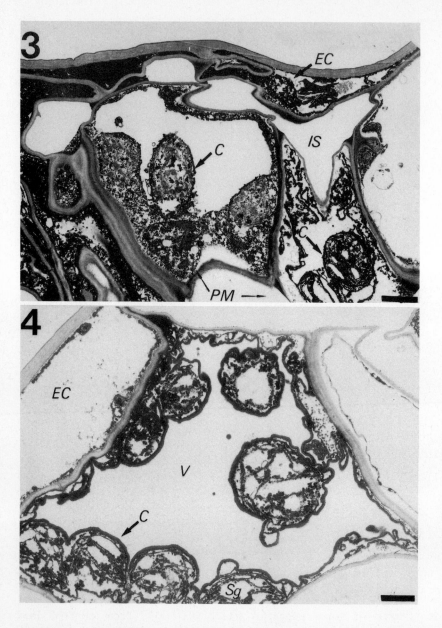

Figs. 3-4. Epidermal and palisade mesophyll tissue from UV-C-irradiated
leaves. In Fig. 3 collapsed adaxial epidermal cells, below which are two
partially collapsed palisade mesophyll cells containing damaged chloroplasts. In Fig. 4 palisade mesophyll cell from the adaxial surface with
denatured chloroplasts containing ovoid starch grains. Fusion of chloroplast membranes and lamellae give chloroplasts an amorphous appearance. C,
chloroplast; EC, epidermal cell; IS, intercellular space; PM, palisade
mesophyll cell. Bar = 1 μm.

203

al. 1983). However, contiguous cells were sometimes found where (except for slight granal fusion, Fig. 5) the one displayed little structural change as compared to an adjacent cell. Figure 6, from a deeper lying area, shows extensively disrupted chloroplast structure, although the damage is less severe than that shown in Figs. 3 and 4.

Seemingly distinctive features for the UV-C-treated material are the numerous membrane-bound crystalloid inclusions in otherwise apparently undamaged cells such as that shown in Fig. 7 (also reported in Bornman et al. 1983). Figure 8, showing sieve tube cells in part of a vascular bundle, gives a good example of thylakoid unstacking or dilation.

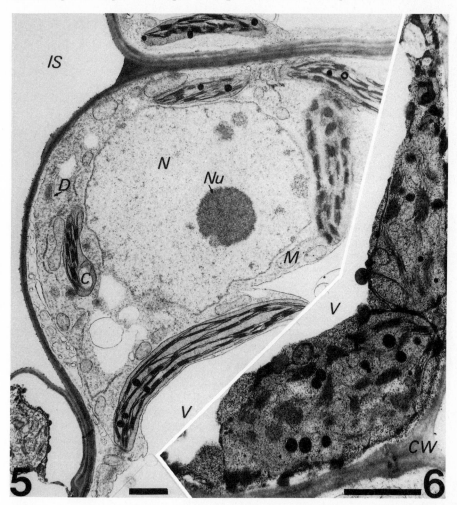

Figs. 5-6. Parts of apparently undamaged (Fig. 5) and damaged (Fig. 6) adaxial spongy mesophyll cells. In Fig. 5 an undamaged cell is contiguous with a severely structurally altered one. Grana in Fig. 6 partially fused and stroma lamellae virtually obliterated. Tissue in both figures is from UV-C-treated leaves. D, dictyosome; CW, cell wall; Nu, nucleolus. Bar = 1 μm.

While fusion of thylakoid membranes (Figs. 5 and 6) or total chloroplast disorganization coupled also with destruction of whole cells (Figs. 3 and 4) seems characteristic of UV-C treated material, structural changes from UV-B radiation not only are less pronounced, but sometimes not visually discernible. Figures 9 and 10 show examples of this. Disruption of other membrane-bound organelles, besides that of the chloroplast, was not observed. From studying about 100 electron micrographs of numerous sections taken from different leaves of UV-B treated material, we infer that the outer and inner chloroplast membranes are major targets. Figure 11 shows distinct breaks in the chloroplast envelope (as indicated by unlabeled arrows).

An interesting and often observed feature in UV-B-treated tissue is indicated in Fig. 12. Here a quasi-crystalline appearance of the stroma (bold arrows) is evident, as is also a marked dilation of the thylakoids (unlabeled arrows). It is possible that the crystalline formation of the stroma is the result of UV-induced water stress. Significant reductions in fresh weight in UV-B-treated material as compared to untreated controls have been found in several crop plants (Tevini et al. 1981; Vu et al. 1981). Crystalline stroma formation was only seen in UV-B-treated material, although in order to further verify this, it would be necessary to subject plants to low doses of UV-C radiation.

The exact molecular or structural target of UV-B radiation has not as yet been established with certainty; moreover there is very likely to be more than one target. Although disruptions of the chloroplast membranes are easily seen, these may be secondary effects with macromolecules such as membrane proteins and lipids as the primary targets. Tevini et al. (1981) reported a reduction in total protein and glycerolipids in a number of crop plants and suggested a possible correlation with the processes for senescence. It seems plausible that UV-B radiation effects on membranes resemble those of advanced aging. Support for this assumption comes from studies by Skokut et al. (1977) based on chlorophyll loss coupled with chloroplast degradation in Nicotiana glutinosa treated with UV-C radiation (254 nm). They concluded that UV radiation induced accelerated aging of leaf tissue, discernible at the physiological and ultrastructural level, and that this aging effect was similar to that of natural senescence. It appears that during senescence endogenously-formed free radicals may remove electrons from polysaturated lipid components resulting in lipid peroxidation upon reaction with molecular oxygen, thus impairing membrane integrity. Berjak (1974) in fact has suggested that damage to subcellular membranes as a result of peroxidation of, for example, lipid molecules is an underlying deteriorative process in cellular aging. Dhindsa et al. (1981) among others have reported increased lipid peroxidation with aging. There have also been reports of damage to lipids by UV radiation as a consequence of peroxidation (e.g., Roshchupkin et al. 1975).

Cathodic protection, i.e., the supply of electrons, to a given system, which can react with endogenously-formed free radicals and thereby reduce deterioration by peroxidation, has been reported in caryopses of Zea mays to counteract the aging process of membranes (Berjak 1974). It would be interesting to apply cathodic protection to leaves under UV treatment to ascertain whether or not some protection against membrane damage is afforded. This could be investigated either physiologically or ultra-structurally.

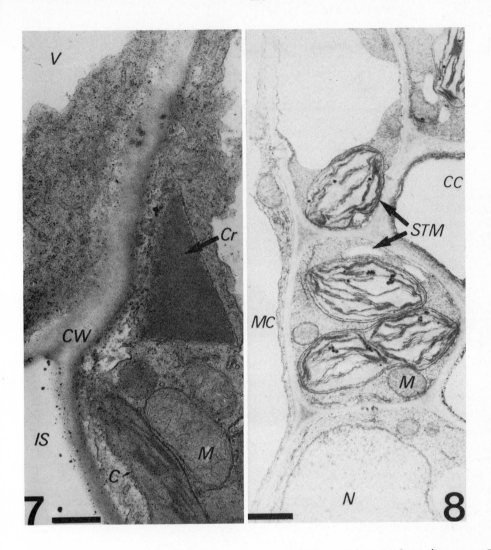

Figs. 7–8. UV–C–treated material from mesophyll and vascular tissues. In Fig. 7 a close–up view of a membrane–bound crystalloid observed in spongy mesophyll cells. In Fig. 8 portion of vascular tissue showing chloroplasts with altered thylakoid structure in phloem cells. Thylakoid dilation can also be seen. CC, companion cell; Cr, membrane–bound crystalloid; MC, spongy mesophyll cell; STM, sieve tube member. Bar = 1 µm.

Based on the findings of both this study and other investigations at the ultrastructural level, it is not surprising that physiological processes are affected by UV radiation, given the structural alterations induced. A correlation of the effects of UV radiation on photosynthesis with ultra-structural changes in spinach chloroplasts was reported by Mantai et al. (1970). On the other hand it is highly probable, as already mentioned, that the structural changes are secondary effects. As suggested by L.A.

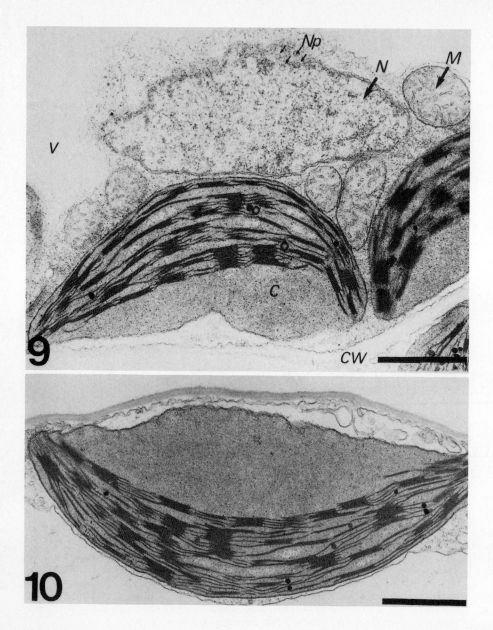

Figs. 9-10. Parts of spongy mesophyll cells of cells from UV-B-treated leaf material. Organelles appear unaffected by UV-B irradiation. Np, nuclear pores. Bar = 1 μm.

Wright, should electrogenic pumps involved in transmembrane ion flux across chloroplast membranes become altered in some way by UV radiation, the subsequent osmotic irregularities could in themselves lead to dilation and rupture of the membrane. At this stage gross ultrastructural changes would in turn further impair physiological processes.

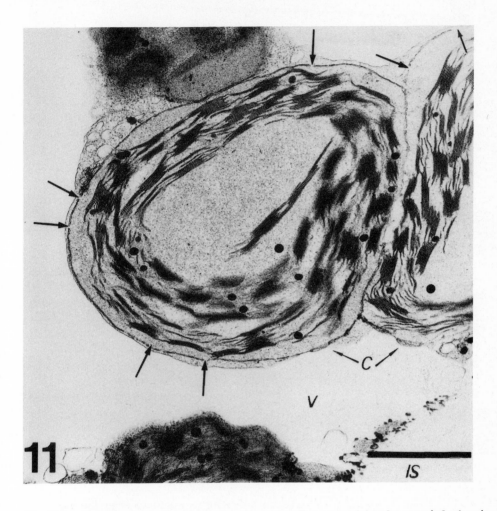

Fig. 11. Spongy mesophyll tissue from UV-B-treated leaf material showing disruptions in the chloroplast membrane (unlabeled arrows). Bar = 1 μm.

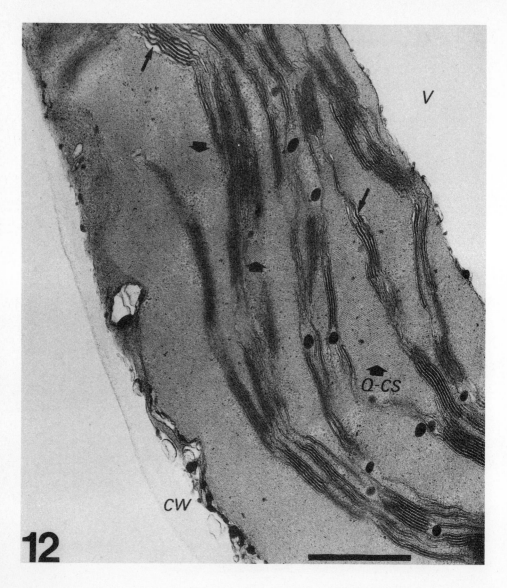

Fig. 12. UV-B-treated leaf material showing a quasi-crystalline stroma of the chloroplast; bold arrows, particularly pronounced areas; thin arrows, thylakoid dilation; Q-CS, quasi-crystalline structure. Bar = 1 μm.

ACKNOWLEDGMENTS

This work was supported by The Carl Trygger Foundation for Scientific Research to J.F.B. and in part by the U.S. National Science Foundation grant PCM 78-03872 to R.F.E.

REFERENCES

Berjak P (1974) Cathodic protection and viability extension: preliminary subcellular studies in plant embryos. Proc Elect Microsc Soc SA 4:31-32

Bornman JF, Evert RF, Mierzwa RJ (1983) The effect of UV-B and UV-C radiation on sugar beet leaves. Protoplasma 117:7-16

Brandle JR, Campbell WF, Sisson WB, Caldwell MM (1977) Net photosynthesis, electron transport capacity, and ultrastructure of Pisum sativum L. exposed to ultraviolet-B radiation. Plant Physiol 60:165-169

Campbell WF, Caldwell MM, Sisson WB (1974) Electron microscopy of ultraviolet-induced changes in mesophyll cells of Glycine max. Am J Bot (Suppl) 61:27

Dhindsa RS, Plumb-Dhindsa P, Thorpe TA (1981) Leaf senescence: correlated with increased levels of membrane permeability and lipid peroxidation, and decreased levels of superoxide dismutase and catalase. J Exp Bot 32:93-101

Doughty CJ, Hope AB (1973) Effects of ultraviolet radiation on the membranes of Chara corallina. J Memb Biol 13:185-198

Levitt J (1980) Responses of plants to environmental stresses, vol 1. Academic Press, New York, ISBN 0-12-445501-8

Mantai KE, Wong J, Bishop NI (1970) Comparison studies on the effects of ultraviolet irradiation on photosynthesis. Biochim Biophys Acta 197:257-266

Murphy TM (1983) Membranes as targets of ultraviolet radiation. Physiol Plant 58:381-388

Palta JP, Li PH (1978) Cell membrane properties in relation to freezing injury. In: Li PH and Sakai A (eds) Plant cold hardiness and freezing stress. Mechanisms and crop implications. Academic Press, New York, p 93, ISBN 0-12-447650-3

Pauls KP, Thompson JE (1982) Effects of cytokinins and antioxidants on the susceptibility of membranes to ozone damage. Plant Cell Physiol 23:821-832

Roshchupkin DI, Pelenitsyn AB, Potapenko A Ya, Talitsky VV, Vladimirov Yu A (1975) Study of the effects of ultraviolet light on biomembranes - IV. The effects of oxygen on UV-induced hemolysis and lipid photoperoxidation in rat erythrocytes and liposomes. Photochem Photobiol 21:63-69

Skokut TA, Wu JH, Daniel RS (1977) Retardation of ultraviolet light accelerated chlorosis by visible light or by benzyladenine in Nicotiana glutinosa leaves: Changes in amino acid content and chloroplast ultrastructure. Photochem Photobiol 25:109-118

Tevini M, Iwanzik W, Thoma U (1981) Some effects of enhanced UV-B irradiation on the growth and composition of plants. Planta 153:388-394

Vu CV, Allen LH, Garrard LA (1981) Effects of supplemental UV-B radiation on growth and leaf photosynthetic reactions of soybean Glycine max. Physiol Plant 52:353-365

Comparative Sensitivity of Binucleate and Trinucleate Pollen to Ultraviolet Radiation: A Theoretical Perspective

S. D. Flint and M. M. Caldwell

Utah State University, Logan, Utah USA

ABSTRACT

Angiosperm pollen shed in the binucleate condition (which is the case for most species) may be more susceptible to UV radiation damage than pollen shed in the trinucleate condition. This speculation is supported by data summarized from the literature which show that trinucleate pollen is more physiologically advanced at the time when it is shed from the anther. The time course of germination, penetration of the stigma, and successful fertilization is thus more rapid in these trinucleate species. This results in less exposure to solar ultraviolet and consequently less damage should be expected. Experiments are needed to test the sensitivity of these pollen types in vivo and to evaluate the efficacy of possible repair mechanisms.

INTRODUCTION

Ultraviolet radiation is capable of damaging angiosperm pollen. For example, chromosome aberrations induced by ultraviolet radiation have been detected by cytological methods (Kirby-Smith and Craig 1957) and by noting the loss of marker genes in the seed produced (Stadler and Sprague 1936). Whether pollen in nature suffers chromosomal aberrations due to solar ultraviolet is not known. Floral parts provide substantial screening from solar UV-B (Flint and Caldwell 1983); thus, the potential for UV-induced chromosome aberrations and other nucleic acid damage exists between the time of dehiscence and when the pollen tube and the generative nucleus penetrate the stigma. Ultraviolet radiation may also decrease pollen germinability. Only limited data from in vitro experiments with pollen of temperate latitude origin are available. In species with binucleate pollen, germination is significantly inhibited only when ultraviolet radiation levels similar to those of equatorial latitudes at high elevations are administered to germinating pollen (Flint and Caldwell 1984). Whether pollen from plants native to low latitudes would show similar sensitivity is not known. Species with trinucleate pollen may be more resistant to ultraviolet damage (Pfahler 1981; Flint and Caldwell 1984), although only two species have been examined.

To investigate pollen sensitivity to UV-B radiation under conditions which are similar to a field situation is difficult and time consuming. Thus, perspective on the potential sensitivity of pollen of different species is useful in selecting species for intensive study. By synthesizing data in

NATO ASI Series, Vol. G8
Stratospheric Ozone Reduction, Solar Ultraviolet Radiation
and Plant Life. Edited by R. C. Worrest and M. M. Caldwell
© Springer-Verlag Berlin Heidelberg 1986

the literature on physiological, developmental and protective characteristics of pollen, we are able to examine, from a theoretical perspective,
the potential differences in germination inhibition between species which
possess binucleate and trinucleate pollen. These data may also be used as
a framework for the examination of mechanisms involved in UV damage to
pollen.

SURVEY METHODS

Reports describing pollen physiology, the time course of pollen life
history, and potential protective mechanisms were classified according to
whether the species had binucleate pollen, trinucleate pollen, or an intermediate condition known as "advanced" binucleate pollen. The extensive
survey of Brewbaker (1967) was used to classify species as to whether the
pollen was binucleate or trinucleate, and the separation of some taxa into
a phylogenetically "advanced" binucleate category was taken from Hoekstra
and Bruinsma (1979).

In determining the time course of pollen life history (germination, stigma
penetration, pollen tube growth, and fertilization), in vivo data were used
wherever possible. The only in vitro data included in this tabulation were
for time to germination and tube growth rate of the "advanced" binucleate
species and for the intervals to protein synthesis. Time to "fertilization" should be considered only as an approximation because some are
reported as time to reach the ovule, others as time to the micropyle, and
others are calculated based on tube growth rate and length of the style.
When more than one report existed for a particular species, the values were
averaged and counted as one entry.

RESULTS AND DISCUSSION

In the extensive survey of Brewbaker (1967), 70% of all taxa shed their
pollen in the binucleate condition, while the remaining 30% shed trinucleate pollen. This characteristic usually applies to an entire plant
family: only 12% of the families surveyed by Brewbaker (1967) contain both
pollen types. No decisive correlations have been discovered between pollen
type and plant phenology, morphology, habit, or distribution. The
"advanced" binucleate category was proposed by Hoekstra and Bruinsma
(1979). As this category is based on physiological, rather than cytological characteristics of pollen, only Tradescantia and Impatiens taxa
have thus far been placed in this group (Hoekstra and Bruinsma 1979). It
is likely that a number of other species with binucleate pollen will also
be placed in this category.

It has also long been known that the two cytological pollen types differ
dramatically in some physiological characteristics. Binucleate pollen
germinates under a broader range of in vitro conditions and retains its
viability longer. The reactions preventing incompatible pollination in
species with these two pollen types occur in different tissues (Brewbaker
1967). Subsequent experimental and physiological work has revealed
potential differences between these two cytological pollen types that may
influence their sensitivity to UV radiation.

At the time of dehiscence, binucleate pollen is typically less physiologically advanced than trinucleate pollen (Table 1). Binucleate pollen has not yet completed its second mitosis, and must develop its mitochondria and also conduct protein synthesis before germination can begin. Trinucleate pollen has completed its second mitosis and is physiologically prepared for germination since its mitochondria are developed and the necessary protein synthesis (Table 2) has been completed before dehiscence. This early completion of protein synthesis accounts for the lack of ribosomes in pollen of many trinucleate species. (Maize, an exception, has numerous ribosomes which may function primarily in the growth of the pollen tube through the exceedingly long style.) Thus, in trinucleate pollen there are fewer processes which may be susceptible to UV radiation damage, which suggests that germination of this pollen type may be less inhibited by UV radiation than binucleate pollen. Whether the presence of different metabolic pathways and respiration rates (Table 1) influences sensitivity to UV radiation is uncertain.

Because of the physiological factors discussed previously, dramatic differences exist between the two cytological pollen types in the time required for development (Table 2). Most species with trinucleate pollen complete the germination-to-fertilization process in a much shorter time interval than species with binucleate pollen. (Had we included binucleate taxa requiring months to effect fertilization, such as Corylus and Quercus [Kramer and Kozlowski 1979], this difference would have been more pronounced.) As germination and stigma penetration usually occur more rapidly in species with trinucleate pollen, there is a shorter period of exposure to sunlight and thus less time for damage to occur. (Trinucleate pollen of Brassica oleracea is unusual in that it germinated slower than any other species in Table 2.)

"Advanced" binucleate pollen shares some, but not all, of the characteristics of trinucleate pollen, thus sensitivity of this group to UV radiation would be expected to be intermediate. Development is rapid as in trinucleate pollen (Table 2), in part because mitochondria are developed at the time of dehiscence (Table 1). Other characteristics are less developed (Table 1). Mitotic division is arrested at dehiscence, as in binucleate pollen. In Tradescantia pollen, polyribosomes are present, supporting the idea that, unlike in trinucleate pollen, protein synthesis is necessary for germination. However, polyribosomes rather than free ribosomes suggest that this will occur more rapidly than in typical binucleate pollen. In Impatiens pollen, a lack of ribosomes suggests that protein synthesis has been completed before dehiscence, as in trinucleate pollen. This category should perhaps include some of the rapidly germinating binucleate species, e.g., Persea americana, 6 min (Sedgley 1977) and Pisum sativum, 5 min (Hoekstra and Bruinsma 1978). Overlap between the lag times for germination of the binucleate and trinucleate categories would then be less. The frequency of occurrence of "advanced" binucleate taxa among the angiosperms needs examination.

As pollen tubes grow for a short distance on the surface of the stigma before penetrating it, potential protective mechanisms may be important at this stage (Table 3). Some species with binucleate pollen have wet stigmas, while species with trinucleate pollen tend to have dry stigmas. As these stigmatic exudates contain UV-absorbing phenolic compounds (Martin 1970), pollen grains may be at least partly shielded from solar UV radiation shortly after landing on the stigma. In addition, binucleate pollen

Table 1. Physiological characteristics of binucleate, "advanced binucleate", and trinucleate pollen which may result in differential sensitivity to UV-B radiation.

Factor	Binucleate	"Advanced Binucleate"	Trinucleate	Reference
Presynthesized RNA	yes	yes	- [a]	3,8,9
Second mitosis	arrested at prophase when shed	arrested at prophase when shed	occurs hours to days before shed	1,2,8
Ribosomes	monoribosomes	polyribosomes or no ribosomes	few ribosomes[a] (e.g. Asteraceae)	6,10
Mitochondrial development at dehiscence	undeveloped	nearly complete	nearly complete	4
Alternative oxidase pathway	no	no	yes	7
Respiration rate	low	low	high	4,5,6

[a] while in trinucleate maize pollen both RNA (Mascarenhas et al. 1984) and polyribosomes (Hoekstra and Bruinsma 1979) are present, they most likely function primarily in the growth of the pollen tube through the exceedingly long style. Maize may not be representative of most trinucleate pollen.

References: 1. Brewbaker 1967; 2. Brewbaker and Emery 1962; 3. Frankis and Mascarenhas 1980; 4. Hoekstra 1979; 5. Hoekstra and Bruinsma 1975; 6. Hoekstra and Bruinsma 1979; 7. Hoekstra and Briunsma 1980; 8. Mascarenhas 1975; 9. Mascarenhas 1978; 10. van Went 1974.

Table 2. Developmental characteristics of binucleate, "advanced binucleate", and trinucleate pollen which may result in differential sensitivity to UV-B radiation.

Factor	Binucleate	"Advanced Binucleate"	Trinucleate	Reference
Interval to germination[a] (min)	6-120 n=8 md=27.5[b]	2-6 n=2	1-360 n=15 md=7	1
Interval to protein synthesis (min)[a]	10 n=2	2 n=1	negligible after dehiscence, n=2	2
Interval to stigma penetration (min)[a]	10-75 n=3 md=10	-	4.75-480 n=9 md=8.5	3
Interval to fertilization (hr)[a]	3.5-168 n=22 md=36	1-6 n=2	0.5-16 n=18 md=4.75	4
Rate of tube growth (mm hr^{-1})	0.1-4.6 n=18 md=0.8	0.65 n=1	0.6-35 n=10 md=4.6	5

[a]Time intervals begin with the arrival of the pollen grain on the stigmatic surface or the germination medium.

[b]md = median

Table 2 (continued)

References:

1. Ameele 1982; Brown and Shands 1957; Cass and Peteya 1979; Chandra and Bhatnagar 1974; Cresti et al. 1980; Dickenson and Lewis 1973; Ferrari et al. 1983; J. Heslop-Harrison 1979; J. Heslop-Harrison 1980; Y. Heslop-Harrison 1977; J. Heslop-Harrison et al. 1975a; Y. Heslop-Harrison et al. 1984; Hoekstra and Bruinsma 1978; Kroh and Linskens 1963; Owens and Horsfield 1982; Randolph 1936; Russell 1982; Sastri and Shivanna 1979; Sedgley 1977; Sedgley and Buttrose 1978; Wilms 1980; Zeven and Van Heemert 1970.

2. Hoekstra and Bruinsma 1979.

3. Brown and Shands 1957; Chandra and Bhatnagar 1974; Ferrari et al. 1983; Y. Heslop-Harrison 1977; J. Heslop-Harrison and Y. Heslop-Harrison 1981; J. Heslop-Harrison et al.1975b; Hoekstra and Bruinsma 1978; McLean and Ivimey-Cook 1956; Mattsson 1983; Owens and Horsfield 1982; Sastri and Shivanna 1979; Wilms 1980.

4. Ascher and Peloquin 1966; Brown and Shands 1957; Chandra and Bhatnagar 1974; Chen and Gibson 1973; Clarke et al. 1977; Herd and Beadle 1980; Y. Heslop-Harrison 1977; Hoekstra and Bruinsma 1978; Hopping and Jerram 1979; Jalani and Moss 1980; Kho et al. 1980; Kramer and Kozlowski 1979; Kroh et al. 1979; Lange and Wojciechowska 1976; Lee 1980; Luxová 1967; Maheshwari 1949; Mulcahy et al.1983; Pope 1937; Pundir 1982; Pundir et al. 1983; Reger and James 1982; Russell 1982; Sastri and Shivanna 1979; Sedgley 1977; Sedgley 1983; Sedgley and Scholefield 1980; Socias i Company 1976; Staudt 1982; Stephens and Quinby 1934; Stott 1972; Thompson and Liu 1973; Vasil 1960; Wilms 1974.

5. Ascher and Peloquin 1966; Buchholz and Blakeslee 1927; Chandra and Bhatnagar 1974; Chen and Gibson 1973; Currah 1981; Facteau et al. 1973; Gillissen and Brantjes 1978; J. Heslop-Harrison 1980; J. Heslop-Harrison 1982; Y. Heslop-Harrison 1977; Hoekstra and Bruinsma 1978; Hopping and Jerram 1979; Jensen and Fisher 1970; Kroh et al. 1979; Kwack 1965; Lee 1980; Maheshwari 1949; Mascarenhas 1978; Mulcahy et al. 1983; Pundir 1982; Pundir et al. 1983; Russell 1982; Sedgley 1983; Sedgley and Buttrose 1978; Staudt 1982; Van Herpen and Linskens 1981; Vasil 1960; Williams and Maier 1977; Wilms 1981.

Table 3. Characteristics possibly conferring differential protection from UV-B radiation: comparison of binucleate, "advanced binucleate", and trinucleate pollen.

Factor	Binucleate	"Advanced Binucleate"	Trinucleate	Reference
Elutable compounds	high	–	low	6
Stigma surface	both wet and dry	dry	usually dry	2
Pollen tube growth on surface of stigma	yes	–	yes	1,2,7,8,9
Photoreactivation of DNA damage	yes	–	yes	3,5
Dark repair of DNA damage	yes	–	–	4

References: 1. Cass and Peteya 1979; 2. Y. Heslop-Harrison and Shivanna 1977; 3. Ikenaga et al. 1974; 4. Jackson and Linskens 1978; 5. Jackson and Linskens 1979; 6. Kirby and Smith 1974; 7. Kroh et al. 1979; 8. Sedgley and Scholefield 1980; 9. Wilms 1980.

contains more readily elutable compounds than trinucleate pollen. These elutable compounds probably include UV-screening flavonoids, since they are localized on the surface of the pollen and are quickly released (Wiermann and Vieth 1983). However, the relative flavonoid contents of binucleate and trinucleate pollens remain to be verified experimentally. These factors may provide greater protection for binucleate pollen and possibly offset the effect of greater UV exposure accumulated during their slow development.

Repair mechanisms may also ameliorate UV-B damage to DNA. Photoreactivation of DNA damage has been shown to occur in both binucleate and trinucleate pollen grains while dark repair has only been demonstrated in binucleate pollen (Table 3). Whether the efficiencies of these two repair systems differ between binucleate and trinucleate pollen has never been examined.

This review of pollen characteristics, and their potential contributions to UV sensitivity, can serve as a useful framework for the selection of test species for further work. Nevertheless, a considerable physiological and morphological diversity exists within each pollen category and diverse response to UV radiation would not be unexpected.

ACKNOWLEDGMENT

This work has been supported by the U.S. Environmental Protection Agency. Although the research described in this chapter has been funded primarily by the U.S. Environmental Protection Agency through Cooperative Agreement CR808670 to Utah State University, it has not been subjected to the Agency's required peer and policy review and therefore does not necessarily reflect the views of the Agency and no official endorsement should be inferred.

REFERENCES

Ameele RJ (1982) The transmitting tract in Gladiolus. I. The stigma and the pollen-stigma interaction. Amer J Bot 69:389-401

Ascher PD, Peloquin SJ (1966) Effect of floral aging on the growth of compatible and incompatible pollen tubes in Lilium longiflorum. Amer J Bot 53:99-102

Brewbaker JL (1967) The distribution and phylogenetic significance of binucleate and trinucleate pollen grains in the angiosperms. Amer J Bot 54:1069-1083

Brewbaker JL, Emery GC (1962) Pollen radiobotany. Radiat Bot 1:101-154

Brown CM, Shands HL (1957) Pollen tube growth, fertilization, and early development in Avena sativa. Agron J 49:286-288

Buchholz JT, Blakeslee AF (1927) Pollen-tube growth at various temperatures. Amer J Bot 14:358-369

Cass DD, Peteya DJ (1979) Growth of barley pollen tubes in vivo I. Ultrastructural aspects of early tube growth in the stigmatic hair. Can J Bot 57:386-396

Chandra S, Bhatnagar SP (1974) Reproductive biology of Triticum. II. Pollen germination, pollen tube growth, and its entry into the ovule. Phytomorphology 24:211-217

Chen C, Gibson PB (1973) Effect of temperature on pollen-tube growth in Trifolium repens after cross- and self-pollinations. Crop Sci 13:563-566

Clarke AE, Considine JA, Ward R, Knox RB (1977) Mechanism of pollination in Gladiolus: roles of the stigma and pollen-tube guide. Ann Bot 41:15-20

Cresti M, Ciampolini F, Sarfatti G (1980) Ultrastructural investigations on Lycopersicum peruvianum pollen activation and pollen tube organization after self- and cross-pollination. Planta 150:211-217

Currah L (1981) Pollen competition in onion (Allium cepa L.). Euphytica 30:687-696

Dickinson HG, Lewis D (1973) Cytochemical and ultrastructural differences between intraspecific compatible and incompatible pollinations in Raphanus. Proc R Soc Lond B 183:21-38

Facteau TJ, Wang SY, Rowe KE (1973) The effect of hydrogen fluoride on pollen germination and pollen tube growth in Prunus avium L. cv. 'Royal Ann'. J Amer Soc Hort Sci 98:234-236

Ferrari TE, Comstock P, More TA, Best V, Lee SS, Wallace DH (1983) Pollen stigma interactions and intercellular recognition in Brassica: pathways for water uptake. In: Mulcahy DL, Ottaviano E (eds) Pollen: Biology and implications for plant breeding. Elsevier, New York, p 243

Flint SD, Caldwell MM (1983) Influence of floral optical properties on the ultraviolet radiation environment of pollen. Amer J Bot 70:1416-1419

Flint SD, Caldwell MM (1984) Partial inhibition of in vitro pollen germination by simulated solar ultraviolet-B radiation. Ecology 65:792-795

Frankis R, Mascarenhas JP (1980) Messenger RNA in the ungerminated pollen grain: a direct demonstration of its presence. Ann Bot 45:595-599

Gillissen LJW, Brantjes NBM (1978) Function of the pollen coat in different stages of the fertilization process. Acta Bot Neerl 27:205-212

Herd YR, Beadle DJ (1980) The site of the self-incompatibility mechanism in Tradescantia pallida. Ann Bot 45:251-256

Heslop-Harrison J (1979) An interpretation of the hydrodynamics of pollen. Amer J Bot 66:737-743

Heslop-Harrison J (1980) Aspects of the structure, cytochemistry, and germination of the pollen of rye (Secale cereale L.). Ann Bot Suppl 1

Heslop-Harrison J (1982) Pollen-stigma interaction and cross-incompatibility in the grasses. Science 215:1358-1364

Heslop-Harrison J, Heslop-Harrison Y (1981) The pollen-stigma interaction in the grasses. 2. Pollen-tube penetration and the stigma response in Secale. Acta Bot Neerl 30:289-307

Heslop-Harrison J, Knox RB, Heslop-Harrison Y, Mattsson O (1975a) Pollen-wall proteins: emission and role in incompatibility responses. In: Duckett JG, Racey PA (eds) The biology of the male gamete. (Supplement 1 to Biol J Linn Soc, vol 7), Academic Press, p 189

Heslop-Harrison J, Heslop-Harrison Y, Barber J (1975b) The stigma surface in incompatibility responses. Proc R Soc Lond B 188:287-297

Heslop-Harrison Y (1977) The pollen-stigma interaction: pollen-tube penetration in Crocus. Ann Bot 41:913-922

Heslop-Harrison Y, Shivanna KR (1977) The receptive surface of angiosperm stigma. Ann Bot 41:1233-1258

Heslop-Harrison Y, Reger BJ, Heslop-Harrison J (1984) The pollen-stigma interaction in the grasses. 6. The stigma ('silk') of Zea mays L. as host to the pollens of Sorghum bicolor (L.) Moench and Pennisetum americanum (L.) Leeke. Acta Bot Neerl 33:205-227

Hoekstra FA (1979) Mitochondrial development and activity of binucleate and trinucleate pollen during germination in vitro. Planta 145:25-36

Hoekstra FA, Bruinsma J (1975) Respiration and vitality of binucleate and trinucleate pollen. Physiol Plant 34:221-225

Hoekstra FA, Bruinsma J (1978) Reduced independence of the male gametophyte in angiosperm evolution. Ann Bot 42:759-762

Hoekstra FA, Bruinsma J (1979) Protein synthesis of binucleate and trinucleate pollen and its relationship to tube emergence and growth. Planta 146:559-566

Hoekstra FA, Bruinsma J (1980) Control of respiration of binucleate and trinucleate pollen under humid conditions. Physiol Plant 48:71-77

Hopping ME, Jerram EM (1979) Pollination of Kiwifruit (Actinidia chinensis Planch.): stigma-style structure and pollen tube growth. N Z J Bot 17:233-240

Ikenaga M, Kondo S, Fujii T (1974) Action spectrum for enzymatic photoreactivation in maize. Photochem Photobiol 19:109-113

Jackson JF, Linskens HF (1978) Evidence for DNA repair after ultraviolet irradiation of Petunia hybrida pollen. Molec Gen Genet 161:117-120

Jackson JF, Linskens HF (1979) Pollen DNA repair after treatment with the mutagens 4-nitroquinoline-1-oxide, ultraviolet and near-ultraviolet irradiation, and boron dependence of repair. Molec Gen Genet 176:11-16

Jalani BS, Moss JP (1980) The site of action of the crossability genes (Kr$_1$, Kr$_2$) between Triticum and Secale. I. Pollen germination, pollen tube growth, and number of pollen tubes. Euphytica 19:571-579

Jensen WA, Fisher DB (1970) Cotton embryogenesis: the pollen tube in the stigma and style. Protoplasma 69:215-235

Kho YO, Den Nijs APM, Franken J (1980) Interspecific hybridization in Cucumis L. II. The crossability of species, an investigation of in vivo pollen tube growth and seed set. Euphytica 29:661-672

Kirby EG, Smith JE (1974) Elutable substances of pollen grain walls. In: Linskens HF (ed) Fertilization in higher plants. North Holland, Amsterdam, p 127

Kirby-Smith JS, Craig DL (1957) The induction of chromosome aberrations in Tradescantia by ultraviolet radiation. Genetics 42:176-187

Kramer PJ, Kozlowski TT (1979) Physiology of woody plants. Academic Press, New York

Kroh M, Linskens HF (1963) Biochemie der Befruchtungs- "Inkompatibilität". Die Ursache für die Unfruchtbarkeit von Blütenpflanzen bie Selbstbestäutbung, I. Umschau in Wissenschaft und Technik 63:266-269

Kroh M, Gorissen MH, Pfahler PL (1979) Ultrastructural studies on styles and pollen tubes of Zea mays L.: general survey of pollen tube growth in vivo . Acta Bot Neerl 28:513-518

Kwack BH (1965) Stylar culture of pollen and physiological studies of self-incompatibility in Oenothera organensis. Physiol Plant 18:297-305

Lange W, Wojciechowska B (1976) The crossing of common wheat (Triticum aestivum L.) with cultivated rye (Secale cereale L.). I. Crossability, pollen grain germination, and pollen tube growth. Euphytica 25:609-620

Lee CL (1980) Pollenkeimung, Pollenschlauchwachstum und Befruchtungsverhältnisse bei Prunus domestica. II. Pollenschlauchwachstum im Griffel. Gartenbauwissenschaft 45:241-248

Luxová M (1967) Fertilization of barley (Hordeum distichum L.). Biol Plant 9:301-307

Maheshwari P (1949) The male gametophyte of the angiosperms. Bot Rev 15:1-75

Martin FW (1970) The ultraviolet absorption profile of stigmatic extracts. New Phytol 69:425-430

Mascarenhas JP (1975) The biochemistry of angiosperm pollen development. Bot Rev 41:259-314

Mascarenhas JP (1978) Ribonucleic acids and proteins in pollen germination. In: Proceedings of the fourth international palynological conference, vol 1. Lucknow, India, p 400

Mascarenhas NT, Bashe D, Eisenberg A, Willing RP, Xiao C-M, Mascarenhas JP (1984) Messenger RNAs in corn pollen and protein synthesis during germination and pollen tube growth. Theor Appl Genet 68:323-326

Mattsson O (1983) The significance of exine oils in the initial interaction between pollen and stigma in Armeria maritima. In: Mulcahy DL, Ottaviano E (eds) Pollen: Biology and implications for plant breeding. Elsevier, New York, p 257

McLean RC, Ivimey-Cook WR (1956) Textbook of theoretical botany, vol 1. Longman, Green and Co, London

Mulcahy DL, Curtis PS, Snow AA (1983) Pollen competition in a natural population. In: Jones CE, Little RJ (eds) Handbook of experimental pollination biology. Scientific and Academic Editions, New York, p 330

Owens SJ, Horsfield NJ (1982) A light and electron microscopic study of stigmas in Aneilema and Commelina species (Commelinaceae). Protoplasma 112:26-36

Pfahler PL (1981) In vitro germination characteristics of maize pollen to detect biological activity of environmental pollutants. Environ Health Perspect 37:125-132

Pope M (1937) The time factor in pollen-tube growth and fertilization in barley. J Agric Res 54:525-529

Pundir NS (1982) Pollen germination and pollen tube growth in Gossypium hirsutum L. in the field and in vitro. Phyton 42:59-62

Pundir NS, Abbas RF, Al-Attar AAA (1983) Pollen germination and pollen tube growth in Raphanus sativus L. following self- and cross-pollinations. Phyton 43:127-130

Randolph LF (1936) Developmental morphology of the caryopsis in maize. J Agric Res 53:881-916

Reger BJ, James J (1982) Pollen germination and pollen tube growth of sorghum when crossed to maize and pearl millet. Crop Sci 22:140-144

Russell SD (1982) Fertilization in Plumbago zeylanica: entry and discharge of the pollen tube in the embryo sac. Can J Bot 60:2219-2230

Sastri DC, Shivanna, KR (1979) Role of pollen-wall proteins in intraspecific incompatibility in Saccharum benegalense. Phytomorphology 29:324-330

Sedgley M (1977) The effect of temperatures on floral behaviour, pollen tube growth, and fruit set in the avocado. J Hort Sci 52:135-141

Sedgley M (1983) Pollen tube growth in Macadamia. Sci Hort 18:333-341

Sedgley M, Buttrose, MS (1978) Some effects of light intensity, daylength, and temperature on flowering and pollen tube growth in the watermelon (Citrullus lanatus). Ann Bot 42:609-616

Sedgley M, Scholefield PB (1980) Stigma secretion in the watermelon before and after pollination. Bot Gaz 141:428-434

Socias i Company R, Kester DE, Bradley MV (1976) Effects of temperature and genotype on pollen tube growth in some self-incompatible and self-compatible almond cultivars. J Amer Soc Hort Sci 101:490-493

Stadler LJ, Sprague GF (1936) Genetic effects of ultra-violet radiation in maize. I. Unfiltered radiation. Proc Nat Acad Sci USA 22:572-578

Staudt G (1982) Pollenkeimung und Pollenschlauchwachstum in vivo bei Vitis und die Abhängigkeit von der Temperatur. Vitis 21:205-216

Stephens JC, Quinby JR (1934) Anthesis, pollination and fertilization in sorghum. J Agric Res 49:123-136

Stott KG (1972) Pollen germination and pollen-tube characteristics in a range of apple cultivars. J Hort Sci 47:191-198

Thompson MM, Liu LJ (1973) Temperature, fruit set and embryo sac development in 'Italian' prune. J Amer Soc Hort Sci 98:193-197

Van Herpen MMA, Linskens HF (1981) Effect of season, plant age and temperature during plant growth on compatible and incompatible pollen tube growth in Petunia hybrida. Acta Bot Neerl 30:209-218

Van Went JL (1974) Ultrastructure of Impatiens pollen. In: Linskens HF (ed) Fertilization in higher plants. North-Holland, Amsterdam, p 81

Vasil IK (1960) Studies on pollen germination of certain Cucurbitaceae. Amer J Bot 47:239-247

Wiermann R, Vieth K (1983) Outer pollen wall, an important accumulation site for flavonoids. Protoplasma 118:230-233

Williams RR, Maier M (1977) Pseudocompatibility after self-pollination of the apple Cox's Orange Pippin. J Hort Sci 52:475-483

Wilms HJ (1974) Branching of pollen tubes in spinach. In: Linskens HF (ed) Fertilization in higher plants. North-Holland, Amsterdam, p 155

Wilms HJ (1980) Ultrastructure of the stigma and style of spinach in relation to pollen germination and pollen tube growth. Acta Bot Neerl 29:33-47

Wilms HJ (1981) Pollen tube penetration and fertilization in spinach. Acta Bot Neerl 30:101-122

Zeven AC, Van Heemert C (1970) Germination of pollen of weed rye (Secale segetale L.) on wheat (Triticum aestivum L.) stigmas and the growth of the pollen tubes. Euphytica 19:175-179

The Effect of Enhanced Solar UV-B Radiation on Motile Microorganisms

D.-P. Häder

Philipps Universität, Lahnberge, Marburg, Federal Republic of Germany

ABSTRACT

Slightly enhanced levels of UV-B radiation which do not immediately harm or kill microorganisms may have a catastrophic effect on the survival of microorganisms since it impairs the development, motility or photoorientation of the organisms. Exposure to UV-B wavelengths $\leq$ 300 nm for as short as 0.5 h inhibits slug and sorocarp development of the slime mold Dictyostelium. Fluences of $\leq$ 100 J m^{-2}, reduce the speed of movement by 50% and inhibit the phototactic orientation. The green flagellate Euglena becomes immotile with moderate doses of UV-B radiation which inhibits the active movement to suitable light conditions necessary for survival. The photosynthetic blue-green alga Phormidium is prevented from entering dark areas by a simple reversal of movement (step-down photophobic response). This response is inhibited by slightly enhanced levels of UV-B radiation which prevents the organism from staying in the suitable niche of its photoenvironment.

INTRODUCTION

Increased doses of UV-B radiation have been shown to have a detrimental effect on growth and development of higher plants (Biggs et al. 1981, Tevini and Iwanzik 1983, Teramura 1983). Several targets of UV-B radiation have been found both in higher plants and in microorganisms. UV-B radiation has been demonstrated to break the cellular DNA in bacteria (Ascenzi and Jagger 1979, Peak and Peak 1983) and to terminate DNA and RNA synthesis in slime molds (Kielman and Deering 1980, Hamer and Cotter 1982). The photosynthetic electron transport chain in chloroplasts is inhibited by increased doses of UV-B radiation (Noorudeen and Kulandaivelu 1982, Iwanzik et al. 1983) and membranes and genetic materials are affected (Murphy 1983, Soyfer 1983).

Since UV-B effects on growth and production in higher plants usually require long term studies, it is desirable to find a model system in which effects of increased levels of UV-B can be determined after only short exposure times. Microorganisms can be used as suitable systems since they have been shown to be extremely sensitive to even only slightly increased levels of UV-B radiation. Furthermore, the effect of UV-B radiation can be evaluated after short exposure times. Three parameters are reviewed in this chapter as indicators for UV-B effects: development, motility, and orientation with respect to light stimuli.

NATO ASI Series, Vol. G8
Stratospheric Ozone Reduction, Solar Ultraviolet Radiation and Plant Life. Edited by R. C. Worrest and M. M. Caldwell
© Springer-Verlag Berlin Heidelberg 1986

MATERIALS AND METHODS

The experiments were carried out on the following organisms for which the culture conditions are given by the associated references: the cellular slime mold Dictyostelium discoideum strain NC4 (Häder et al. 1983), the filamentous blue-green alga Phormidium uncinatum (Nultsch 1962) and the green flagellate Euglena gracilis (Häder et al. 1981).

The UV-B effect on the development of Dictyostelium was assayed by following the aggregation of amoebae and differentiation through the slug stage to mature sorocarps. Twenty µl of a dense amoebae suspension was placed in the center of 1% water agar plates (Plexiglas petri dishes, 10 cm in diameter). The UV irradiation entered the plate through a window in the lid covered with a cutoff filter (WG 230, 1-mm thick, Schott & Gen.) and retarded or inhibited the further development of the amoebae (Häder 1983a).

The speed of movement was determined in Dictyostelium by averaging the distance slugs had moved in a constant time interval proportional to UV-B irradiation. In Euglena, motility was assayed by calculating the number of motile cells as a percentage of the total population after being irradiated with a given UV-B dose.

The phototactic orientation of the organism was quantified using a mathematical approach based on directional statistics (Batschelet 1965). The data acquisition was facilitated by a computer-aided input via a bit pad (Bit Pad I Summagraphics) (Lipson and Häder 1984). The computer programs for the data analysis have been described previously (Häder 1981). Photophobic responses of blue-green algae were assayed using light traps. This population technique allows one to integrate over the responses of a large number of organisms (Nultsch 1962).

White actinic light was produced with slide projectors equipped with quartz halogen bulbs (Leitz, Prado). The infrared radiation was cut off by inserting a KG1 heat absorbing filter (Schott & Gen.). The fluence rate was primarily adjusted by inserting neutral density filters and fine-tuned by varying the voltage with a triac dimmer.

The UV radiation was produced with 1-kW xenon arc lamps (Kratos, type 102) equipped with a 145-mm path length quartz cuvette filled with boiled distilled water to remove the infrared component. Monochromatic light was produced by a Ebert-type monochromator (Kratos GM 250) with a grating (1180 lines/mm blazed at 300 nm) and 4-mm slits. Fluence rates of the actinic light were measured with a thermopile (CA1, Kipp & Zonen) connected to a microvoltmeter (Keithley, type 155). UV-B irradiance levels were determined with a UV sensitive photomultiplier (PM 271 F) connected to a radiometer (IL 1500, International Light).

RESULTS

UV-B effects on sorocarp development in Dictyostelium

Aggregation competent amoebae of the slime mold Dictyostelium discoideum were exposed to UV-B radiation at an irradiance of 0.125 W m^{-2} at the beginning of the developmental cycle. Radiation with wavelengths $\leq$ 305 nm effectively prevented aggregation of amoebae and development of pseudoplasmodia (slugs), while in the uninhibited control, slugs were formed which started to leave the inoculation spot in less than 24 h (Fig. 1a). Forty-eight h after the onset of the experiment, the control organisms had formed mature sorocarps, and organisms exposed to UV-B radiation with wavelengths $\leq$ 305 nm were drastically retarded, although they had been exposed to the UV-B radiation for only between 0.5 and 4 h at the beginning of the experiment (Fig. 1b). UV-B irradiation for 1 h was most effective when given during the first 14 h (Fig. 2). After that time, it barely retarded further development, though the organism had only started to aggregate by then (Häder 1983a). At wavelengths $\leq$ 300 nm an exposure for 1 h at a fluence rate of < 0.1 W m^{-2} was sufficient to retard the development by a factor of two.

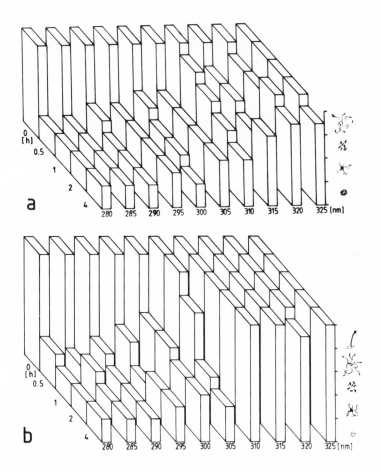

Fig. 1. Effect of increasing exposure times to an irradiance of 0.125 W m^{-2} at various UV-B wavelengths given at the beginning of the developmental cycle of Dictyostelium discoideum. The developmental stage reached after 24 h (Fig. 1a) and 48 h (Fig. 1b) is indicated by symbols in ascending order: no development, aggregating amoebae, formation of slugs, motile slugs outside the inoculation spot and mature sorocarps.

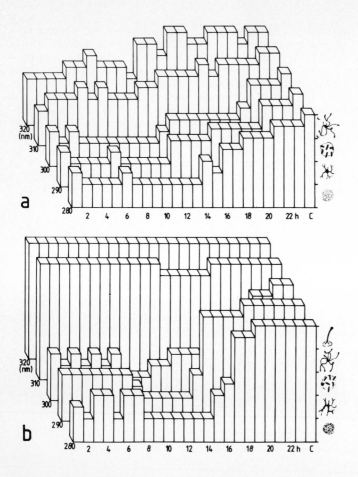

Fig. 2. Development of Dictyostelium affected by 1-h UV-B radiation (0.125 W m^{-2}) of various wavelengths given at various times during the developmental cycle, estimated 24 h (Fig. 2a) and 48 h (Fig. 2b) after the onset of the experiment. Symbols as in Fig. 1.

UV-B inhibition of motility

The speed of movement of Dictyostelium slugs was decreased by about 50% during a 24-h period of lateral white light exposure (Fig. 3) when exposed to 0.5 h of UV-B radiation with wavelengths $\leq$ 305 nm at 0.1 W m^{-2} at the beginning of the experiment (Häder 1983b). Larger doses completely inhibited motility. In the green flagellate Euglena, exposure to a dose of $\geq$1600 J m^{-2} almost completely inhibited motility. Less than 10% of the cells were motile (Fig. 4). In the filamentous blue-green alga Phormidium, the effect of UV-B radiation is cumulative. When assayed over 5-min periods every 0.5 h the percentage of motile filaments in the population was found to decrease steadily (Fig. 5). Wavelengths $\leq$ 290 nm were extremely effective and wavelengths of 295 nm and 300 nm had an intermediate effect. Irradiation with longer wavelengths hardly impaired motility.

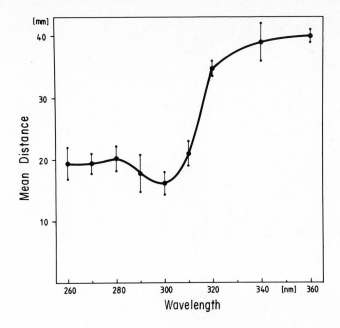

Fig. 3. Effect of 0.5-h supplemental UV-B irradiation of 0.1 W m^{-2} (abscissa, wavelength in nm) given at the beginning of the actinic irradiation (100 lx) on the mean distance (ordinate in mm) <u>Dictyostelium</u> slugs had travelled during 24 h in lateral white light.

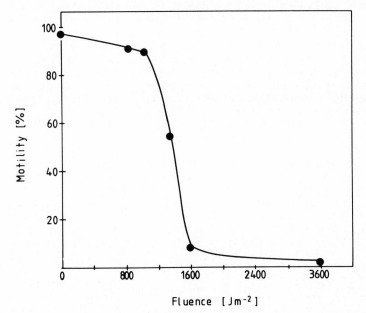

Fig. 4. Inhibition of movement in <u>Euglena gracilis</u> (in percentage of the total population) by increasing fluences of UV-B radiation at 290 nm.

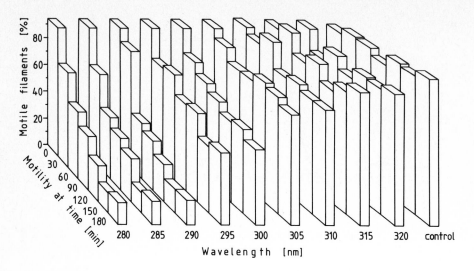

Fig. 5. Percentage of motile filaments in a population of Phormidium measured during 5-min periods every 30 min at various UV-B wavelengths compared with the uninhibited control.

Effect of UV-B radiation on photomovement

In most organisms studied, the orientation with respect to light stimuli is impaired by much lower UV-B doses than development or motility (Häder 1983b). When exposed to lateral actinic white light Dictyostelium pseudoplasmodia show a pronounced positive phototaxis (movement toward the light source) (Fig. 6a). After an additional exposure to UV-B radiation of 0.1 W m^{-2} at 280 nm during the first hour of the experiment, the organism moved almost randomly in the lateral white light (Fig. 6b). The fluence-response curves at various wavelengths demonstrate the sensitivity of the response (Fig. 7). At wavelengths $\leq$ 280 nm a fluence of < 50 J m^{-2} reduces the degree of phototactic orientation to 50%.

The photophobic response of Phormidium, which consists of a reversal of movement elicited by a sudden decrease in light intensity, is impaired drastically with increasing exposure time to UV-B radiation (Fig. 8). The percentage of responding organisms decreases with the wavelength of the inhibitory UV-B radiation, which was adjusted to a fluence rate of 20 mW m^{-2}.

DISCUSSION

A slight increase in the solar UV-B radiation reaching the surface of the earth may not be immediately hazardous to the survival of many micro-organisms since expected doses could be well below those needed for a substantial killing. However, even slightly increased levels of UV-B radiation affect development, motility and photoorientation of many micro-organisms, which would have a catastrophic effect on survival.

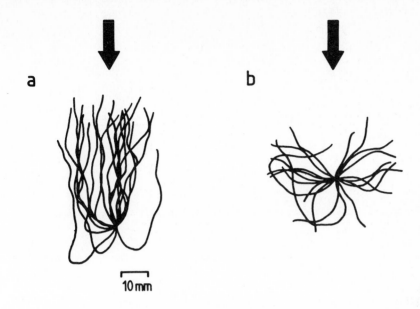

Fig. 6. Patterns of Dictyostelium discoideum slugs moving for 24 h on a water agar surface in lateral white light (100 lx, arrows) without (a) and with (b) additional UV-B irradiation (0.1 W m^{-2} at 280 nm) during the first 1 h.

The development of the slime mold Dictyostelium is completely and irreversibly inhibited at fluences of about 100 J m^{-2} at wavelengths of $\leq$ 300 nm. We compare this result to the actual global irradiance measured over a daily cycle: Measurements at Gainesville (Florida) on September 15, 1973, showed a fluence rate at 300 nm of about 10^{-4} W m^{-2} nm^{-1} at 8:00 AM, about 3 x 10^{-3} W m^{-2} nm^{-1} at 10:00 AM and about 8 x 10^{-3} W m^{-2} nm^{-1} at noon (Billen and Green 1974). Dictyostelium would not develop when exposed for 3.5 h to a fluence measured at noon. A shift of 5 nm to lower wavelengths in the atmospheric transmission would reduce the exposure time to 20 min, which essentially blocks any development. Shielding by a canopy of leaves in a forest would reduce the UV-B impact and prolong the tolerated exposure time.

Similar calculations can be carried out for the inhibition of motility and photoorientation. Dictyostelium amoebae are positively phototactic in low light intensities and negative in higher light intensities (Häder and Poff 1979a,b) and thus tend to populate layers deeper down in the forest mulch. Slugs are exclusively positively phototactic (Bonner et al. 1950) and thus migrate to the surface. This is of an advantage to the organisms since the spores will be produced in a location where they can be easily distributed.

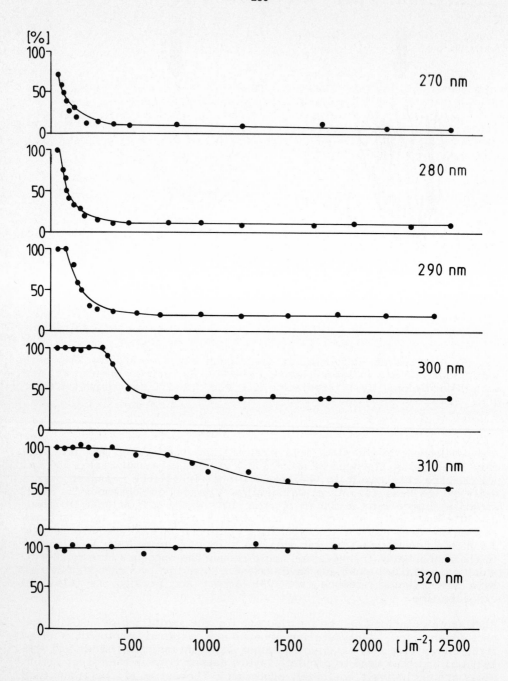

Fig. 7. Fluence response curves of the phototactic orientation (ordinate) to lateral white light (100 lx) for <u>Dictyostelium</u> slugs at various UV-B wavelengths.

When the motility and the phototactic orientation are impaired, the migration into an ecologically advantageous position is inhibited.

A similar situation, with likewise catastrophic results, occurs for photosynthetic flagellates. These organisms depend on the availability of photosynthetically active light. Euglena relies on positive phototaxis to swim towards the light source and on negative phototaxis to escape from harmful high light intensities. This light-induced movement ensures a delicate position between excessively dim and excessively bright light. The active movement is impaired by hazardous UV-B radiation; therefore the organisms will not survive for an extended period of time during UV-B exposure.

The filamentous blue-green alga Phormidium lives on the bottom of shallow bodies of water, where it glides over the surface of the substrate. It is prevented from moving into the sediment by a step-down photophobic response. Each time the tip of a filament moves into a darker area, a reversal of movement is induced. The phobic reversal is cancelled within a short time by low doses of UV-B radiation so that the organisms will not be prevented from moving into the sediment where they die due to the lack of photophosphorylation.

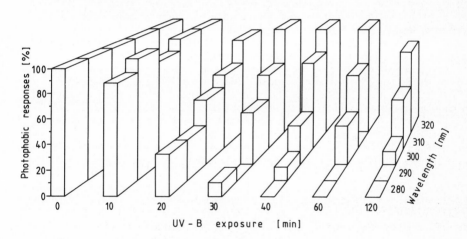

Fig. 8. Inhibition of photoaccumulation (in percentage of the control) of Phormidium in white light traps (1000 lx) by increasing exposure times to UV-B radiation (0.1 W m^{-2}).

In summary, slightly enhanced UV-B irradiation seriously impairs development, motility and photoorientation of microorganisms. This would have a catastrophic effect on their survival even at levels which do not immediately kill the cells.

ACKNOWLEDGMENTS

This work was supported by the Bundesministerium für Forschung und Technologie (KBF 57). The author thanks I. Herrmann, R. Möller and G. Traxler for excellent technical assistance.

REFERENCES

Ascenzi JM, Jagger J (1979) Ultraviolet action spectrum (238-405 nm) for inhibition of glycine uptake in E. coli. Photochem Photobiol 30:661-666

Batschelet E (1965) Statistical methods for the analysis of problems in animal orientation and certain biological rhythms. In: Galles SR, Schmidt-Koenig K, Jacobs GJ, Belleville RF (eds) Animal orientation and navigation. NASA, Washington, Amer Inst Biol Sci, p 1

Biggs RH, Kossuth SV, Teramura AH (1981) Response of 19 cultivars of soybeans to ultraviolet-B irradiance. Physiol Plant 53:19-26

Bonner JT, Clarke WW Jr, Neely CL Jr, Slifkin MK (1950) The orientation to light and the extremely sensitive orientation to temperature gradients in the slime mold Dictyostelium discoideum. J Cell Comp Physiol 36:149-158

Billen D, Green AES (1974) Comparison of germicidal activity of sunlight with the response of a sun-burning meter. Photochem Photobiol 21:449-481

Häder D-P (1981) Computer-based evaluation of phototactic orientation in microorganisms. EDV Med Biol 12:27-30

Häder D-P (1983a) Effects of UV-B irradiation on the development of Dictyostelium discoideum. Photochem Photobiol 38:551-555

Häder D-P (1983b) Inhibition of phototaxis and motility by UV-B irradiation in Dictyostelium discoideum slugs. Plant Cell Physiol 24:1545-1552

Häder D-P, Poff KL (1979a) Light-induced accumulations of Dictyostelium discoideum amoebae. Photochem Photobiol 29:1157-1162

Häder D-P, Poff KL (1979b) Photodispersal from light traps by amoebas of Dictyostelium discoideum. Exp Mycol 3:121-131

Häder D-P, Colombeti G, Lenci F, Quaglia M (1981) Phototaxis in the flagellates, Euglena gracilis and Ochromonas danica. Arch Microbiol 130:78-82

Häder D-P, Williams KL, Fisher PR (1983) Phototactic orientation by amoebae of Dictyostelium discoideum slug phototaxis mutants. J Gen Microbiol 129:1617-1621

Hamer JE, Cotter DA (1982) Ultraviolet light-induced termination of RNA synthesis during Dictyostelium discoideum spore germination. Exp Mycol 6:353-363

Iwanzik W, Tevini M, Dohnt G, Voss M, Weiss W, Gräber P, Renger G (1983) Action of UV-B radiation on photosynthetic primary reactions in spinach chloroplasts. Physiol Plant 58:401-407

Kielman JK, Deering RA (1980) Ultraviolet light-induced inhibition of cell division and DNA synthesis in axenically grown repair mutants of Dictyostelium discoideum. Photochem Photobiol 32:149-156

Lipson ED, Häder D-P (1984) Video data acquisition for movement responses in individual organisms. Photochem Photobiol 39:437-441

Murphy TM (1983) Membranes as targets of ultraviolet radiation. Physiol Plant 58:381-388

Noorudeen AM, Kulandaivelu G (1982) On the possible site of inhibition of photosynthetic electron transport by ultraviolet-B radiation. Physiol Plant 55:161-166

Nultsch W (1962) Der Einfluß des Lichtes auf die Bewegung der Cyanophyceen. III. Mitteilung: Photophobotaxis bei *Phormidium* *uncinatum*. Planta 58:647–663

Peak MJ, Peak JG (1983) Use of action spectra for identifying molecular targets and mechanisms of action of solar ultraviolet light. Physiol Plant 58:367–372

Soyfer VN (1983) Influence of physiological conditions on DNA repair and mutagenesis in higher plants. Physiol Plant 58:373–380

Teramura AH (1983) Effects of ultraviolet-B radiation on the growth and yield of crop plants. Physiol Plant 58:415–427

Tevini M, Iwanzik W (1983) Inhibition of photosynthetic activity by UV-B radiation in radish seedlings. Physiol Plant 58:395–400

UV-B Radiation and Adaptive Mechanisms in Plants

C. J. Beggs, U. Schneider-Ziebert and E. Wellmann

Universität Freiburg, Freiburg im Breisgau, Federal Republic of Germany

ABSTRACT

Further examples of UV-B dependent formation of UV-B screening pigments have been investigated. This active protective mechanism also has been found to be present in the leaf epidermis of various lightgrown vegetables (greenhouse conditions excluding UV-B radiation). Typical action spectra for this kind of positive UV-B effect are interpreted in terms of effectiveness under natural growth conditions and also with respect to the role of interaction with longer wavelengths. The importance of photorepair of UV-B damage is demonstrated for two systems, bean leaves and mustard cotyledons. Photorepair phenomena are further characterized and their importance for adaptation to an increased UV-B environment are pointed out. UV-B effects on growth and the hypersensitivity reaction (isoflavonoids, phytoalexins) in bean are discussed both from the viewpoint of damaging and beneficial consequences for the plant.

INTRODUCTION

This chapter seeks to review the present state of knowledge with respect to possible protective mechanisms in plants. Protective mechanisms may be conveniently divided into three main classes: (i) those whereby UV damage is repaired or its effects negated, (ii) those whereby the amount of UV radiation actually reaching sensitive targets is reduced and (iii) those in which the plant, by some appropriate response, minimizes the negative effects of damage that has taken place.

One of the most important and sensitive targets of UV radiation is certainly DNA. Action spectra for UV damage in most organisms suggest strongly that DNA is involved as a target (Caldwell 1971; Setlow 1974). Although proteins and RNA also absorb at the relevant wavelengths and are almost certainly also subject to UV damage, the special role of DNA as the storehouse of genetic information makes damage to it of particular consequence to the organism.

Because of this special role, organisms have developed several mechanisms for repairing DNA damaged by UV radiation or other factors. Experiments with microorganisms have shown that the significant products formed as a consequence of absorption of UV radiation by DNA are cyclobutane-type dimers formed between pyrimidine bases in the DNA (Setlow et al. 1965;

NATO ASI Series, Vol. G 8
Stratospheric Ozone Reduction, Solar Ultraviolet Radiation
and Plant Life. Edited by R. C. Worrest and M. M. Caldwell
© Springer-Verlag Berlin Heidelberg 1986

Setlow 1966). These are not only the most common photoproducts, but almost certainly the cause of most DNA-mediated damaging effects (Setlow and Setlow 1962).

To combat such damage organisms have developed three important repair processes. In the case of photoreactivation or photoenzymatic repair (see Sutherland 1981) the dimers are monomerized in situ by an enzymatic process using light energy. The photoreactivating enzyme binds to dimer sites and using light energy (usually UV-A and blue radiation) splits the dimer. This type of repair appears to be widespread and, in plants, photo-reversible negative effects of UV radiation have often been described (Bawden and Kleczkowski 1952; Tanada and Hendricks 1953; Hadwiger and Schwochau 1971; Bridge and Klarman 1973). Also, the presence of the photoreactivating enzyme has been demonstrated (Saito and Werbin 1969). Some further results concerning this system in plants are described later and the role of the system will be discussed.

Excision repair of DNA involves the removal of the UV photoproduct from the DNA molecule and its replacement by a new resynthesized correct sequence. It is a repair process requiring no light energy, contrary to the case for photoreactivation. The information for resynthesis is obtained from the complementary undamaged strand of DNA. This repair method is also very widespread and is found in plants (Harm 1980), but although considerable work has been done on bacterial systems little is known of its role and importance in plants.

Finally, postreplication repair involves replication and combination of undamaged strands to form a complete new whole. This form of repair occurs in bacteria, but as yet no information is available concerning its role, if any, in plants.

It has often been suggested that UV damage may also involve photo-oxidation of cell structures and the production of (or cause of further damage by) free radicals. Indeed there is evidence that some UV damage may occur in this way (Moss and Smith 1981; Thomas et al. 1981; Peak and Peak 1983). On the other hand it seems that excited oxygen radicals are more important in photodamage caused by UV-A and visible radiation than in the case of UV-B damage. For example, in E. coli oxygen dependence of lethal effects is only found for wavelengths above 325 nm (Peak and Peak 1983), and for Neurospora Thomas et al. (1981) suggest that visible radiation alone causes inactivation of conidia via singlet oxygen. In at least some of these cases DNA is believed to be damaged by these radicals, leading to an indirect effect of the radiation (UV-B or otherwise) on the DNA (Peak et al. 1973; Peak and Peak 1982). On the other hand UV-B radiation can induce isoflavonoid phytoalexins. This effect can be photoreversed by subsequent UV-A/blue irradiation and therefore is most probably mediated via DNA dimer production (see Bridge and Klarman 1973). These phytoalexins are in some cases potent sources of free radicals, which again can cause cellular damage (Bakker et al. 1983).

Plants, and other organisms, protect themselves from photo-oxidation via quenching molecules and radical scavenging enzymes such as superoxide dismutases and glutathione peroxidase. Surveys of the role and mode of action of these enzymes have been given by Halliwell (1974) and Fridovich (1975). There is no evidence as yet for any radiation effects on the levels of such enzymes whereby they increased on exposure to UV or visible radiation. It may be that the plant possesses enough enzyme capacity to deal with all but severe photo-oxidation or that a protective response in particular cells (e.g., epidermal) cannot be detected against the high background of permanent dismutases involved in protecting the highly oxidation-prone photosynthetic system. It is, however, well known (Wellmann 1974) that UV-B radiation of wavelengths around 300 nm induces enzymes involved in the production of plant phenolics and that some of these phenolics are efficient quenchers of photo-oxidation (Tajima et al. 1983). It is, therefore, not to be excluded that this response, in part, also has a function in protecting against the effects of free radicals.

The second class of protective mechanisms are those whereby the amount of UV radiation actually reaching the sensitive targets is reduced. These may be structural aspects of the plant or screening molecules which absorb the UV radiation. They may be static, that is always present, or appear as the plant develops normally without any special outside stimulus or they may appear after a special stimulus such as radiation treatment.

Structural attenuation of UV radiation appears to play little role in most plants and has recently been reviewed by Caldwell et al. (1983). The cuticle and cell walls as such do not absorb UV radiation although in some cases external secretions or powders may contain large quantities of flavonoids which absorb UV radiation efficiently. These secretions may be the thick waxes of desert scrub plants (Johnson 1983) or the oils and waxes found around the buds of some trees (Egger et al. 1970; Wollenweber and Egger 1971). In general however flavonoids are found in the cell vacuoles, although they also do occur in chloroplasts (McLure 1975).

Often it has been suggested (Caldwell 1981; Wellmann 1983) that flavonoids and anthocyanins have as one of their major functions the absorption of UV radiation that might otherwise cause damage to the plant. These theories are based on the observations that, in addition to the fact that the pigments absorb strongly in the relevant waveband and are usually accumulated in the outer cell layers of the plant, flavonoid production is very often stimulated by radiation. Although often induced by visible radiation (presumably via phytochrome and/or the blue light receptor), there is also a separate and widespread UV-B induction of flavonoids and anthocyanins. In certain other systems UV-B radiation is mandatory, although further regulation via phytochrome or the blue light receptor can occur (see Wellmann 1983). The view that this represents a protective mechanism is also supported by the observations that (i) the highest quantum efficiency occurs at the most damaging UV wavelengths occurring in solar radiation at the earth's surface (about 300 nm), (ii) there is a linear fluence-effect relationship, and (iii) the response after UV induction is fast (Wellmann 1976, 1983). Since UV induction of flavonoids in cell cultures of parsley was first described (Wellmann 1971), further experiments have shown that pigment induction is an exceedingly complicated process, often involving confusing interactions of various wavebands of radiation. Some examples of recent work are described subsequently in order to give a picture of the complexity of the system.

A further question with respect to screening pigments is that of whether, in spite of the good arguments mentioned previously, they actually do form a functional, effective protection against UV-radiation damage. Some experiments indicating that this is indeed the case also will be discussed.

The final group of protective mechanisms are those involved in minimizing the effects of damage that has occurred in spite of the various other protective mechanisms. In plants an important mechanism is growth delay (inhibition). It has been argued that UV-induced growth delay represents an active protective reaction whereby, for example, cell division does not occur or is reduced when the DNA is exposed to potentially damaging radiation, and that repair mechanisms are given a chance to act before cell division (DNA replication) begins again. Discussion of the role of growth delay as a protective mechanism is complicated by two factors. Firstly, a growth inhibition due to UV radiation, which is certainly a direct damaging effect, can and does occur in plants and it is difficult to distinguish between a damaging and a "positive" growth delay. Secondly, many other wavebands also affect plant growth with various effects on different organs, such as stimulation or inhibition of growth.

The idea of growth delay as a protective response is analogous to the theories proposed regarding bacterial systems. In E. coli a UV-induced growth delay is found and it has been argued that this protects the bacterium in the way previously described. UV-induced growth delay in E. coli shows an action spectrum peaking at about 340 nm and the photoreceptor is believed to be the nucleoside 4-thiouridine found in certain transfer RNA's (see Jagger 1960; Jagger et al. 1964; Ramabhadran and Jagger 1976).

The significance of growth inhibition in plants is also complicated by the fact that most plant growth involves both a cell division and a cell enlargement component. Which of these two components is responsible for growth, varies both with time (age of the organ) and from organ to organ. The argument that growth delay protects the replicating DNA is only tenable if indeed the organ growth that is being delayed is at least to a significant extent due to cell division. In most plant organs, cell division occurs predominantly at the beginning of organ development (growth) and most later growth is due to cell enlargement (see e.g., Häcker et al. 1964; Gaba and Black 1983). Thus it seems unlikely that one is dealing here with a protective mechanism of the type described previously. In roots however cell division occurs continuously throughout growth. UV-induced root-growth inhibition has been studied by Duell-Pfaff (1980, Ph.D. Thesis, University of Freiburg, West Germany). The results show that growth inhibition correlates well with a decrease in DNA production after UV treatment. This and the fact that cell division is important in root growth suggest that, in roots at least, a protective role should be considered. In the presence of potentially damaging UV radiation the UV-sensitive processes involving nucleic acids and proteins would be reduced and the repair and protective reactions as discussed previously would have time to act. Simultaneously UV-screening pigments are formed (Wellmann 1974) allowing the roots to continue their growth under an increased UV environment. That UV-induced root-growth inhibition can be partially modulated by phytochrome (Wellmann 1974) also supports the idea that it is not a simple damage effect. Both the results of Wellmann and those of Duell-Pfaff imply that the response cannot occur via the same system as for E. coli because the most effective wavelengths lie far below the 340 nm quoted by Jagger et al. (1964).

MATERIALS AND METHODS

Seeds of the plant species used were obtained from G. Vath, Samenfachgeschaft, Freiburg, West Germany, with the exception of Rumex patientia, which were a gift from Dr. M. M. Caldwell, Utah State University, Logan, Utah USA. Except where otherwise stated in the figure legends they were germinated on paper with distilled water in the dark at 25°C. Cell suspension cultures were produced as described for Petroselinum by Wellman (1974).

The standard broad band sources used had the following properties:

Far red: maximum 740 nm; half-bandwidth (hbw), 123 nm; energy fluence rate, 3.5 W m^{-2}.

Red: maximum 658 nm, hbw 15 nm; energy fluence rate, 0.67 or 6.7 W m^{-2}.

Blue: maximum 434 nm, hbw 43 nm; energy fluence rate, 7 W m^{-2}.

UV-350: maximum 350 nm, hbw 40 nm; energy fluence rate, 7.8 W m^{-2}.

UV-310: maximum 310 nm, hbw 40 nm.

The construction of these light sources is as described in Schäfer (1977) with the exception of the UV sources. The UV sources consisted of a bank of Osram L 40W/73 fluorescent tube for UV-350, and a single Philips TL 40W/73 fluorescent tube for UV-310. Lamps of the Osram L 40W/73 type, although having a maximum at 350 nm, still emit a significant proportion as UV-B radiation (measurements of Osram GmbH and see Diffey 1983).

Monochromatic radiation was obtained using radiation from a Leitz Wetzlar projector fitted with Osram XBO 450 W xenon arc lamps and quartz optics (see Mohr and Schoser 1960) filtered through Schott UV-M-IL interference filters (hbw about 10 nm) from Schott and Gen. Cutoff filters used were also from Schott (type WG). Fluence-rate measurements were made using a thermopile or, for wavelengths above 400 nm, a Gamma Scientific spectroradiometer alone, both calibrated against a standard lamp (see Beggs et al. 1980 for details).

For the experiments in Table 1, anthocyanins were extracted in methanol, HCl (100:1). The samples were left to extract in the dark at -25°C for 24 h. Anthocyanin was measured as absorbance at 546 nm using an Eppendorf photometer 1101M (Eppendorf Geratebau GmbH). Sinapis anthocyanin was extracted and measured as described by Lange et al. (1971).

Flavonoids were extracted in 70% ethanol, 1% acetic acid at 85°C for 10 min. They were then hydrolyzed by heating at 100°C for 1 h in 1:1 HCl, methanol. After hydrolysis the extracts were chromatographed on paper with 15% acetic acid and 3:1:1 tert. butanol, acetic acid, water (two dimensional). The flavonoids were eluted from the paper with methanol and characterized by means of their UV spectra using the published spectra of Mabry et al. (1970) as reference. For the results of Fig. 3, raw extracts in 70% ethanol were used. The flavonoid content of these extracts had previously been checked according to the method given above. Absorption spectra were measured using a Perkin-Elmer 554 spectrophotometer.

RESULTS AND DISCUSSION

Photoreactivation

This section describes the results of experiments to determine the role of photoreactivation in a plant response to UV which is clearly a damage response. Cotyledons of Sinapis alba L. produced anthocyanins in the lower epidermis as a response to red light pulses operating through a phytochrome system (see Lange et al. 1971). If, after a 5-min red light pulse, a 5-min pulse of UV-B radiation is given, the amount of anthocyanin produced, after a further 24 h in darkness, was increased. The amount produced depended on the fluence rate and wavelength of the UV-B radiation used. An action spectrum for this effect showed a peak at about 280 nm and below, with little effectiveness above 300 nm (Wellmann et al. 1984). Table 1 shows the results obtained when, after the red and UV-B pulses, the cotyledons were irradiated for a further hour with various light sources and the anthocyanin then measured after a further 24 h of darkness. It can be seen that the effect of the damaging UV pulse can be at least partially reversed by subsequent exposure to sunlight. UV-A radiation is also very effective in reversing the damage effect, blue and red are not effective. It seems reasonable to conclude that one is dealing here with photoreactivation.

Table 1. Effect of light of various qualities in reactivating UV-B damage to phytochrome-induced anthocyanin formation in Sinapis alba L. cotyledons. 40-h old cotyledons were excised and irradiated on two layers of filter paper irrigated with distilled water. After the irradiation program the cotyledons were returned to darkness for 24 h at 25°C at which point anthocyanin was extracted and measured. Radiation sources (except sunlight) were standard sources. For UV-B the TL 40W/12 source was used, for red the 6.7 W m^{-2} source. The results are those of three independent experiments. (From Wellmann et al. 1984).

Irradiation Program	Anthocyanin (Mean $A_{546} \pm$ SD)
5 min Red + 10 min UV-B + 1 h Blue	0.022 ± 0.021
5 min Red + 10 min UV-B + 1 h Red	0.023 ± 0.012
5 min Red + 10 min UV-B + 1 h UV-A	0.183 ± 0.020
5 min Red + 10 min UV-B + 1 h Sunlight	0.243 ± 0.081
5 min Red + 10 min UV-B + 1 h Dark	0.018 ± 0.008

The effectiveness of the light sources tested fits the known action spectra for photoreactivation (see Harm 1980) and light of longer wavelengths is not effective, thereby excluding an effect mediated by phytochrome or photosynthesis. The young etiolated cotyledons have, in any case, no capacity for photosynthesis.

Figure 1 shows the results of preliminary experiments on UV-mediated growth reduction in Phaseolus vulgaris L. secondary leaves. UV radiation of 291 nm (fluence rate = 2.6 W m^{-2}) applied for 10 min leads to a decrease in the growth rate with respect to that of a nonirradiated control leaflet. However a difference in the effectiveness of UV-B radiation was noted depending on whether the plants were, after UV irradiation, returned immediately to white light or kept for 12 h in darkness before being

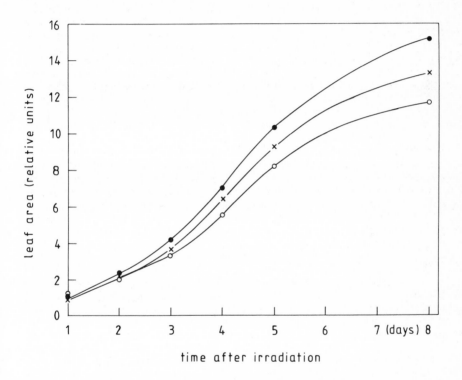

time after irradiation

Fig. 1. Effect of white light on UV-B induced secondary leaf growth in Phaseolus vulgaris L. The plants were grown in soil in a growth chamber at 25°C under 12-h white light : 12-h dark until secondary leaves were just present (about two weeks). For each secondary leaf, one side leaflet was irradiated for 10 min with UV-B radiation, the other was shielded (●). The plants were then kept in the dark for 12 h and then returned to white light (○) or returned immediately to white light (×). The UV-B source was mono-chromatic radiation using a 291-nm interference filter (fluence rate 2.6 W m^{-2}). (Mean values of three independent experiments.)

returned to white light. In the case of those returned to white light immediately, the effect of the UV irradiation was smaller. It remains, in this case, to be determined which wavebands are responsible for the effects. As yet, photosynthesis cannot be excluded.

These experiments provide further evidence that a functioning photo-reactivation system exists in plants. It remains, however, important to determine the capacity of this system in the sense of how much damage can be caused by UV radiation before the photoreactivation system can no longer cope. Regarding the prediction of increased levels of UV radiation in the natural environment it seems that plants probably will have sufficient capacity to deal with any likely increases in solar UV reaching the earth's surface, particularly when one considers that the plants will always be subject to polychromatic radiation containing photoreactivating wave-lengths.

Screening Pigments

A survey of UV-B induced pigment formation in various crop plants: Table 2 shows the results of a survey of UV-B induced flavonoid and anthocyanin formation in a variety of plants, mostly crop plants. The Table makes it clear that UV-B induced pigment formation is not just an isolated response of Petroselinum cell cultures (Wellmann 1971), the system in which it was first detected and characterized, but occurs widely throughout the plant kingdom. In certain cases, e.g., Zea where no effect on flavonoids other than anthocyanins is seen, it is found that the flavonoid content is already very high even in fully dark grown material. In many cases UV-B radiation is not the only waveband which effects pigment formation. Blue, red, and far-red light may also be effective either alone, or in conjunc-tion with UV-B radiation. These interactions may be exceedingly compli-cated as is demonstrated in the sample case history given in the next section.

Interactions between various wavebands in radiation-induced flavonoid formation: As mentioned in the previous section the interactions between various wavebands can be very complicated when one is studying radiation-induced flavonoid formation. Figure 2 shows some results obtained using Petroselinum crispum cell cultures as an example of some typical interac-tions. The dark-grown cell cultures produce no flavonoids at all. A UV pulse leads to flavonoid formation and is a necessary prerequisite for their formation. Blue light alone with up to 5-h irradiation does not lead to any synthesis, nor does a red light pulse alone. On the other hand if blue light is given in conjunction with a UV pulse considerably more flavo-noids are formed than when UV radiation is given alone. The magnitude of the response is also dependent on when and how the blue light is given. A 5-h exposure to blue light given before the UV irradiation is more effec-tive than if given after the UV irradiation. Maximum response however is obtained if over the 5-h period the UV radiation is given as short pulses during continuous blue irradiation. Earlier experiments (Duell-Pfaff and Wellmann 1982) have suggested that a separate blue light receptor is responsible for much of this effect of blue light and that it is not due to the phytochrome system.

Table 2.   UV-B induced pigment formation in various plant species.   Column 2 shows the pigment type formed.   Column 3 shows the approximate irradiation time required to induced significant pigment formation when irradiated with monochromatic UV radiation (298 nm) at a fluence rate of 0.1 W m^{-2}. Pigments were extracted 24 h after irradiation began.   Between irradiation and extraction plants were kept in the dark at 25°C.

System	Effect	Irradiation Time
a. Hypocotyl		
Petroselinum crispum	flav.	min
Antheum gravolans	flav.	h
Daucus carota	flav.	h
Rumex patientia	anth.	h
Sinapis alba[a]	flav., anth.	h, min
b. Seedling Root		
Petroselinum	flav.	min
Antheum	flav.	min
Daucus	flav.	min
Sinapis	flav.	min
Lepidium sativum	flav.	min
c. Coleoptile		
Zea mays	anth.	s
Triticum aestivum	anth., flav.	h, min
Secale cereale	anth., flav.	min
d. Cell Cultures		
Sinapis	flav.	h
Anethum	flav.	min
Petroselinum	flav.	s

[a]With white light background

Figure 2 also shows that the UV effect can also be manipulated to a certain extent by phytochrome.   A certain proportion of the effect can be reversed in all cases if far-red light is given at the end rather than red light. This far-red reversible proportion of the response makes up the same proportion of the total regardless of which pretreatment has been given (UV alone, UV + blue, or blue + UV).

It can be seen that the control of the response is complicated, involving at least three different photoreceptors with a complex synergism between the UV and blue receptors and further control by phytochrome imposed on this.   For consideration of plants in the natural environment it must be remembered that blue and red light will always be present. Blue and near-UV radiation play a particularly complicated role, because it not only induces a synergism with the UV-B receptor but it is also responsible for photoreactivation, and furthermore has some effect on the phytochrome system. The mode of action of these interactions is not as yet well understood and a discussion of it would be outside the scope of this chapter.

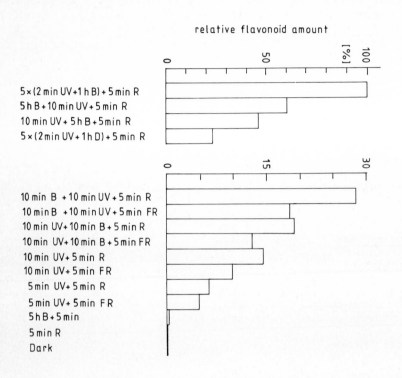

Fig. 2. Radiation control of flavonoid formation in cell suspension cultures of <u>Petroselinum crispum</u>. Flavonoids were extracted 40 h after onset of the irradiation treatment. From the end of the treatment until extraction they were kept in the dark at 25°C. For extraction cells were frozen in liquid nitrogen and then extracted in 1:1 ethanol, borate buffer (pH 8.8) at 85°C for 10 min. Results are means from 3 separate experiments of 5 parallels each. Standard error was less than 5%.

<u>Examples of the effect of UV radiation on the absorption of plants</u>: This section shows the results of some experiments to demonstrate the effect of UV treatment on the absorption properties of various plants. The seedlings were exposed to sunlight through a series of different cutoff filters to vary the quality of the UV radiation reaching the plant. After irradiation the plants were returned to darkness until 24 h after the beginning of the irradiation period. At this point the relevant organ was excised and

extracted. Chromatography was used to show that the main pigment increase was in flavonoids (see Materials and Methods section).

Figure 3 shows some examples of the type of response obtained. Figure 3a shows the response of <u>Triticum</u> after 1 h UV-B exposure and Fig. 3b shows the same for <u>Secale</u> after 5 h of sunlight. In both cases it can be seen that the absorption between 300 and 400 nm for extracts from irradiated plants is considerably higher than for the dark controls. Furthermore the use of cutoff filters shows clearly that UV-B radiation is the most important waveband in inducing this response. In <u>Triticum</u> the use of a WG 365 cutoff filter almost abolishes the response, and in <u>Secale</u> the response is considerably lower when such a filter is used compared to responses when using WG 280 or 305 filters.

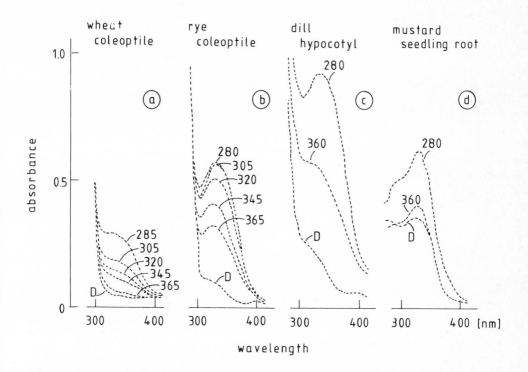

Fig. 3. Absorption of extracts from various plants after irradiation treatment. The plants were germinated in the dark and irradiated one day after germination. The irradiation source was either UV-B radiation (TL 40W/12) or sunlight. The plants were covered with various WG cutoff filters as shown. The sunlight treatments took place in August on a sunny day with few clouds. (a) wheat coleoptile <u>Triticum</u>, 1 h UV-B (TL 40W/12); (b) rye coleoptile <u>Secale</u>, 5 h sunlight; (c) dill hypocotyl <u>Anethum</u>, 9 h sunlight; (d) mustard seedling root <u>Sinapis</u>, 9 h sunlight.

Figures 3c and 3d show the results of similar experiments for two dicotyle-donous organs, namely hypocotyls of Anethum and roots of Sinapis. Again, a considerable increase in absorption is found after irradiation using a WG 280 cutoff filter, which approximates unfiltered extraterrestrial sunlight; whereas the effect is much less when a WG 365 cutoff filter is used.

The experiments demonstrate that the plants respond to UV-B radiation by rapid production of UV-B absorbing flavonoids and that UV-B absorbance in the tissue extracts is increased considerably by this response. Thus an important requirement for a protective role for flavonoids is fulfilled although proof is still required to show that the flavonoids not only can, but do provide protection.

Evidence that UV-B induced flavonoid production does increase protection against UV damage: Experiments are described here to demonstrate that flavonoid production can indeed increase the resistance of plant organs to UV radiation damage. In these experiments Petroselinum roots were irra-diated with the UV-350 standard source (which also emits some UV-B radiation, see Materials and Methods) for a short period of time in order to stimulate flavonoid production. After sufficient flavonoids had been accumulated the seedlings were further irradiated with UV-B radiation. The damaging effect of the UV-B radiation was then compared with that of UV-B radiation on plants which had received no UV-350 pretreatment. Earlier experiments (Gasser and Wellmann unpublished) had shown that if the UV-350 irradiation took place under an atmosphere of N_2 the quantity of flavonoids produced was approximately half that occurring when the radiation was performed under a normal atmosphere. On the other hand the use of the N_2 atmosphere had no effect on root growth. Accordingly a second control was carried out where the UV-350 irradiation took place under an N_2 atmosphere.

Figure 4 shows the results of these experiments. The experimental program was:

$$5 \text{ days dark} \Big\langle \begin{array}{l} 1 \text{ h UV-350} \\ 1 \text{ h UV-350} + N_2 \\ 1 \text{ h dark} \end{array} \Big\rangle 9 \text{ h dark} \rightarrow 5 \text{ min UV-B} \rightarrow 60 \text{ h dark}$$

After the 60-h dark period the roots were examined and analyzed for damage according to the criteria shown in the figure legend. In roots, which had received the same pretreaments, the pigments present at the beginning of the UV-B treatment were extracted and the quantity determined spectrophoto-metrically. This quantity is also shown in the figure.

It can be seen that a definite correlation occurs between freedom from damage and pre-existence of flavonoids in the tissue. Those roots which had received a UV-350 pretreatment contained more flavonoids and survived the UV-B irradiation to almost 100%; whereas those receiving no pre-irradiation contained few flavonoids and were almost all damaged. Those pre-irradiated under an N_2 atmosphere had an intermediate flavonoid content and also occupied an intermediate position with respect to damage. These results therefore strongly suggest that flavonoids do indeed act to protect plants from damage due to excessive UV-B radiation.

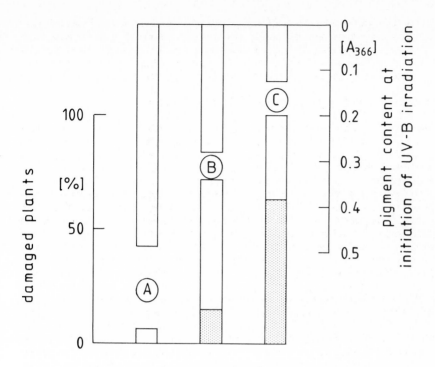

Fig. 4. UV-B damage to <u>Petroselinum</u> roots and its correlation with root pigment content. The irradiation program is as given in the text. Radiation sources were UV-350 (L 40W/73), and UV-B (TL 40W/12). Column A represents plants receiving a pretreatment of 1 h UV-350, column B those receiving UV-350 under an N_2 atmosphere and column C those receiving no pretreatment. Damage was analyzed after 60 h as (i) dead roots (shaded part of the column) and (ii) roots which showed no sign of damage but did not grow further. In each case 100 seedlings were analyzed in 3 separate experiments. Standard error was less than 5%. Pigments were measured as absorbance at 366 nm.

CONCLUSION

The survey and results presented in this chapter suggest that plants possess at least two important mechanisms whereby they are able to adapt to excess UV-B radiation or minimize its damaging effects. It is probable that other mechanisms also exist (e.g., dark repair), but the data here are as yet too scanty to allow one to speculate on how widespread or significant they are. It seems reasonable to assume that photoreactivation is widespread and effective. It also appears as if screening by flavonoids and other pigments, and their production in response to UV radiation, is a widespread and effective protection.

In both cases the importance of longer wavelengths in the plant's response to UV-B radiation is evident and therefore studies of UV-B effects must always consider the polychromatic nature of the irradiation in the natural environment. The blue/UV-A waveband appears here to be of particular importance. This waveband is not only the one responsible for photoreactivation but can also show a strong synergistic interaction with UV-B radiation in stimulating flavonoid production. Furthermore this waveband is, through the blue light receptor, responsible for many independent radiation responses of its own. Red and far-red light operating via the phytochrome system also interact with the UV-B system.

Effects of UV-B radiation itself can probably be divided into damaging effects, adaptive (protective) responses, and those which can be considered a part of normal photomorphogenesis, equivalent to the independent responses mediated by the blue light receptor and phytochrome. It is often difficult to determine which of these response types one is dealing with, because on the one hand the exact significance of particular photomorphogenic responses is often unclear, and on the other hand a mild growth inhibition due to damage is often indistinguishable from a nondamaging growth inhibition. It appears reasonable to assume that effects that can be photoreactivated must be damage effects, at least in so far as pyrimidine dimer formation must be involved in some way. It is not, however, always easy to determine whether or not photoreactivation occurs without actually isolating dimers, because the effect of subsequent blue/UV-A radiation, via photoreactivation, is often disguised, due to overlapping with separate effects resulting from the blue light receptor.

Finally, with respect to the possibility of increased solar UV-B radiation reaching the earth's surface due to a decrease in stratospheric ozone levels, we feel that plants possess effective protective mechanisms of a capacity probably sufficient to cope with the UV increase forecast. However it must be kept in mind that increased solar UV-B radiation will occur as a small fluence-rate increase over a long period of time and, of necessity, laboratory experiments take place over a short time period using relatively high fluence rates that are most unlikely for the natural environment. How far one can extrapolate from the one situation to the other remains, as yet, uncertain.

ACKNOWLEDGMENT

This work was supported by the Bundesministerium für Forschung und Technologie (Grant Nr. KBF 55).

REFERENCES

Bakker J, Gommers FJ, Smits L, Fuchs A, de Vries FW (1983) Photoactivation of isoflavonoid phytoalexins: involvement of free radicals. Photochem Photobiol 38:323-330

Bawden FC, Kleczkowski A (1952) Ultraviolet injury to higher plants counteracted by visible light. Nature 169:90-93

Beggs CJ, Holmes MG, Jabben M, Schäfer E (1980) Action spectra for inhibition of hypocotyl growth by continuous irradiation in light and dark grown Sinapis alba L. seedlings. Plant Physiol 66:615-618

Bridge MA, Klarman WL (1973) Soybean phytoalexin, hydroxyphaseollin induced by ultraviolet radiation. Phytopathology 63:606-609

Caldwell MM (1971) Solar UV irradiation and the growth and development of higher plants. In: Giese AC (ed) Photophysiology, vol 6. Academic Press, New York, p 131, ISBN 0-12-282606-X

Caldwell MM (1981) Plant response to solar ultraviolet radiation. In: Lange OL, Nobel PS, Osmond CB, Ziegler H (eds) Encyclopedia of plant physiology, new series, physiological plant ecology I, vol 12A. Springer-Verlag, Berlin Heidelberg New York, p 169, ISBN 3-540-10763-0

Caldwell MM, Robberecht R, Flint SD (1983) Internal filters: Prospects for UV acclimation in higher plants. Physiol Plant 58:445-450

Diffey BL (1983) The UV-B content of "UV-A fluorescent lamps" and its erythemal effectiveness in human skin. Physics in Biology and Medicine 28:351-360

Duell-Pfaff N, Wellmann E (1982) Involvement of phytochrome and a blue light photoreceptor in UV-B induced flavonoid synthesis in parsley (Petroselinum hortense Hoffm.) cell suspension cultures. Planta 156:213-217

Egger K, Wollenweber E, Tissot M (1970) Freie Flavonol-Aglykone im Knospensekret von Aesculus Arten. Z Pflanzenphysiol 62:464-466

Fridovich I (1975) Superoxide dismutases. Ann Rev Biochem 44:147-159

Gaba V, Black M (1983) The control of cell growth by light. In: Shropshire W Jr, Mohr H (eds) Encyclopedia of plant physiology, new series, photomorphogenesis vol 16A. Springer-Verlag, Berlin Heidelberg New York, p 385, ISBN 3-540-12143-9.

Hadwiger LA, Schwochau ME (1971) Ultraviolet light-induced formation of pisatin and phenylalanine ammonia lyase. Plant Physiol 47:588-590

Häcker M, Hartmann KM, Mohr H (1964) Zellteilung und Zellwachstum im Hypokotyl von Lactuca sativa L. unter dem Einfluss des Lichtes. Planta 63:253-268

Halliwell B (1974) Superoxide dismutase, catalase, and glutathione peroxidase: Solutions to the problems of living with oxygen. New Phytol 73:1075-1080

Harm W (1980) Biological effects of ultraviolet radiation. Cambridge University Press, Cambridge, ISBN 0-521-22121-8

Jagger J (1960) Photoprotection from ultraviolet killing in Escherichia coli B. Radiat Res 13:521-525

Jagger J, Wise WC, Stafford RS (1964) Delay in growth and division induced by near UV radiation in E. coli B and its role in photoprotection and liquid holding recovery. Photochem Photobiol 3:11-24

Johnson ND (1983) Flavonoid aglycones from Eriodyction californicum resin and their implication for herbivory and UV screening. Biochemical Systematics and Ecology 11:211-215

Lange H, Shropshire Jr W, Mohr H (1971) An analysis of phytochrome mediated anthocyanin synthesis. Plant Physiol 47:649-655

Mabry TJ, Markham KR, Thomas MB (1970) The systematic identification of flavonoids. Springer-Verlag, Berlin Heidelberg New York

McLure JW (1975) Physiology and function of flavonoids. In: Harbone JB, Mabry TJ, Mabry H (eds) The flavonoids. Academic Press, New York, p 970, ISBN 0-412-11960-9

Mohr H, Schoser G (1960) Eine mit Xenonbogen austerüstete Interferenzfilter Monohcromatoranlage für kurzwellige sichtbare und langwellige ultraviolette Strahlung. Planta 55:143-152

Moss SH, Smith KC (1981) Membrane damage can be a significant factor in the inactivation of E. coli by near UV radiation. Photochem Photobiol 33:203-210

250

Peak MJ, Peak JG (1982) Single strand breaks induced by Bacillus subtilis DNA by ultraviolet light: action spectrum and properties. Photochem Photobiol 35:675-680

Peak MJ, Peak JG (1983) Use of action spectra for identifying molecular targets and mechanisms of action of solar ultraviolet light. Physiol Plant 58:367-372

Peak MJ, Peak JG, Webb RB (1973) Inactivation of transforming deoxyribonucleic acid by ultraviolet light III. Further observations on the effect of 365 nm radiation. Mutation Res 20:143-148

Ramabhadran TV, Jagger J (1976) Mechanism of growth delay induced in E. coli by near UV radiation. Proc Natl Acad Sci USA 73:59-63

Saito N, Werbin H (1969) Evidence for a DNA photoreactivating enzyme in plants. Photochem Photobiol 9:389-393

Schäfer E (1977) Kunstlicht und Pflanzenzucht. In: Albrecht H (ed) Optische Strahlung Quellen. Lexika-Verlag, Grafenau, West Germany, p 249, ISBN 3-88146-112-4

Setlow RB (1966) Cyclobutane type pyrimidine dimers in polynucleotides. Science 153:379-386

Setlow RB (1974) The wavelengths in sunlight effective in producing skin cancer: a theoretical approach. Proc Natl Acad Sci USA 71:3363-3366

Setlow RB, Setlow JK (1962) Evidence that ultraviolet induced thymine dimers in DNA cause biological damage. Proc Natl Acad Sci USA 48:1250-1257

Setlow JK, Boling ME, Bollom FJ (1965) The chemical nature of photoreactivable lesions in DNA. Proc Natl Acad Sci USA 53:1430-1436

Sutherland BM (1981) Photoreactivation. BioScience 31:439-444

Tanada T, Hendricks SB (1953) Photoreversal of ultraviolet effects in soybean leaves. Amer J Bot 40:634-637

Tajima K, Sakamoto M, Okuda K, Mukai K, Ishizu K, Sakurai H, Mori H (1983) Reaction of biological phenolic antioxidants with super-oxide generated by a cytochrome P-450 model system. Biochem Biophys Res Commun 115:1002-1008

Thomas SA, Sargent ML, Tuveson RW (1981) Inactivation of normal and mutant Neurospora crassa conidia by visible light and near UV. Role of $^1O_2$, carotenoid composition and sensitizer location. Photochem Photobiol 33:345-349

Wellmann E (1971) Phytochrome mediated flavone glycoside synthesis in cell suspension cultures of Petroselinum hortense after preirradiation with ultraviolet light. Planta 101:283-286

Wellmann E (1974) Regulation der Flavonoidbiosynthese durch ultraviolettes Licht und Phytochrom in Zellkulturen und Keimlingen von Petersilie (Petroselinum hortense Hoffm.). Ber Dtsch Bot Ges 87:267-273

Wellmann E (1976) Specific ultraviolet effects in plant morphogenesis. Photochem Photobiol 24:659-660

Wellmann E (1983) UV radiation in photomorphogenesis. In: Shropshire Jr W, Mohr H (eds) Encyclopedia of plant physiology, new series, photomorphogenesis, vol 16B. Springer-Verlag, Berlin Heidelberg New York, p 745, ISBN 3-540-10763-0

Wellmann E, Schneider-Ziebert Y, Beggs CJ (1984) UV-B inhibition of phytochrome mediated anthocyanin formation in Sinapis alba L. cotyledons. Action spectrum and the role of photoreactivation. Plant Physiol 75:997-1000

Wollenweber E, Egger K (1971) Die lipophilen Flavonoide des Knospenols von Populus nigra. Phytochemistry 10:225-226

Leaf UV optical properties of Rumex patientia L. and Rumex obtusifolius L. in regard to a protective mechanism against solar UV-B radiation injury

R. Robberecht and M. M. Caldwell

University of Idaho, Moscow, Idaho USA
and Utah State University, Logan, Utah USA

ABSTRACT

Effective UV attenuation in the outer leaf layers may represent an important protective mechanism against potentially damaging solar UV-B radiation. Epidermal optical properties for Rumex patientia and Rumex obtusifolius were examined on field collected and greenhouse grown plants. Rumex patientia, a relatively UV-B sensitive plant, has substantially higher epidermal UV transmittance than Rumex obtusifolius, which indicated that the UV-B flux at the mesophyll layer for Rumex patientia would also be higher. Attenuation of UV-B radiation increased in Rumex obtusifolius by 27% after exposure to solar UV-B radiation. Flavonoid extract absorbance also increased in whole leaves of both species after solar UV-B irradiation. The epidermis is not only an effective filter for UV-B radiation, but is wavelength selective, and shows a degree of plasticity in this attenuation.

INTRODUCTION

Leaves are among the primary organs of higher plants for the interception of visible radiation. Morphological characteristics such as leaf arrangement on the stem, orientation, and structure are adaptations that often maximize the interception of light and its penetration to the mesophyll for photosynthesis. These same characteristics may not always afford effective protection from potentially injurious solar ultraviolet-B radiation (290-320 nm). Leaf optical properties that allow high transmittance of visible radiation, while selectively filtering out UV-B radiation before it penetrates to sensitive physiological targets in the leaf, present a viable protective mechanism for reducing UV-B injury.

Plant sensitivity to UV-B-induced damage varies considerably among species (Biggs et al. 1975; Van and Garrard 1975; Teramura 1983). These differences may, in part, be related to the effectiveness of factors in leaf tissue that affect UV-B attenuation. Several studies have shown that leaf surface reflectance of UV-B radiation and the capacity for UV-B attenuation in the outer leaf layers differs considerably among species (Lautenschlager-Fleury 1955; Gausman et al. 1975; Robberecht and Caldwell 1978; Robberecht et al. 1980). A mechanism that combines attributes such as effective UV-B attenuation, wavelength selectivity, and a degree of plasticity to respond to different UV-B irradiances would be of significant adaptive value. The role of UV-absorbing pigments such as flavonoids and related phenolic compounds as a mechanism for plant acclimation to solar UV-B radiation has been suggested (Caldwell 1968; McCree and Keener 1974; Gausman et al. 1975; Wellmann 1975; Lee and Lowry 1980; Robberecht and Caldwell 1983). Plasticity in such a mechanism has significance not only for species that colonize high UV-B irradiance habitats, but also for plant growth in the intensified UV-B irradiance environment expected as a consequence of reduced stratospheric ozone (National Academy of Sciences 1982). The present study examined the leaf optical properties of two Rumex species in regard to a possible protective mechanism against damage induced by UV-B radiation. Rumex patientia L. has been shown to be a relatively UV-B-sensitive species (Sisson and Caldwell 1976), whereas R. obtusifolius L. appears relatively less sensitive (Sisson unpublished data). The two species differ significantly in epidermal UV optical properties for both field and greenhouse-grown plants.

MATERIALS AND METHODS

Epidermal UV transmittance was determined for fresh tissue obtained from greenhouse-grown plants and from plants collected in foothill habitats of northern Utah. The epidermis was mechanically removed. Transmittance and reflectance were measured with the integrating sphere/spectroradiometer system described in detail by Robberecht and Caldwell (1978). Absorptance was calculated by subtracting the sum of transmittance and reflectance from 100%.

A methanol-water-HCl solution (70:29:1 v/v) was used to examine the contribution of UV-absorbing pigments, such as flavonoids and related phenolic compounds, to the UV attenuating capability of the epidermis. This method effectively removes such compounds from leaf tissue (see Caldwell 1968; Ribereau-Gayon 1972). Fresh epidermal tissue was immersed in this solution for 15 minutes and subsequently re-examined for transmittance. Whole-leaf extracts were obtained for analysis of possible changes in flavonoid absorbance after exposure to solar UV-B radiation. Leaf tissue was ground in 5 ml of the extraction solution and centrifuged for 10 minutes. The resulting supernatant was then examined spectrophotometrically (Beckman Model 35) for absorbance.

Greenhouse cultivated plants of both species were transferred to the field and placed beneath either a 0.08-mm Aclar or 0.13-mm Mylar plastic film (DuPont Co.). Mylar excludes UV-B radiation, whereas Aclar is relatively transparent in this waveband (Fig. 1). The species were examined successively. Ultraviolet irradiance was weighted with the generalized plant spectrum normalized at 280 nm (Caldwell 1971). Mean biologically effective UV-B irradiation beneath the Aclar filter was 1125 J m^{-2} d^{-1} for Rumex patientia and 1151 J m^{-2} d^{-1} for R. obtusifolius. Determination of epidermal transmittance and whole-leaf extract absorbance were made at the end of the UV-B exposure period.

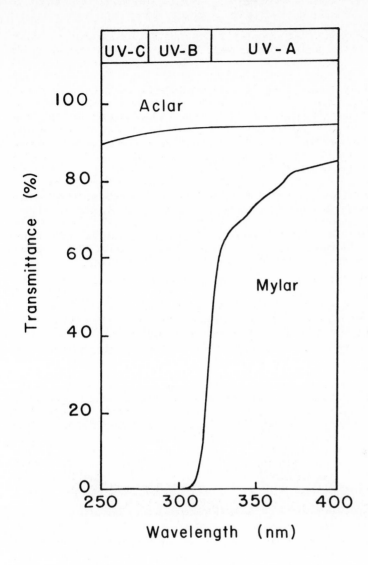

Fig. 1. Transmittance spectra of Aclar (0.08 mm) and Mylar (0.13 mm) plastic films.

RESULTS

Optical Properties

The two Rumex species differed most significantly in the degree of UV-B
radiation transmitted through the epidermis to the mesophyll (Fig. 2).
Mean epidermal UV-B transmittance for greenhouse-grown plants was 18% for
R. patientia and 10% for R. obtusifolius. Ultraviolet-B reflectance of 4%
was similarly low for both species, and showed little variation with wavelength in the UV waveband. The epidermis absorbed more than 80% of the
incident UV-B radiation on the leaf.

Examination of field-grown plants also showed epidermal transmittance for
Rumex patientia to be substantially greater than that for R. obtusifolius.
After UV-absorbing compounds had been extracted from the tissue, UV-B
transmittance was similar for both species (Fig. 3). Removal of UV-B-
absorbing compounds increased mean UV-B transmittance from 8% for fresh
tissue to 51% after extraction for R. obtusifolius, and similarly from 24%
to 52% for R. patientia.

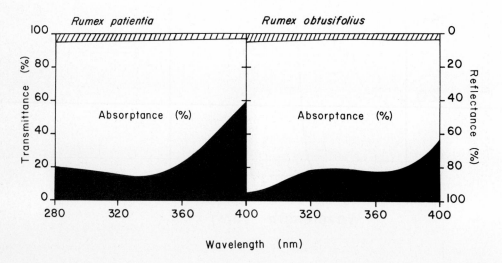

Fig. 2. Epidermal UV optical properties of two Rumex species cultivated
under greenhouse conditions. Transmittance is indicated by the solid area
and reflectance by the hatched area. Each spectrum represents the mean of
three samples.

Influence of Solar UV-B Radiation

Epidermal UV-B transmittance for Rumex obtusifolius was significantly
reduced (p $\leq$ 0.05) by 27% after eight days of exposure to solar UV-B
radiation (Fig. 4). This occurred in fully expanded leaves on plants that
were transferred from the greenhouse to Aclar-covered frames in the field.
After five days of solar UV-B irradiation, the epidermis of R. patientia

became brittle and prone to disintegration upon removal. Since epidermal samples of sufficient size for examination could not be obtained, no transmittance measurements were possible for this species.

Ultraviolet-B absorbance in extracts from whole-leaf tissue showed significant increases (p $\leqq$ 0.05) in response to solar UV-B exposure (Table 1). Absorbance increased an average of 11% for R. patientia and 92% for R. obtusifolius.

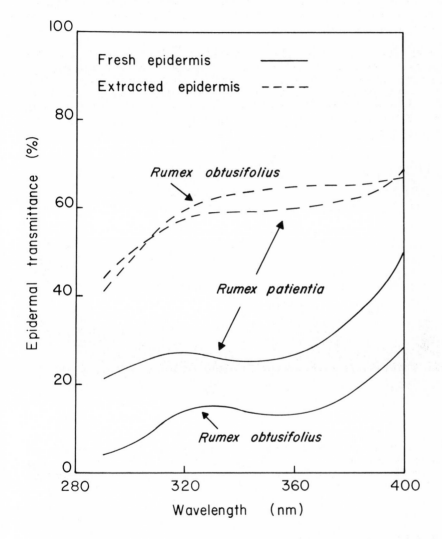

Fig. 3.   Epidermal UV transmittance of fresh tissue and of tissue from which UV-absorbing pigments were extracted. The species were collected from the field. Mean transmittance of 5 and 10 samples per spectrum are shown for Rumex obtusifolius and R. patientia, respectively.

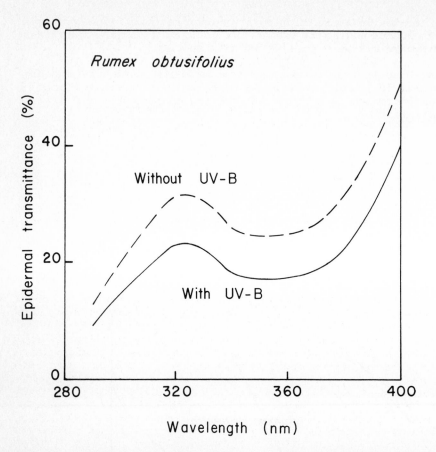

Fig. 4. Epidermal UV transmittance of <u>Rumex</u> <u>obtusifolius</u> leaves with or without solar UV-B irradiation. The plants were exposed to solar UV-B radiation at a mean biologically effective dose rate of 1151 J m^{-2} d^{-1} for eight days. These spectra represent the mean of nine samples per treatment. Epidermal transmittance of UV-B-irradiated leaves was significantly lower ($p \leqq 0.05$) than that of nonirradiated leaves.

Table 1. Extract absorbance of whole-leaf tissue for two Rumex species exposed to solar UV-B radiation. The mean of three to five samples per treatment for each species is presented as absorbance per mg dry weight. Significant increases (p $\leqq$ 0.05) in absorbance are indicated by an asterisk.

Species	Days of solar UV-B irradiation	Mean UV-B daily dose (Biologically effective J m^{-2} d^{-1})	Absorbance (at 305 nm)	
			No UV-B	With UV-B
R. patientia	5	1125	0.27	0.30*
R. obtusifolius	8	1151	0.26	0.50*

DISCUSSION

Reflectance of UV-B radiation from the leaf surface appears to be of limited value for preventing the penetration of this radiation into the leaf. The relatively low and nonwavelength-selective reflectance of 4%, found for the two Rumex species in this study, is typical for glabrous leaves (Gausman et al. 1975; Robberecht et al. 1980). Leaf surfaces covered with certain types of heavy pubescence or cuticular waxes can reflect more than 40% of the UV-B radiation incident on the leaf surface. Notable examples of high UV-B reflectance are Dudleya brittonia (Mulroy 1979) and Argyroxiphium sandwicense (Robberecht et al. 1980). Reflectance in these rather uncommon cases is relatively nonselective for wavelength, and is generally high in visible as well as UV wavebands. No significant increase or decrease in reflectance has been found in response to UV-B irradiation (Gausman and Escobar 1982).

The epidermis is a highly effective filter for UV-B radiation for both Rumex species examined in this study. Similar levels of UV attenuation have been observed for other species (Lautenschlager-Fleury 1955; McCree and Keener 1974; Gausman et al. 1975), though UV attenuation of up to 99% has been reported (Robberecht et al. 1980). For Rumex patientia and R. obtusifolius, as well as for Oenothera stricta (Robberecht and Caldwell 1983), nearly all of the UV-B radiation that penetrates through the epidermis is absorbed in the mesophyll layer. Thus, whole leaves appear to be opaque to UV-B radiation.

Increased epidermal attenuation in Rumex obtusifolius after solar UV-B irradiation indicates a degree of plasticity in epidermal optical properties. Such plasticity has been documented in other species (Lautenschlager-Fleury 1955; Robberecht and Caldwell 1983). The increased epidermal UV-attenuation capacity found for Rumex obtusifolius was most likely due to an increase in UV-absorbing pigments. Flavonoid and related phenolic compounds can be induced by UV irradiation (Wellmann 1974, 1975). Further, flavonoid extracts from Rumex leaves showed significant increases in absorbance after UV-B irradiation. This has also been reported for other species (Sisson 1981; Robberecht and Caldwell 1983). The substantial difference in fresh epidermal transmittance between the two Rumex species

258

is most likely related to the concentration of UV-absorbing compounds in the tissue, since after extraction of these compounds the spectra were similar (see Fig. 3).

The results of this study, along with the investigations of Caldwell (1968), Wellmann (1974, 1975), Sisson (1981), and Robberecht and Caldwell (1983), strongly suggest that the epidermis has a significant protective role against UV-B induced damage to sensitive physiological targets in the mesophyll. Furthermore, flavonoid and related phenolic compounds are most likely the major sources of UV-B absorption in the epidermis, and to some extent in the mesophyll. Since these compounds generally do not absorb visible radiation (Mabry et al. 1970; McClure 1976), their presence in leaf tissue provides a viable protective mechanism against potential damage from UV-B radiation. A mechanism involving the induction of flavonoids in response to UV-B irradiation results in a leaf epidermis that is an effective filter for UV-B radiation, has the plasticity for increased attenuation, and is highly selective for wavelength. In an intensified UV-B radiation environment expected to occur with stratospheric ozone depletion, such a protective mechanism could have particular adaptive significance. Predictions of plant sensitivity in the intensified UV-B environment, however, are difficult because the UV-attenuation capacity and prevalence of the protective mechanism in species of different plant communities is not known (Caldwell et al. 1983).

ACKNOWLEDGMENTS

This research was supported by grants from the U.S. National Aeronautics and Space Administration (NAS-9-14871) and the U.S. Environmental Protection Agency. Although the research described in this chapter has been funded partly by the U.S. Environmental Protection Agency through EPA-CR808167 to Utah State University, it has not been subjected to the Agency's required peer and policy review and therefore does not necessarily reflect the views of the Agency and no official endorsement should be inferred. We thank S.D. Flint for his assistance with the reflectance spectra of Rumex.

REFERENCES

Biggs RH, Sisson WB, Caldwell MM (1975) Response of higher terrestrial plants to elevated UV-B irradiance. In: Nachtwey DS, Caldwell MM, Biggs RH (eds) Impacts of climatic change on the biosphere, part I, ultraviolet radiation effects, monograph 5. Climatic Impact Assessment Program, US Dept Transportation Report No DOT-TST-75-55, NTIS, Springfield, Virginia, p (4)34
Caldwell MM (1968) Solar ultraviolet radiation as an ecological factor for alpine plants. Ecol Monogr 38:243-268
Caldwell MM (1971) Solar UV irradiation and the growth and development of higher plants. In: Giese AC (ed) Photophysiology, vol 6. Academic Press, New York, p 131
Caldwell MM, Robberecht R, Flint SD (1983) Internal filters: prospects for UV-acclimation in higher plants. Physiol Plant 58:445-450

Gausman HW, Escobar DE (1982) Reflectance measurement of artificially induced ultraviolet radiation stress on cotton leaves. Rem Sens Environ 12:485–490

Gausman HW, Rodriquez RR, Escobar DE (1975) Ultraviolet radiation reflectance, transmittance, and absorptance by plant epidermises. Agron J 67:719–724

Lautenschlager-Fleury D (1955) Über die Ultraviolettdurchlässigkeit von Blattepidermen. Ber Schweiz Bot Ges 65:343–386

Lee DW, Lowry JB (1980) Solar ultraviolet on tropical mountains: can it affect plant speciation? Am Nat 115:880–883

Mabry TJ, Markham KR, Thomas MB (1970) The systemic identification of flavonoids. Springer, Berlin Heidelberg New York Tokyo

McClure JW (1976) Progress toward a biological rationale for the flavonoids. Nova Acta Leopoldina, Suppl 7:463–496

McCree KJ, Keener ME (1974) Effect of atmospheric turbidity on the photosynthetic rates of leaves. Agric Met 13:349–357

Mulroy TW (1979) Spectral properties of heavily glaucous and non-glaucous leaves of a succulent rosette-plant. Oecologia (Berl) 38:349–357

National Academy of Sciences (NAS) (1982) Causes and effects of stratospheric ozone reduction: an update. Washington DC, p 145, ISBN 0-309-03248-2

Ribereau-Gayon P (1972) Plant phenolics. Hafner, New York, ISBN 0-05-0025-120

Robberecht R, Caldwell MM (1978) Leaf epidermal transmittance of ultraviolet violet radiation and its implications for plant sensitivity to ultraviolet-radiation induced injury. Oecologia (Berl) 32:277–287

Robberecht R, Caldwell MM (1983) Protective mechanisms and acclimation to solar ultraviolet-B radiation in Oenothera stricta. Plant Cell Environ 6:477–485

Robberecht R, Caldwell MM, Billings WD (1980) Leaf ultraviolet optical properties along a latitudinal gradient in the arctic-alpine life zone. Ecology 61:612–619

Sisson WB (1981) Photosynthesis, growth, and ultraviolet irradiance absorbance of Cucurbita pepo L. leaves exposed to ultraviolet-B radiation (280–315 nm). Plant Physiol 67:120–124

Sisson WB, Caldwell MM (1976) Photosynthesis, dark respiration, and growth of Rumex patienta L. exposed to ultraviolet irradiance (288 to 315 nanometers) simulating a reduced atmospheric ozone column. Plant Physiol 58:563–568

Teramura AH (1983) Effects of ultraviolet-B radiation on the growth and yield of crop plants. Physiol Plant 58:415–427

Van TK, Garrard LA (1975) Effect of UV-B radiation on net photosynthesis of some C_3 and C_4 crop plants. Soil Crop Sci Soc Florida Proc 35:1–3

Wellmann E (1974) Regulation der Flavonoidbiosynthese durch ultraviolettes Licht und Phytochrom in Zellkulturen und Keimlingen von Petersilie (Petroselinum hortense Hoffm.). Ber Dtsch Bot Ges 87:267–273

Wellmann E (1975) UV dose-dependent induction of enzymes related to flavonoid biosynthesis in cell suspension cultures of parsley. FEBS Lett 51:105–107

UV-B-Induced Effects upon Cuticular Waxes of Cucumber, Bean, and Barley Leaves

D. Steinmüller and M. Tevini

University of Karlsruhe, Federal Republic of Germany

ABSTRACT

Barley, bean, and cucumber seedlings were grown in a growth chamber with white light and low levels of UV-B radiation. Cuticular waxes found in barley leaves were five times greater than in bean or cucumber leaves based on leaf area. Predominant compounds identified were primary alcohols in barley wax, primary alcohols, and wax monoesters in bean wax, and alkanes in wax from cucumber leaves. Irradiation with enhanced levels of UV-B radiation caused an increase of total wax by about 25% in all plant species investigated. The wax composition was not markedly affected with respect to these wax classes. However, aldehyde content, detected as a minor constituent of cucumber and barley wax, increased twofold with enhanced UV-B radiation. The distribution pattern of the homologs within each wax class was different at both low and enhanced UV-B levels. In general, the distribution of the homologs was shifted towards shorter acyl chain lengths in wax classes of leaves exposed to enhanced UV-B levels. This effect was most apparent in cucumber wax, and less in bean or barley wax.

It is assumed that increased amounts of epicuticular waxes may have a function in attenuating enhanced UV radiation via UV reflectance and scattering.

INTRODUCTION

Epicuticular waxes minimize cuticular transpiration and play an important role as barriers to water diffusion (Hall and Jones 1961; Martin and Juniper 1970). The water diffusion coefficient increases up to 500-fold upon extraction of cuticular waxes (Schönherr 1976). Amount, composition, and structural arrangement of plant surface waxes vary greatly among plant species and depend on leaf age. Furthermore, cuticular waxes are influenced by environmental factors such as light intensity, photoperiod, humidity, temperature, and others (Hull et al. 1975; Macey 1970; Baker 1974). Although the cuticular membrane represents the barrier between the plant and its environment, and coincidently is the first target of all radiation incident on the plant surface, less is known about the function of surface lipids. The aim of this chapter is to describe the effects of UV-B radiation on wax deposition and composition in three plant species with differing UV-B sensitivities.

NATO ASI Series, Vol. G8
Stratospheric Ozone Reduction, Solar Ultraviolet Radiation
and Plant Life. Edited by R. C. Worrest and M. M. Caldwell
© Springer-Verlag Berlin Heidelberg 1986

MATERIALS AND METHODS

Plant Material and Growth Conditions

Cucumber (Cucumis sativus L. cv. Delikatess), bean (Phaseolus vulgaris L. cv. Favorit), and barley (Hordeum vulgare L. cv. Villa) were grown in plastic pots containing standard greenhouse soil. Plants were placed into a growth chamber with an environment described previously (Tevini et al. 1983a; Teramura et al. 1983) and were harvested 14 days after germination. Leaf areas were determined by Li-Cor Model LI-3000 area meter. The control UV-B and enhanced UV-B treatments in the chamber were 3412 and 7928 effective $J m^{-2} d^{-1}$ (using the generalized plant action spectrum normalized to 300 nm).

Wax Extraction, Separation, and Identification

Redistilled solvents were used for all extractions. The extraction was done according to Köhlen and Gülz (1976). Crude wax was separated into single wax classes by thin layer chromatography (TLC) on silica (Merck Kieselgel H, 0.25 mm) and developed with benzene. Lipids were stained with iodine vapor or with bromothymol blue in alkaline ethanol. Wax classes were identified by gas-liquid chromatography (GLC) and TLC by (i) comparing the R_f-values and the retention times resulting from GLC with known standards, and (ii) chemical modification of the fractions comparing the altered R_f-values and altered retention times with the R_f-values and retention times of standards modified as follows. Wax esters were transesterified with concentrated H_2SO_4 in ethanol and refluxed for 12 h at 75°C. The resulting fatty ethyl esters and free fatty alcohols were separated by TLC as described above. Free primary and secondary alcohols were acetylated with acetic anhydride in pyridine, purified, and separated by TLC. Aldehydes and ketones were reduced with potassium boron hydride in aqueous dioxan and the corresponding alcohols and purified by TLC. Free fatty acids were converted to methyl esters with boron trifluoride in methanol and purified by TLC.

The R_f-values (at room temperature) of the most frequent cuticular lipids were as follows: alkanes, 0.76; wax esters, 0.58; ketones, 0.52; aldehydes, 0.44; secondary alcohols, 0.20; primary alcohols, 0.09; free fatty acids, 0.03.

Test samples of alkenes and wax monoesters were supplied by Dr. P. Gülz, University of Köln, FRG. Free fatty acids were from Fluka, Switzerland. All others were prepared from natural sources rich in some wax classes, such as cabbage, lemon, paraffin candles, bees wax.

Gas Liquid Chromatography (GLC)

The lipid fractions were recovered from silica gel with hot chloroform. After solvent evaporation fractions were resolved in n-octane (alkanes, fatty methyl- and ethyl-esters) or in butyl acetate (others). All wax classes were separated on a 12 m x 0.31 mm fused silica capillary column, OV-1, and quantified via GLC using a 5880A Hewlett-Packard model equipped with flame ionization detection. Oven temperatures were programmed from 90°C to 200°C at 20°C min^{-1} and from 200°C to 350°C at 6°C min^{-1}. Helium was used as a carrier gas at a linear gas velocity of 1.8 m s^{-1} at the initial temperature. All samples were applied to GLC by the on-column injection technique. Peak area was determined by electron integration in proportion to weighted standards. The detector response depended on the number of carbon atoms and on the type of lipid class analyzed.

RESULTS

The amount of wax varied considerably among different types of leaves and different plant species (Table 1). Cotyledons of cucumber exhibited thinner wax sheets than true leaves. Barley contained a 5-fold greater amount of cuticular waxes compared to cucumber and bean. The wax composition also differed among these species. Alkanes represented the main compound of cucumber epicuticular wax (Fig. 1), the primary alcohols dominated in barley wax (Fig. 2), whereas primary alcohols and wax mono-esters mainly occurred in the wax obtained from bean leaves (Fig. 3). Enhanced UV-B radiation produced a 20-28% increase in total wax in all plants investigated when based on leaf surface area, but a 8-23% increase when based on leaf dry weight (Fig. 4).

Table 1. Amount of cuticular wax obtained from cucumber, bean, and barley leaves based on 100 cm^2 leaf area (I) and on 1 g dry weight (II). Values are expressed in µg $\pm$ SD.

	I	II
Cucumber cotyledons	273 $\pm$ 11	666 $\pm$ 26
Cucumber 1st - 3rd leaf	384 $\pm$ 14	878 $\pm$ 32
Bean leaves	538 $\pm$ 18	1559 $\pm$ 51
Barley leaves	2584 $\pm$ 88	15289 $\pm$ 521

Aldehydes, a minor compound in waxes from cucumber and barley, was substantially affected by UV-B radiation. On a leaf area basis the aldehyde content rose from 15.0 µg in controls to 40.4 µg in cucumber wax from plants exposed to an enhanced UV-B flux and from 56.6 µg to 136.7 µg in barley wax. The free fatty acid content was reduced in cucumber cotyledon wax, but increased in wax from the primary leaves of cucumber and barley. Epicuticular wax of bean leaves showed no significant change in this respect.

The distribution of homologs within each wax class was studied by GLC. Odd numbered acyl chains were predominantly found in all alkanes and secondary alcohols, but even numbered species in fatty acids, primary alcohols, aldehydes, and wax esters. The even numbered alkanes constituted 18% of

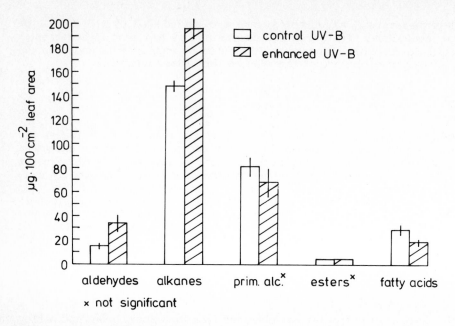

Fig. 1. Amount and composition of cuticular wax obtained from cucumber cotyledons irradiated with low (control UV-B) and enhanced UV-B doses. Values are given as the mean of five independent determinations expressed in µg ± SD.

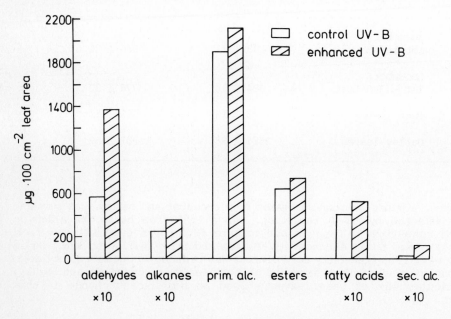

Fig. 2. Amount and composition of cuticular wax obtained from barley leaves. Conditions as in Fig. 1. Alkanes, aldehydes, fatty acids and secondary alcohols were enlarged 10-fold. Values are the mean of three independent determinations.

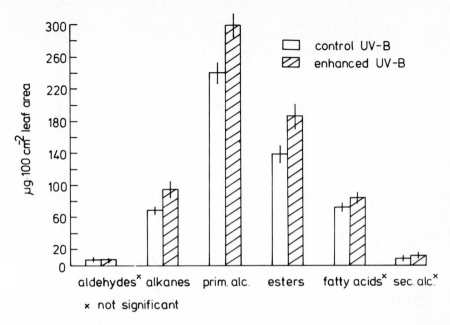

Fig. 3. Amount and composition of cuticular wax obtained from bean leaves. Conditions as in Fig. 1. Values are the mean of four independent determinations.

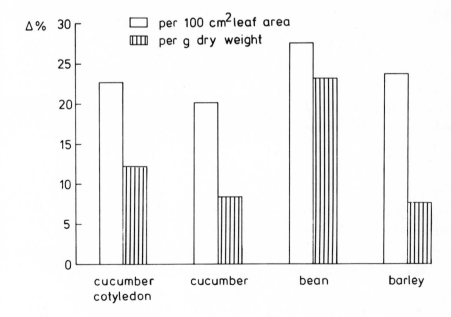

Fig. 4. Relative increase of cuticular wax isolated from plants irradiated with enhanced UV-B radiation compared to control plants.

the total alkanes in cucumber wax. Cucumber had 15% odd-numbered aldehydes; whereas, the odd-numbered aldehydes were less than 1.5% in the aldehydes and primary alcohols of barley wax. Heptacosane, nonacosane, and hentriacontane were the major hydrocarbons of cucumber wax (Fig. 5); whereas hexacosanol, octacosanol, triacontanol, and the corresponding aldehydes comprised 80% of total alcohols or aldehydes (Fig. 6). In alkanes derived from bean leaves nonacosane and hentriacontane dominated and octacosanol represented 45% of total alcohols. Aldehydes and alcohols isolated from barley were characterized by one predominant homolog, hexacosanal (Fig. 7) and hexacosanol, respectively, which had up 90% of these lipids. In all plants investigated, the free fatty acids exhibited a broader distribution pattern ranging from C-12 to C-34 homologs. In wax esters obtained from barley C-38 to C-52 homologs have been identified and C-40 to C-54 esters were found in bean wax. Enhanced UV-B radiation primarily affected these distribution patterns. In fractions of alkanes, aldehydes, and primary alcohols of cucumber leaves, a shift in distribution toward shorter acyl chain lengths was clearly demonstrated (Figs. 5 and 6). Similar results had been observed in bean and barley alkanes, but the effect was less pronounced in these plants. Since the pattern of the free fatty acids and that of the primary alcohols was not significantly affected in bean and barley epicuticular wax, the pattern of the wax esters composed of both fatty acids and fatty alcohols was also not affected.

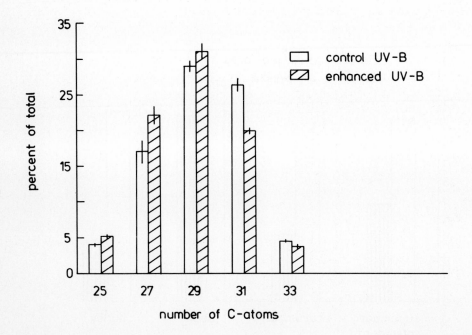

Fig. 5. Distribution of the odd-numbered alkane homologs obtained from cucumber cotyledons. Values are means of five measurements expressed as a percentage of total alkanes ± SD. Total amount, both for controls and for enhanced UV-B conditions, was set to 100%.

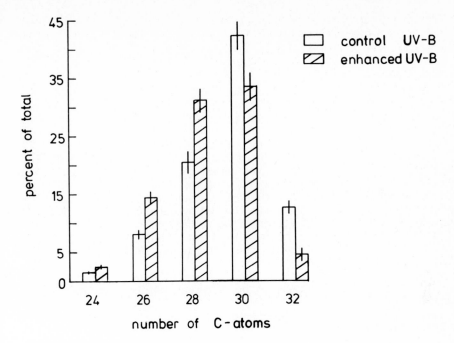

Fig. 6. Distribution of the even-numbered aldehyde homologs obtained from cucumber cotyledons. Conditions as in Fig. 5.

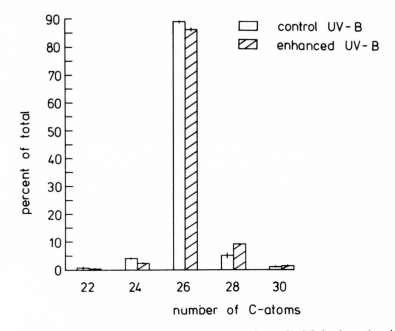

Fig. 7. Distribution of the even-numbered aldehydes obtained from barley leaves. Conditions as in Fig. 5. Values are the mean of three measurements.

DISCUSSION

Most plants exhibit reduced leaf areas when irradiated with enhanced levels of UV-B radiation. Cucumber leaves are very sensitive and barley leaves are more resistant to such changes (Tevini et al. 1981, 1983b). Since cucumber leaves have a higher wax content under enhanced UV-B conditions, the question arose whether UV-B radiation induces wax synthesis or whether the wax sheets merely become thicker as a consequence of smaller leaf areas and a constant endogenous wax production. An increase in total wax was found when based on a leaf area, dry weight, or on the total number of leaves. Although the leaf dry weight of barley increased when irradiated with enhanced UV-B levels, the total wax based on dry weight also increased with enhanced UV-B doses. The composition of the main wax constituents was not markedly affected; however the aldehyde content doubled. The distributional pattern of most wax classes was altered by enhanced UV fluxes. All this suggests that the mode of action of UV-B radiation is on wax biosynthesis. It is well known that increasing light intensity stimulates wax biosynthesis (Macey 1970; for review see Kolattukudy 1980). Although the mechanism of light or UV action cannot be answered at present, based on data obtained from cucumber leaves, the sites of biosynthesis affected by light and by UV-B radiation appear to be different (Steinmüller and Tevini, unpublished).

What is the ecological importance of UV induced wax synthesis of plants? It has been demonstrated that epicuticular wax sheets on greenhouse grown plants were less pronounced and exhibited a lower degree of crystallization than those found on the same plants outdoors (Hull et al. 1975), and this was thought to be a specific UV response. Although most of the cuticular waxes do not absorb within the visible and near UV-waveband (including UV-B), the structural arrangement of waxes on the plant surfaces may enhance light reflectance and light scattering (Cameron 1970). The total reflectance of leaves with mechanically removed wax sheets was found to be much less than those of leaves with normal wax sheets (Sinclair and Thomas 1970; Eller and Willi 1977). In general, reflectance within the UV waveband seems to be lower compared to the visible waveband, but dependent on the morphological characteristics of the leaves. Glabrous green leaves normally reflect up to 10% of the total UV (Gausman et al. 1975). The reflectance rises to 20% in more glaucous leaves (Robberecht and Caldwell 1980), whereas in the conifer blue spruce, reflectance was 3.5-fold greater at 280 nm when compared to 540 nm (Clark and Lister 1975). This was due to the arrangement of cuticular waxes on the conifer needles. As found for epidermal flavonoids (Wellmann 1975, Robberecht and Caldwell 1978) epicuticular waxes might have an additional function in attenuating damage effects of enhanced UV-B radiation.

REFERENCES

Baker EA (1974) The influence of environment on leaf wax development in Brassica oleracea var. Gemmnifera. New Phytol 73:955-966
Cameron RJ (1970) Light intensity and growth of Eucalyptus seedlings. II. The effect of cuticular waxes on light absorption in leaves of Eucalyptus species. Austr J Bot 18:275-284

Clark JB, Lister GR (1975) Photosynthetic action spectra of trees. II. The relationship of cuticle structure to the visible and ultraviolet spectral properties of needles from four coniferous species. Plant Physiol 55:407–413

Eller BW, Willi P (1977) Die Bedeutung der Wachsausblühungen auf Blättern von Kalanchoe pumila Baker für die Absorption der Globalstrahlung. Flora 166:461–473

Hall D, Jones RL (1961) Physiological significance of surface wax on leaves. Nature 191:95–96

Hull HM, Morton HL, Wharrie JR (1975) Environmental influences on cuticle development and resultant foliar penetration. Bot Rev 41:421–452

Gausman HW, Rodriguez RR, Escobar DR (1975) Ultraviolet radiation reflectance, transmittance, and absorptance by plant leaf epidermises. Agronomy J 67:720–724

Kolattukudy P (1980) Cutin, suberin, and waxes. In: Stumpf PK (ed) The biochemistry of plants, vol 4. Academic Press, New York, p 571

Köhlen L, Gülz P-G (1976) Untersuchungen über die Kutikularwachse in der Gattung Cistus L. (Cistaceae). I. Die Zusammensetzung der Alkane in den Blattwachsen. Z Pflanzenphysiol 77:99–106

Macey MJK (1970) The effect of light on wax synthesis in leaves of Brassica oleracea. Phytochemistry 9:757–761

Martin JT, Juniper BE (1970) The cuticles of plants. Edward Arnold, Edinburgh

Robberecht R, Caldwell MM (1978) Leaf epidermal transmittance of ultraviolet radiation and its implications for plant sensitivity to ultraviolet-radiation induced injury. Oecologia 32:277–287

Robberecht R, Caldwell MM (1980) Leaf ultraviolet optical properties along a latitudinal gradient in the arctic-alpine life zone. Ecology 61:612–619

Schönherr J (1976) Water permeability of isolated cuticular membranes: The effect of cuticular waxes on diffusion of water. Planta 131:159–164

Sinclair R, Thomas DA (1970) Optical properties of leaves of some species in arid South Australia. Aust J Bot 18:275–284

Teramura AH, Tevini M, Iwanzik W (1983) Effects of ultraviolet-B irradiation on plants during mild water stress. I. Effects on diurnal stomatal resistance. Physiol Plant 57:175–180

Tevini M, Iwanzik W, Thoma U (1981) Some effects of enhanced UV-B radiation on the growth and composition of plants. Planta 58:395–400

Tevini M, Iwanzik W, Teramura AH (1983a) Effects of UV-B irradiation on plants during mild water stress. II. Effects on growth, protein, and flavonoid contents. Z Pflanzenphysiol 110:459–467

Tevini M, Thoma U, Iwanzik W (1983b) Effects of enhanced UV-B radiation on germination, seedling growth, leaf anatomy and pigments of some crop plants. Z Pflanzenphysiol 109:435–448

Wellmann E (1975) UV-dose-dependent induction of enzymes related to flavonoid biosynthesis in cell suspensions of parsley. FEBS Lett 51:105–107

Effects of UV-B Radiation on Growth and Development of Cucumber Seedlings

M. Tevini and W. Iwanzik

University of Karlsruhe, Federal Republic of Germany

ABSTRACT

Cucumber seedlings (Cucumis sativus, cv. Delikatess) were grown for three weeks in a growth chamber at different UV-B levels and high white light conditions (800 μmol m^{-2} s^{-1}). In one set of experiments, different cutoff filters (Schott series WG 345, 320, 305, 295, and 280), and in another set, different thicknesses of one filter type (WG 305, 2-6 mm) were used to provide the plants with different UV levels from a 6000-W xenon lamp and additional UV-B lamps (Philips TL 40/12). Growth parameters were measured in terms of hypocotyl length and cotyledon length and width during the development of the seedlings. After three weeks, cotyledon area and fresh and dry weight were determined. All growth parameters were reduced, depending on the wavelength and intensity of UV radiation. Reduced growth was due to lower absolute growth rates at higher UV-B levels. Cell division of cotyledons decreased in relation to UV-B intensity, as demonstrated by reduced surface cell and stomate numbers on either side of the cotyledons. Furthermore, the results show that even moderate enhancements of UV-B radiation, such as would follow small depletions of ozone in the stratosphere, cause significant reductions in the growth of UV-B sensitive cucumber seedlings.

INTRODUCTION

In the earliest studies on the effects of near-UV radiation (300-400 nm) on plant growth and development, Stoklasa (1911) and Popp (1926) described "better" growth in plants not exposed to radiation in that waveband. Pfeiffer (1928), and later Popp and McIlvaine (1937), particularized the "better" growth, and reported longer shoots, but thinner leaves and poor root development in those plants. These effects, however, are typical for plants grown in the dark or at low white light intensities. At that time it was not clear whether the effects were due to low visible light intensities or to the exclusion of near-UV radiation.

Pirschle and von Wettstein (1941) showed that alpine plants grew better in their biotope when UV radiation was excluded. Later, Brodführer (1956) noted that plants grown under near-UV radiation exhibited the same characteristics as alpine plants growing under relatively high UV levels at high elevations. From these results it was evident that UV radiation had some influence on plant growth. The ecological and photomorphogenic background of further UV research, as described by Caldwell (1971) and others, changed rapidly to include more economic interests, because of the ozone reduction problem. In this context, screening tests with large numbers of plant species were performed in growth chambers and in the field with supplementary UV radiation in order to examine the effects on growth, photosynthesis, and yield of plants grown under enhanced levels of UV-B radiation. Studies in growth chambers, greenhouses, and in the field showed differences in UV-B sensitivity between plant species and even between cultivars with respect to growth, photosynthesis, and dry matter production (Klein et al. 1965; Caldwell et al. 1975; Krizek 1975; Sisson and Caldwell 1976; Van et al. 1976; Brandle et al. 1977; Biggs and Kossuth 1978; Lindoo and Caldwell 1978; Vu et al. 1978; Teramura et al. 1980; Sisson 1981; Biggs et al. 1981; Teramura and Caldwell 1981; Tevini et al. 1981).

Teramura (1983) summarized the effects of enhanced UV-B radiation on more than 40 plant species examined in growth chambers and in the field. With respect to plant growth it was shown that sensitive plants usually exhibited reduced leaf expansion and stem elongation, and that quantitative differences between plants grown in growth chambers depended, in addition to the UV-B intensity, on the white light intensity (Teramura 1980). UV-B effects were quantitatively higher in plants grown at low levels of photosynthetically active radiation (PAR) than under high PAR. Stem elongation especially was reduced much more with the same UV-B radiation at low PAR than at high PAR. These differences were also obvious between plants grown in the field and in growth chambers, provided low PAR intensities were used. The large quantitative differences might easily be explained by the fact that low white light intensities induce the typical characteristics of an etiolated plant: for example, a taller stem. At high light intensities, such as those found in the field, photomorphogenic reduction of stem elongation occurs and therefore the quantitative differences due to UV-B radiation become smaller. The comparison of growth chamber and field grown plants is therefore not very valuable because there are two competing effects: namely the photomorphogenic effect of visible light (with phytochrome as the responsible receptor) and the effect of UV-B radiation, with an unknown photoreceptor. The real contribution of UV-B radiation to photomorphogenesis, especially to early seedling growth, is not well established in plants grown at high white light intensities. This important question was studied in cucumber seedlings, which were shown to be very sensitive to UV-B radiation with respect to growth and development (Tevini et al. 1983).

Thus, in one set of experiments, described later in this chapter, different cutoff filters were used to increase UV radiation stepwise to relatively high levels. The second aim of the study was to show that ambient and enhanced levels of UV-B radiation equivalent to UV-B levels at different latitudes have photomorphogenic effects on the growth and development of plants grown under relatively high white light intensities. Different levels of UV-B radiation equivalent to ambient UV-B levels at different latitudes were provided by different thicknesses of one filter type.

MATERIALS AND METHODS

Cucumber seeds (Cucumis sativus L., cv. Delikatess) were sown into plastic pots (15 x 15 x 4 cm) containing greenhouse soil (TSK1). After 3-days growth, plants were thinned to 15 plants per pot to obtain seedlings of equal height and cotyledon development at the beginning of growth analysis.

Two series of experiments with different UV irradiation conditions were performed. In one set of experiments, a series of cutoff filters (Schott series WG 345, 320, 305, 295, and 280, each 2-mm thick) was employed to modify the radiation from the light sources in the growth chamber. These filters have rapidly increasing transmittances in the UV waveband. With the WG 345 filter only UV-A radiation and white light were transmitted, and no UV-B radiation. With all other filters, increasing amounts of UV-B radiation (and, with WG 280, very small amounts of UV-C) were transmitted. Before exposure to UV-B radiation seeds grew for three days in a chamber with continuous white light (Philips TL 40/29, 100 μmol m^{-2} s^{-1}) at 22°C until the opening of the hypocotyl hook. At day 3 after sowing, seedlings were transferred into a humidity- and temperature-controlled growth chamber equipped with UV radiation sources. Temperature regime was 20°C/12°C at a relative humidity of 86/88%. The pots were rotated every day to minimize possible effects from small light gradients within the filter-covered pots.

In a second set of experiments, Schott WG 305 filters of different thicknesses (2, 3, 4, 5, 6 mm) were used to simulate ambient UV-B radiation levels which are found at lower and higher latitudes, compared to Karlsruhe, Federal Republic of Germany (49°N). Seedlings were grown from the time of sowing in the growth chamber under these radiation conditions, as would happen under natural conditions. The lower night temperatures during the first three days resulted in a slight retardation of germination when compared to plants from the first set of experiments, which were germinated at 22°C.

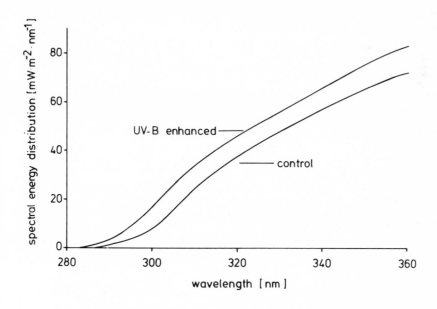

Fig. 1. Spectral energy distribution (unweighted) of UV radiation (280-360 nm) in the growth chamber. Control is with 4-mm Schott filters, enhanced UV-B is with 2-mm Schott filters (WG 305).

Radiation was supplied by a combination of a 6000-W xenon lamp (Osram) and
two UV-B lamps (Philips TL 40/12), one on either side of the xenon lamp.
The radiation sources provided 800 μmol m^{-2} s^{-1} PAR for 6 h during the
middle of the day, and two 3.5-h periods of gradually increasing and
decreasing light intensity, in the morning and afternoon, respectively.
Spectral energy distribution of visible light was measured with a spectroradiometer (EG&G 550) at a 10-nm effective bandwidth resolution. Radiation
in the UV waveband was measured with a radiometer equipped with a double
monochromator (Tevini et al. 1983). As an example, the spectral energy
distributions of the radiation under 2- and 4-mm thick WG 305 filters are
shown in Fig. 1.

For comparative purposes, radiation was weighted for biological effectiveness according to a generalized plant response function (Caldwell 1971) and
normalized to 300 nm (Rupert 1982). The daily effective doses of UV radiation in the two series of experiments, and for comparative purposes, some
ambient UV-B doses at different latitudes, are summarized in Table 1.

Table 1. Daily effective doses of the different irradiation conditions,
and ambient solar radiation at different latitudes (calculated according to
Green et al. 1980), weighted according to the generalized plant response
function (Caldwell 1971). Normalization wavelength is 300 nm.

Filter	UV-B (BE) J m^{-2}	
WG 280	30,282	
WG 295	12,855	
WG 305	8,151	
WG 320	914	
WG 345	----	
WG 305, 2 mm	8,151	8,070 (= 35°N)[a]
WG 305, 3 mm	4,841	5,044 (= 50°N)[b]
WG 305, 4 mm	3,018	3,354 (= 60°N)[b]
WG 305, 5 mm	1,743	1,649 (= 60°N)[c]
WG 305, 6 mm	1,237	1,271 (= 55°N)[d]

[a] Day 127, altitude 0.25 km, albedo 0.1, aerosol scaling coefficient 1.0,
ozone 0.30 atm-cm.
[b] The same as under (a) except ozone 0.32 atm-cm.
[c] The same as under (b) except day 100.
[d] The same as under (b) except day 82.

Beginning with day 5 (series with WG 280-345 filters) or day 7 (series with
WG 305 filters, 2-6 mm) plant height (hypocotyl length) and cotyledon size
(in terms of length and width) were measured every other day throughout the
entire experiment. Fresh weights and dry weights of cotyledons and roots,
as well as cotyledon leaf areas, were determined at the end of the experiments. Leaf area was measured using a LiCor Model LI-3000 area meter.

The number of stomata and cells in the epidermal layer was determined by light microscopy, using polystyrene replicas of the leaf surface. Both series of experiments were performed three times. The mean values were evaluated with Student's t-test and were statistically significant at the 95% level.

RESULTS

Experiments with Different Filter Types

In this set of experiments cucumber seeds germinated for three days under continuous white light. Seedlings were then transferred to a growth chamber with a day/night cycle, in which the day cycle consisted of white light and different radiation levels: UV-A (WG 345); low (WG 320); moderate (WG 305); or high UV-B levels with small amounts of UV-C (WG 295 and 280). The different UV radiation conditions produced quantitative differences in growth responses of hypocotyls and cotyledons during development (Fig. 2).

Hypocotyl length at every stage of seedling development was greatest in plants grown at the lowest UV-B level (WG 320) and smallest in plants grown at the highest UV-B level (WG 280, Fig. 2a). However, plants grown without UV-B irradiation (WG 345) had significantly shorter hypocotyls than plants grown at the lowest UV-B level. The absolute growth rate of the hypocotyls showed two characteristic maxima during the time course of seedling development, as demonstrated in Fig. 2b. The first maximum was found between days 8 and 10 in plants under all irradiation conditions, but to a lesser extent in plants grown at higher UV-B levels. Between days 11 and 12 growth rates were very low, and continued to decrease in plants irradiated with the highest UV-B levels (WG 295 and 280); however, growth rates soon increased very rapidly in plants grown without UV-B or at the lowest UV-B level. In plants grown under WG 305, the second growth-rate increase was delayed two days. Towards the end of the experiments the growth rate was highest in plants receiving no UV-B radiation.

Length (Fig. 2c) and width (Fig. 2e) of cucumber cotyledons increased most in plants grown at the lowest UV-B level. Increasing amounts of UV-B radiation gradually reduced the growth of cotyledons, and after 17 days the leaf area of plants grown at the highest UV-B level was only 25% that of the leaf area of plants grown with the lowest UV-B level (Fig. 3). Cotyledon dry weight showed a similar relationship to enhanced UV-B radiation (Fig. 4). The reduction in cotyledon dry weight was paralleled by a similar decrease in total root dry weight (Fig. 5). Reduced leaf expansion was due to a reduction of the absolute growth rates for leaf length and width following enhanced UV-B irradiation (Figs. 2d and 2f). Growth rates for leaf length and width were closely correlated and growth curves showed similar time courses. Highest absolute growth rates for cotyledon length and width were found during the early seedling development of all plants, but gradually decreased, depending on the UV-B intensity. Growth of cotyledons was drastically reduced after day 7, when hypocotyls started with maximal growth. In plants exposed to the highest UV-B radiation levels leaf growth in length ceased 13 days after sowing, and in width 15 days after sowing; whereas in all other plants leaf growth continued at rates inversely proportional to UV-B radiation levels.

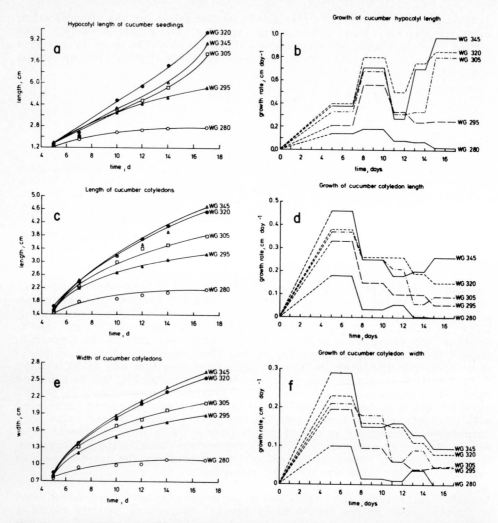

Fig. 2. Effects of different UV-B irradiation conditions (under Schott filters WG 280, 295, 305, 320, and 345) on the growth of cucumber seedlings. 2a, c, e: absolute growth of hypocotyl in length (a); cotyledon in length (c); and width (e). 2b, d, f: growth rates of hypocotyls (b); cotyledon length (d); and width (f).

The reduction of leaf area following UV-B irradiation was correlated to a reduction in cell numbers on the abaxial and adaxial surface of the cotyledons (Fig. 6). After 17 days the number of cells on the abaxial surface of plants which received the highest UV-B doses was four times lower than in plants which received UV-A, or only low UV-B, irradiation. The number of cells, generally lower on the adaxial side than on the abaxial side, was significantly decreased in plants which received moderate or high levels of UV-B radiation. The adaxial cell numbers of plants under highest UV-B irradiation were only roughly estimated because the epidermis was heavily damaged by radiation.

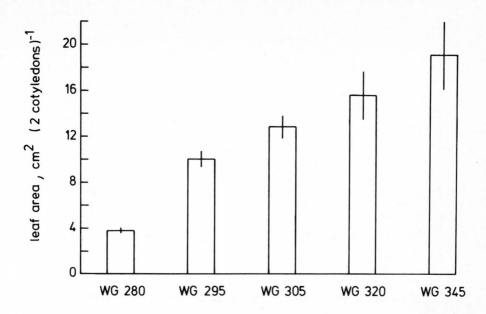

Fig. 3. Leaf area of cucumber cotyledons grown for 17 days under different levels of UV-B radiation.

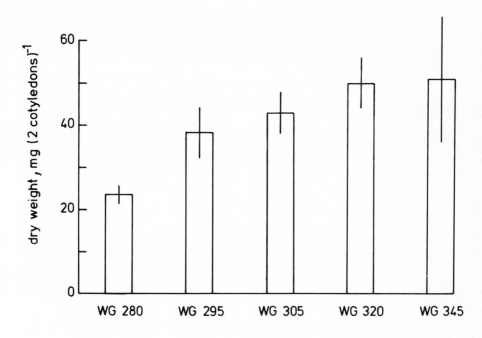

Fig. 4. Dry weight of cucumber cotyledons grown for 17 days under different levels of UV-B radiation.

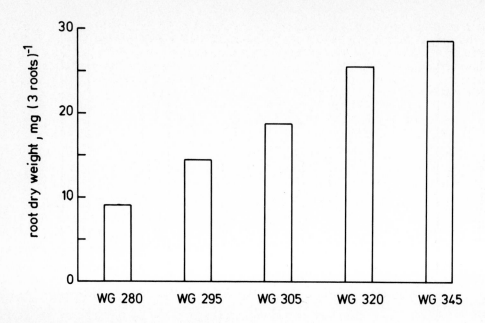

Fig. 5. Dry weight of total cucumber roots grown for 17 days under different levels of UV-B radiation.

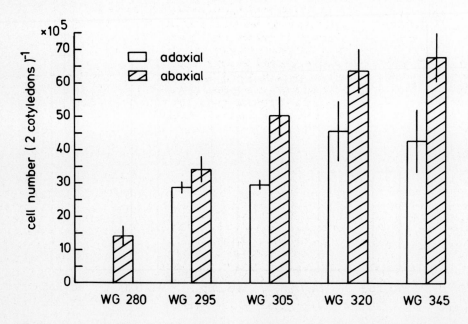

Fig. 6. Cell numbers on cucumber cotyledon abaxial and adaxial surfaces. Plants were grown for 17 days under different levels of UV-B radiation.

The number of stomata of the cucumber cotyledons, about 2-3 times greater in the upper than in the lower epidermis, showed similar responses to UV radiation on both leaf sides. The stomate number was reduced with increasing amounts of UV-B irradiation (Fig. 7). Seedlings exposed to UV-A radiation had fewer stomata in the upper epidermis, but more in the lower epidermis, than plants exposed to the lowest UV-B dose.

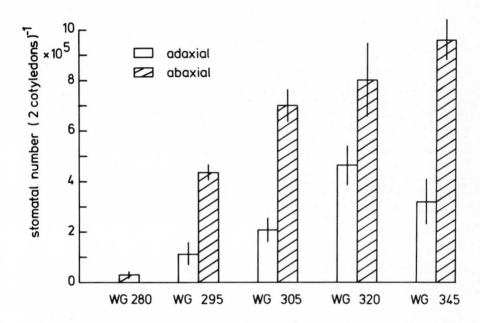

Fig. 7. Stomate numbers on cucumber cotyledon abaxial and adaxial surfaces. Plants were grown for 17 days under different levels of UV-B radiation.

Experiments with Different Thicknesses of One Filter Type

In this set of experiments plants were germinated and grown in a growth chamber under the same white light conditions as in the previously described experiments, but under only one type of cutoff filter (WG 305) of different thicknesses ranging from 2 to 6 mm. With this filter set it was possible to increase the radiation in the UV-B waveband gradually, creating equivalents to ambient UV-B levels at different latitudes. The weighted daily doses ranged from 1238 J m^{-2} d^{-1} (WG 305, 6 mm) [equivalent to ambient daily UV-B doses during sunny days in the early spring (day 82) at northern latitudes of around 55°N] to 8151 J m^{-2} d^{-1} (WG 305, 2 mm) [equivalent to ambient UV-B levels at southern latitudes around 35°N at day 127] (Table 1).

The effects on the growth parameters of cucumber seedlings grown for 21 days under these different UV-B conditions are summarized in Table 2. Leaf area, fresh weight, and dry weight were clearly inversely proportional to the daily UV-B dose and were reduced by about 40%-50% in plants receiving the highest UV-B dose (8151 J m^{-2} d^{-1}) compared to plants grown at the lowest UV-B dose (1238 J m^{-2} d^{-1}). This percentage difference was smaller when it was based on a comparison to the plants which received medium daily UV-B doses, as for example under WG 305, 4 mm (3019 J m^{-2} d^{-1}).

Table 2. Growth parameters of cucumber seedlings grown for 21 days under WG 305 (2-6 mm).

Filter	Daily Dose[a]	Leaf Area[b]	%	Fresh Weight[c]	%	Dry Weight[d]	%
WG 305, 6 mm	1237	17.10 ±1.40	100	0.66 ±0.03	100	44.20 ±3.70	100
WG 305, 5 mm	1743	15.13 ±1.90	88.5	0.67 ±0.13	101.5	42.60 ±8.20	96.4
WG 305, 4 mm	3018	14.57 ±1.60	85.2	0.61 ±0.02	92.4	35.80 ±2.30	81.0
WG 305, 3 mm	4841	13.89 ±1.30	81.2	0.50 ±0.09	75.6	35.60 ±1.30	80.5
WG 305, 2 mm	8151	9.71 ±2.10	56.8	0.37 ±0.05	56.1	23.60 ±0.20	53.4

[a]UV-B (BE) J m^{-2}.
[b]cm^2/2 cotyledons.
[c]g/2 cotyledons.
[d]mg/2 cotyledons.

The reductions in biomass were due to lower growth rates of hypocotyls and cotyledons under enhanced UV-B radiation (Figs. 8b, d, f). The time courses of absolute growth rates were similar to those in the plants grown under different filter types, but somewhat delayed. This delay resulted from the fact that these plants had been grown under UV conditions from the time they were sown. The lowest absolute growth rates of hypocotyls and cotyledons were mostly found at the highest UV-B levels throughout seedling development. The lower growth rates with enhanced UV-B irradiation consequently produced shorter hypocotyls and smaller cotyledons (Figs. 8a, c, e). Significant differences in morphology, however, appeared at the earliest after 14- to 16-days irradiation. By that time growth in plants exposed to the highest UV-B level had slowed down (hypocotyls) or already stopped (cotyledons).

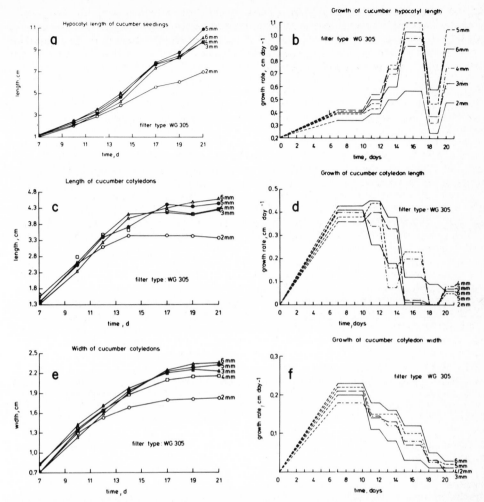

Fig. 8. Effects of different UV-B irradiation conditions (Schott filters WG 305, 2-6 mm) on the growth of cucumber seedlings. 8a, c, e: absolute growth of hypocotyl in length (a); cotyledon in length (c); and width (e). 8b, d, f: growth rates of hypocotyls (b); cotyledon length (d); and width (f).

DISCUSSION

Plant growth is regulated by light qualities, of which the red and blue wavelengths have been intensively studied and described in connection with the responsible photoreceptor, phytochrome (e.g., Gaba and Black 1983). Induction phenomena initiated by short irradiation at low visible light intensities have been distinguished from high irradiance reactions (HIR) which occur at long irradiation times and high light intensities (Jabben and Holmes 1983). Most of the information on the role of phytochrome in the natural environment comes from studies in which photosynthetically active radiation (PAR) and phytochrome photoequilibrium were varied. The extent of growth response is highly dependent on light intensity and light quality. It has been shown that blue light controls plant growth, but we have little information on the contribution of UV radiation to growth responses. In cucumber it is well known that far-red radiation promotes hypocotyl growth, while red and blue light inhibit it (Gaba and Black 1979; Thomas and Dickinson 1979; Cosgrove 1981). Our data show that UV radiation also affects hypocotyl growth in cucumber seedlings. The influence is wavelength and intensity dependent. Hypocotyl growth in plants grown under UV-A radiation is reduced when compared to that in plants grown under long-wavelength UV-B radiation. Short-wavelength UV-B radiation, as encountered by the plants grown under WG 305 or 295, reduced hypocotyl length as well as cotyledon expansion considerably. This confirms results found in several crop plants grown in the field with supplementary UV-B radiation. Stunting of wheat and soybean seedlings was reported under moderate UV-B irradiances, but only at low PAR (Teramura 1980). In contrast, Cucumis sativus, which has been shown to be very UV-B sensitive (Tevini et al. 1983), also exhibits decreased growth under low UV-B radiation and high white light conditions.

During development of cucumber seedlings, growth-rate curves of hypocotyls and cotyledons exhibited characteristic time courses. Expansion of cotyledons occurred first, enabling increased photosynthesis. High growth rates in the hypocotyl followed the cotyledon expansion, and reached a second maximum when the primary leaves expanded. UV-B irradiation, and UV-C as well, decreased growth rate and retarded seedling growth at two points during development. Retardation of seedling growth was observed during early development when plants were germinated and grown in UV-B radiation, in comparison to plants germinated under white light for three days and then transferred to UV-B exposure. Furthermore, plants exposed to moderate or high UV-B doses, such as the plants under WG 305 and 295 filters, showed either a retarded second growth period of hypocotyls (WG 305), or no second growth period at all by the end of seedling development (WG 295). This might be due to inhibition of primary leaf initiation by UV-B or UV-C radiation.

Two mechanisms might be involved in the inhibition and retardation of cotyledon and hypocotyl growth of cucumber seedlings. Firstly, UV-B radiation may have inhibitory effects on cell division. Reduced cell and stomate numbers on each side of the cotyledons showed an influence of UV-B radiation on cell division. This confirms results with Rumex patientia, also irradiated with enhanced UV-B radiation (Dickson and Caldwell 1978). Secondly, reduced cell elongation in either cotyledons or hypocotyls can be due to destruction of phytohormones, especially of auxins responsible for cell growth. Popp and McIlvaine (1937) demonstrated as early as 1937 that plants irradiated with near-UV radiation had shorter stems and less extractable auxin than plants grown without near-UV radiation. Fridborg and Eriksson (1975) concluded that auxins may be destroyed by UV radiation. IAA (indoleacetic acid), which is considered to be responsible for cell wall extension, has an absorption band in the UV-B waveband at 290 nm. Destruction of IAA by longer UV-B wavelengths might be possible through photodynamic action, as demonstrated by Galston (1950). Destruction of phytohormones and inhibition of cell division by UV-B radiation may both contribute to reduced leaf area. Reduced hypocotyl growth seems to be due more to IAA destruction in the tip of the stem.

Cucumber seedlings seem to be good indicators for enhanced UV-B radiation in the field because they are very UV-B sensitive even at high white light intensities. They are, therefore, also suitable for study of the effects of small differences in daily UV-B doses on growth of plant crops. This is of importance if the daily UV-B dose is enhanced by a possible ozone layer depletion. The results clearly showed that an enhanced UV-B dose of 8151 J m^{-2} d^{-1}, which is equivalent to approximately 12% ozone depletion at 49°N (Karlsruhe, Federal Republic of Germany) during the growing season, reduced growth in terms of hypocotyl elongation and cotyledon expansion by about 30%-40% compared to effective UV-B doses as would occur at more northerly latitudes (55° to 60°N) in the very early spring. Lower daily doses are biologically less effective but still show significant effects on growth of cucumber seedlings.

ACKNOWLEDGMENT

The current research was supported by the Bundesministerium für Forschung und Technologie (BMFT).

REFERENCES

Biggs RH, Kossuth SV (1978) Effects of ultraviolet-B radiation enhancements under field conditions. In: Final report on UV-B biological and climate effects research, non-human organisms. US Environmental Protection Agency, Report No EPA-IAG-D6-0618, Washington, DC, p (IV)1-63

Biggs RH, Kossuth SV, Teramura AH (1981) Response of 19 cultivars of soybeans to ultraviolet-B irradiance. Physiol Plant 53:19-26

Brandle JR, Campbell WF, Sisson WB, Caldwell MM (1977) Net photosynthesis, electron transport capacity, and ultrastructure of Pisum sativum L. exposed to ultraviolet-B radiation. Plant Physiol 60:165-168

Brodführer U (1956) Der Einfluß des roten und blauen Spektralbereiches auf ultraviolett bestrahlte Pflanzen. Unpublished, abstracted in Lockhart and Brodführer-Franzgrote (1961)

Caldwell MM (1971) Solar UV irradiation and the growth and development of higher plants. In: Giese AC (ed) Photophysiology, vol 6. Academic Press, New York, p 131

Caldwell MM, Sisson WB, Fox FM, Brandle JR (1975) Growth response to simulated UV irradiation under field and greenhouse conditions. In: Nachtwey DS, Caldwell MM, Biggs RH (eds) Impacts of climatic change on the biosphere, part I, ultraviolet radiation effects, monograph 5. Climatic Impact Assessment Program, US Dept Transportation Report No DOT-TST-75-55, NTIS, Springfield, Virginia, p (4)253

Cosgrove D (1981) Rapid succession of growth by blue light. Occurrence, time course, and general characteristics. Plant Physiol 67:584-590

Dickson JG, Caldwell MM (1978) Leaf development of Rumex patientia L. (Polygonaceae) exposed to UV irradiation (280-320 nm). Am J Bot 65:587-863

Fridborg G, Eriksson T (1975) Partial reversal by cytokinin and (2-chloro-ethyl)-trimethylammonium chloride of near-ultraviolet inhibited growth and morphogenesis in callus cultures. Physiol Plant 34:162-166

Gaba V, Black M (1979) Two separate photoreceptors control hypocotyl elongation in green seedlings. Nature 278:51-54

Gaba V, Black M (1983) The control of cell growth by light. In: Shropshire W Jr, Mohr H (eds) Encyclopedia of plant physiology. New series vol. 16A, Photomorphogenesis. Springer-Verlag, Berlin Heidelberg New York, p 358

Galston HW (1950) Riboflavin, light, and the growth of plants. Science 111:619–624

Green AES, Cross KR, Smith LA (1980) Improved analytical characterization of ultraviolet skylight. Photochem Photobiol 31:59–65

Jabben M, Holmes MG (1983) Phytochrome in light-grown plants. In: Shropshire W Jr, Mohr H (eds) Encyclopedia of plant physiology. New series vol. 16B, Photomorphogenesis. Springer-Verlag, Berlin Heidelberg New York, p 704

Klein RM, Edsall PC, Gentile AC (1965) Effects of near ultraviolet and green radiations on plant growth. Plant Physiol 40:903–906

Krizek DT (1975) Influence of ultraviolet radiation on germination and early seedling growth. Physiol Plant 34:182–186

Lindoo SJ, Caldwell MM (1978) Ultraviolet-B radiation-induced inhibition of leaf expansion and promotion of anthocyanin production. Plant Physiol 61:278–282

Lockhart JA, Brodführer-Franzgrote U (1961) The effect of ultraviolet radiation on plants. In: Ruhland W (ed) Handbuch der Pflanzenphysiologie, vol 16B. Springer, Berlin Heidelberg New York Tokyo, p 532

Pfeiffer NF (1928) Anatomical study of plants grown under glasses transmitting light of various ranges of wavelengths. Bot Gaz 85:427–436

Pirschle K, von Wettstein F (1941) Weitere Beobachtungen über den Einfluß von langwelliger und mittelwelliger UV-Strahlung auf höhere Pflanzen besonders polyploide und hochalpine Formen. Biol Zblt 61:425–436

Popp HW (1926) A physiological study of the effect of light of various ranges of wavelength on the growth of plants. Am J Bot 13:706–736

Popp HW, McIlvaine HRC (1937) Growth substances in relation to the mechanism of the action of radiation on plants. J Agr Res 66:931–936

Rupert CS (1982) Choice of the normalization wavelength for action spectra in polychromatic radiation dosimetry of UV-B. In: Biological effects of UV-B radiation. Gesellschaft für Strahlen- und Umweltforschung mbH, München, p 38, ISSN 0721–1694

Sisson WB (1981) Photosynthesis, growth, and ultraviolet irradiation absorbance of Cucurbita pepo L. leaves exposed to ultraviolet-B radiation (280–315 nm). Plant Physiol 67:120–124

Sisson WB, Caldwell MM (1976) Photosynthesis, dark respiration, and growth of Rumex patientia L. exposed to ultraviolet irradiance (288 to 315 nanometers) simulating a reduced atmospheric ozone column. Plant Physiol 58:563–568

Stoklasa J (1911) Über den Einfluß der ultravioletten Strahlen auf die Vegetation. Centbl Bakt Parasit Infekts Abt 2 31:477–495

Teramura AH (1980) Effects of ultraviolet-B irradiance on soybean. I. Importance of photosynthetically active radiation in evaluating ultraviolet-B irradiance effects on soybean and wheat growth. Plant Physiol 48:333–339

Teramura AH (1983) Effects of ultraviolet-B radiation on the growth and yield of crop plants. Physiol Plant 58:415–427

Teramura AH, Caldwell MM (1981) Effects of ultraviolet-B irradiances on soybean. IV. Leaf ontogeny as a factor in evaluating ultraviolet-B irradiance effects on net photosynthesis. Amer J Bot 68:934–941

Teramura AH, Biggs RH, Kossuth S (1980) Effects of ultraviolet-B irradiances on soybean. II. Interaction between ultraviolet-B and photosynthetically active radiation on net photosynthesis, dark respiration, and transpiration. Plant Physiol 65:483-488

Tevini M, Iwanzik W, Thoma U (1981) Some effects of enhanced UV-B irradiation on the growth and composition of plants. Planta 153:388-394

Tevini M, Iwanzik W, Teramura AH (1983) Effect of UV-B radiation on plants during mild water stress. II. Effects on growth, protein, and flavonoid content. Z Pflanzenphysiol 110:459-467

Thomas B, Dickinson HG (1979) Evidence for two photoreceptors controlling growth in de-etiolated seedlings. Planta 135:249-255

Van TK, Garrard LA, West SH (1976) Effects of UV-B radiation on net photosynthesis of some crop plants. Crop Sci 16:715-718

Vu CV, Allan LH, Garrard LA (1978) Effects of supplemental ultraviolet radiation on growth of some agronomic crops. Soil Crop Sci Soc Fla Proc 38:59-63

Interaction of UV-A, UV-B, and Visible Radiation on Growth, Composition, and Photosynthetic Activity in Radish Seedlings

W. Iwanzik

Universität Karlsruhe, Federal Republic of Germany

ABSTRACT

Radish (Raphanus sativus L., cv. Saxa Treib) seedlings were continuously irradiated in a factorial design with UV-A, UV-B, and visible radiation. Plant growth, measured in terms of changes in fresh weight, was markedly reduced in all treatments where UV-B radiation was present, and especially when unfiltered radiation was employed, allowing small amounts of UV-C radiation to reach the plants. Simultaneously applied UV-A radiation did not improve growth; whereas the addition of white light to UV-A and UV-B radiation increased fresh weight. Anthocyanin and flavonoid production were affected in an opposite manner to fresh weight. UV-A radiation was not effective in inducing anthocyanin and flavonoid biosynthesis; whereas UV-B radiation produced increasing amounts of these pigments with increasing UV-B irradiance. The induction of flavonoid synthesis seemed to be more sensitive to UV-B radiation than that of anthocyanins. Soluble proteins were similarly affected.

Photosynthetic activity, measured in terms of variable fluorescence, was decreased by UV-B radiation; whereas UV-A radiation had no effect. The concentration of photosynthetic pigments (chlorophylls and carotenoids), however, was not affected by any of the treatments with UV-B radiation, suggesting that the primary photochemistry of chloroplasts was influenced by UV-B radiation.

It is concluded from the results in this study that UV-A radiation neither attenuates nor amplifies the UV-B-induced effects.

INTRODUCTION

Many investigations from different laboratories have shown that UV-B radiation can affect plants and plant organelles in various ways. For example, reductions in plant fresh weight (Sisson and Caldwell 1976; Basiouny et al. 1978; Vu et al. 1979, 1981; Tevini et al. 1981, 1982, 1983a,b) and in the photosynthetic activity of intact plants (Teramura et al. 1980; Teramura and Caldwell 1981; Iwanzik and Tevini 1982; Tevini and Iwanzik in this volume) and isolated chloroplasts have been observed (Brandle et al. 1977; Garrard et al. 1977; Van et al. 1977; Noorudeen and Kulandaivelu 1982).

UV-B radiation also induces the biosynthesis of secondary plant substances

NATO ASI Series, Vol. G 8
Stratospheric Ozone Reduction, Solar Ultraviolet Radiation and Plant Life. Edited by R. C. Worrest and M. M. Caldwell
© Springer-Verlag Berlin Heidelberg 1986

like anthocyanins and flavonoids (Wellmann 1971, 1975, 1976, 1982; Lindoo and Caldwell 1978; Hashimoto and Tajima 1980; Drumm-Herrel and Mohr 1981; Chassagne et al. 1981; Tevini et al. 1981, 1983a) which may function as protective screens by absorbing damaging UV-B wavelengths of radiation (Caldwell et al. 1983).

There is less information, however, concerning what effect UV-A radiation has upon UV-B-induced effects in intact plants, i.e., whether they are attenuated or amplified. Furthermore, the importance of photosynthetically active radiation (PAR) on the extent of the UV-B-induced effects has been well established (Teramura 1980; Teramura et al. 1980). Therefore, experiments were also conducted in the presence and absence of white light.

The aim of this investigation was to study the interaction of UV-A, UV-B and visible wavebands of radiation, either applied singly or in combination with each other, on growth, pigment composition, soluble proteins and on photosynthetic activity in radish seedlings.

MATERIALS AND METHODS

Growth and Irradiation Conditions

Radish (Raphanus sativus L., cv. Saxa Treib) seedlings were grown in a temperature-controlled room at 22°C in the same soil for up to 5 days and continuously irradiated immediately upon germination. Radiation was supplied by UV-A (Sylvania F 40/BL), UV-B (Philips TL 40/12) and white light fluorescent lamps (Philips TL 40/29). Six different irradiation programs were used:

1. UV-A
2. UV-A + WL
3. UV-B
4. UV-B + WL
5. UV-A + UV-B
6. UV-A + UV-B + WL

Two different filter types were used: (i) Schott filters (WG 305, 2 mm), transparent in the UV-B waveband, but completely absorbing the small amounts of UV-C radiation which are also emitted by the UV-B lamps, and (ii) window glass with only little transmission in the UV-B waveband. One pot in each series was irradiated without a filter, therefore small amounts of UV-C radiation were included when UV-B lamps were used.

Radiation in the UV-B waveband was measured using a double monochromator spectroradiometer as described elsewhere (Tevini et al. 1983b) and in the visible spectral range using a single monochromator spectroradiometer (EG&G, type 550) with a 10-nm effective bandwidth resolution. The spectral energy distribution of the radiation in the UV waveband is depicted in Figs. 1-3. The resulting UV irradiances are summarized in Table 1. For comparative purposes the UV irradiances were weighted according to the generalized plant response weighting function (Caldwell 1971). Normalization wavelength was 300 nm. The weighted irradiances are also given in Table 1. Photosynthetically active radiation (PAR, 400-700 nm) was 50 $\mu mol\ m^{-2}\ s^{-1}$ when white light was simultaneously applied. Irradiation solely with UV-B lamps resulted in a PAR of 2.8 $\mu mol\ m^{-2}\ s^{-1}$, mainly emitted between 400 and 450 nm as well as between 530 and 590 nm (Fig. 4). No measurable emission in the visible spectral range was obtained when UV-A lamps were used.

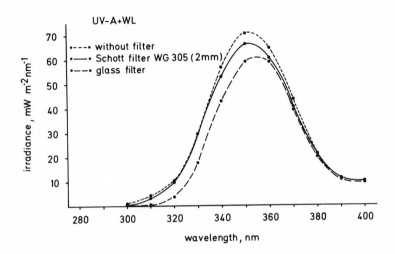

Fig. 1. Spectral energy distribution of the radiation in the UV waveband supplied by UV-A (Sylvania F 40/BL) and white light fluorescent lamps (Philips TL 40/29) filtered by Schott filters (WG 305, 2 mm), window glass and unfiltered.

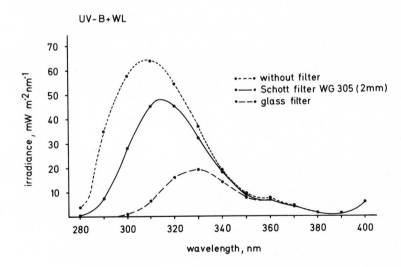

Fig. 2. Spectral energy distribution of the radiation in the UV waveband supplied by UV-B (Philips TL 40/12) and white light fluorescent lamps with and without filters.

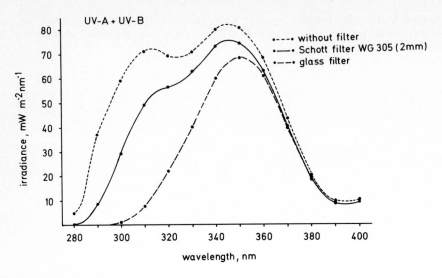

Fig. 3. Spectral energy distribution of the radiation in the UV waveband supplied by UV-A and UV-B fluorescent lamps with and without filters.

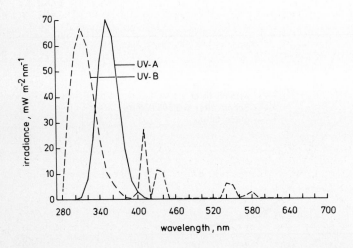

Fig. 4. Spectral energy distribution of the radiation in the UV and visible waveband supplied by unfiltered UV-A and UV-B fluorescent lamps.

Table 1. Unweighted irradiances in the UV-A and UV-B waveband. Weighted irradiances (UV_{BE}) are calculated according to the generalized plant response weighting function (Caldwell 1971). Normalization wavelength is 300 nm.

Irradiation condition		UV-B 280–320 nm	UV-A 320–400 nm	UV_{BE}
UV-A	–glass filter	0.03[a]	2.72[a]	
	–WG 305 2 mm	0.07	3.01	
	–without filter	0.09	3.60	
UV-A + WL	–glass filter	0.02	2.59	
	–WG 305 2 mm	0.06	2.97	
	–without filter	0.08	3.17	
UV-B	–glass filter	0.23	1.08	23.58[b]
	–WG 305 2 mm	1.20	1.86	604.16
	–without filter	2.08	2.17	1826.05
UV-B + WL	–glass filter	0.15	0.73	20.91
	–WG 305 2 mm	1.04	1.23	545.33
	–without filter	1.87	1.39	1774.48
UV-A + UV-B	–glass filter	0.19	3.27	26.75
	–WG 305 2 mm	1.13	4.08	598.92
	–without filter	2.03	4.53	1820.30
UV-A + UV-B + WL	–glass filter	0.17	2.98	25.10
	–WG 305 2 mm	1.08	3.86	560.17
	–without filter	1.94	4.31	1656.90

[a] unweighted irradiances in $W\ m^{-2}$
[b] weighted irradiances in $mW\ m^{-2}$

Analytical Methods

Anthocyanins of whole plants (hypocotyls and cotyledons) were extracted in a mixture of HCl in methanol (5% v/v) and spectrophotometrically measured at 530 nm. Absolute amounts were calculated from a delphinidin calibration curve. Flavonoids of cotyledons were extracted and determined as described elsewhere (Tevini et al. 1981). Soluble proteins were extracted and determined according to the Lowry method (Lowry et al. 1951). Photosynthetic pigments (chlorophylls and carotenoids) were extracted in acetone and separated by HPLC as described earlier (Tevini et al. 1981). Chlorophyll fluorescence of dark adapted cotyledons (5 min.) was excited by a red light laser (Polytec 750, 5 mW, λ = 633 nm) detected by a photodiode and recorded by a transient recorder (Siemens D 1100). The data are mean values of two independent experiments with at least five determinations within each series.

RESULTS

Growth

Changes in cotyledon fresh weight were chosen as representative of seedling growth. Similar results were obtained for cotyledon area (data not shown). The absence of white light did not reduce fresh weight when only UV-A or UV-B radiation were applied (Fig. 5). Plants additionally illuminated with white light exhibited only a slight increase of fresh weight. Plants under unfiltered UV-A radiation had a greater fresh weight than those kept under filters. In contrast, under UV-B radiation only or in combination with white light (UV-B + WL) plant growth improved when the radiation was filtered. Consequently, the lowest fresh weight was found in plants which were not protected by a filter. Filtered UV-B radiation (Schott filter) resulted in an intermediate reduction in fresh weight as compared to plants grown under glass filter. When UV-A radiation was simultaneously employed with UV-B (UV-A + UV-B) there was no change in fresh weight between plants under Schott-filtered and unfiltered radiation. Under the glass filter, a lower fresh weight was found compared to any other treatment where the glass filter was used. When white light was applied together with UV-A and UV-B radiation (UV-A + UV-B + WL), the highest fresh weight was obtained in comparison to plants grown under any other irradiation condition where UV-B was present, even under unfiltered radiation.

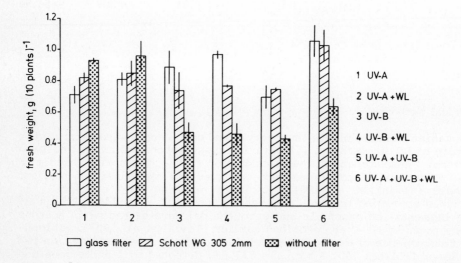

Fig. 5. Fresh weight of radish cotyledons after 5 days of irradiation. Different irradiation conditions were obtained using various radiation sources and filters. Mean of 90 plants $\pm$ SD.

Pigment Content

These compounds can be separated into the water soluble anthocyanins and flavonoids and the lipophilic photosynthetic pigments (chlorophylls and carotenoids). The two groups of pigments were affected very differently by the irradiation conditions.

There was no induction of anthocyanin synthesis by UV-A radiation when applied solely or in combination with white light (Fig. 6). Glass filtered UV-B radiation produced an amount of anthocyanins similar to that as for UV-A radiation. Under unfiltered radiation, the greatest amounts of anthocyanins were produced when UV-B radiation was present. The concentration of anthocyanins declined with decreasing UV-B irradiance. The UV-B irradiance was slightly less in treatments where UV-B lamps were used in combination with UV-A and white light lamps (Table 1) under filtered and unfiltered conditions; whereas the spectral energy distribution was not altered. Flavonoid content was affected similarly to anthocyanins; however, even with UV-A radiation there was a large flavonoid production (Fig. 7). The difference in flavonoid content between UV-A and UV-B irradiated plants was not as pronounced as in anthocyanin content.

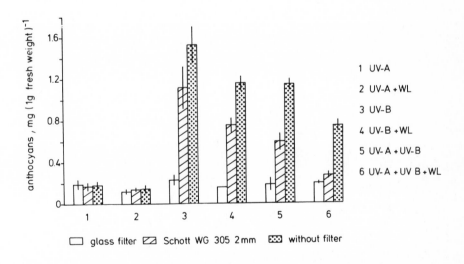

Fig. 6. Amount of anthocyanins in differently irradiated radish seedlings (hypocotyls and cotyledons). The values are given in mg (g fresh weight)$^{-1}$. Mean of 60 plants $\pm$ SD.

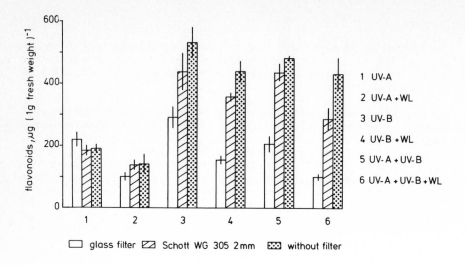

Fig. 7. Amount of flavonoids in differently irradiated radish cotyledons based on 1 g fresh weight. Mean of 60 plants ± SD.

Photosynthetic pigments were not markedly changed by any irradiation condition. An increase in the concentration of chlorophyll a and b was observed when white light was present compared to plants which were solely irradiated with UV-A or UV-B radiation (Figs. 8 and 9). In contrast, the total amount of carotenoids was not significantly different among irradiation treatments (Fig. 10).

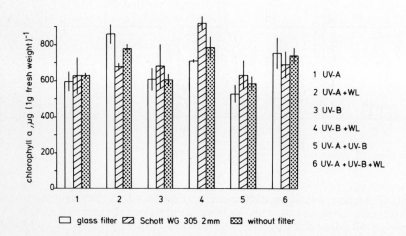

Fig. 8. Amount of chlorophyll a in radish cotyledons grown for 5 days under different irradiation conditions. Values are given in µg (g fresh weight)$^{-1}$. Mean of 60 plants ± SD.

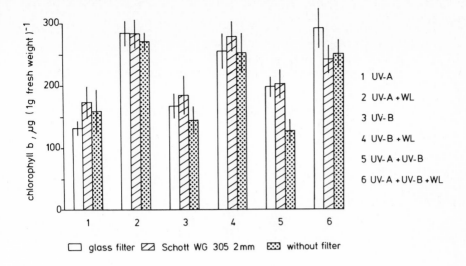

Fig. 9. Effect of different irradiation conditions on the amount of chlorophyll b in radish cotyledons. Mean values of 60 plants in µg (g fresh weight)$^{-1}$ ± SD.

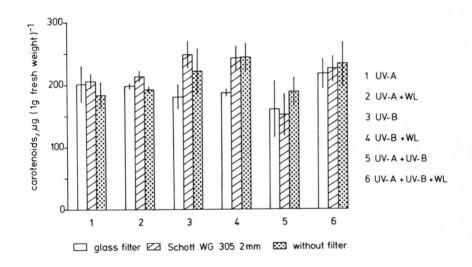

Fig. 10. Effect of different irradiation conditions on the total amount of carotenoids in radish cotyledons. Mean values of 60 plants in µg (g fresh weight)$^{-1}$ ± SD.

Protein Content

Soluble proteins were affected similarly as flavonoids. Plants under UV-A radiation and glass filtered UV-B radiation had the lowest protein content; whereas the greatest amounts of proteins were produced under unfiltered UV-B radiation (Fig. 11).

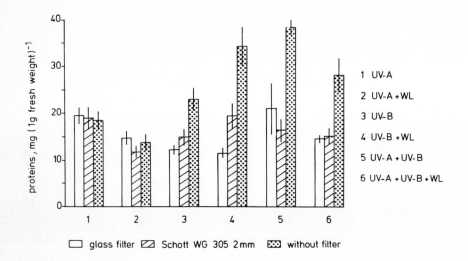

Fig. 11. Effect of different irradiation conditions on the amount of soluble proteins in radish cotyledons. Mean values of 60 plants in mg (g fresh weight)$^{-1} \pm$ SD.

Variable Fluorescence

Within 5 days of irradiation the initial value of fluorescence, F_o, was not altered. Thus, the rise to the maximum fluorescence level, F_{max}, can be taken as a representative of variable fluorescence. Furthermore, the area between the induction curve and the curve obtained in the presence of DCMU gives a relative measure of the pool size of the primary and secondary electron acceptors in photosystem II (Malkin and Kok 1966; Murata et al. 1966; Forbush and Kok 1968). UV-A radiation had no influence on F_{max} and thus on the variable part of fluorescence (Fig. 12). Similarly, there was no decrease of F_{max} when glass filtered UV-B radiation was applied. Variable fluorescence of cotyledons under Schott-filtered UV-B radiation was decreased significantly compared to the value of cotyledons under glass filter. There was no improvement when UV-A radiation was simultaneously employed. Similar, but more pronounced, effects were obtained when the area above the induction curve was taken as a measure (Fig. 13). This indicates that a smaller pool of primary and secondary acceptors was available, suggesting that primary photochemistry of photosystem II was affected by UV-B radiation.

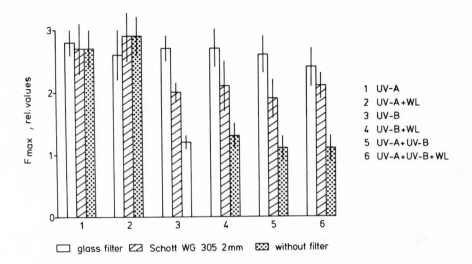

Fig. 12. Effect of different irradiation conditions on the maximum fluorescence level in radish cotyledons. The initial level of fluorescence, F_0, was not changed by any treatment.

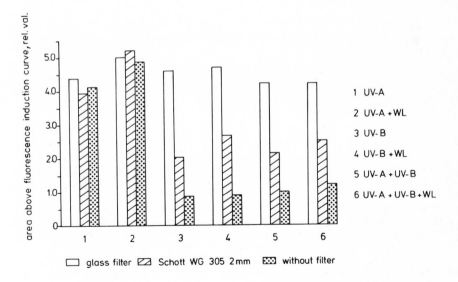

Fig. 13. Effect of different irradiation conditions on the area between the fluorescence induction curves in the presence and in the absence of DCMU (10^{-3} M).

DISCUSSION

Radish seedlings have been irradiated with UV-A, UV-B and visible radiation in different combinations in order to study the interaction of these wavelengths on growth, composition, and photosynthetic activity. The effects on growth, measured in terms of cotyledon fresh weight, are similar to earlier findings that UV-B radiation reduces fresh weight when the radiation was either filtered or unfiltered (e.g., Tevini et al. 1981, 1983a). The addition of white light caused no significant increase compared to plants grown solely under UV-B radiation. This does not support the notion of the great importance of PAR reported by Teramura (1980) for soybean growth and Klein (1963) for Ginkgo pollen tissue growth. Radiation in the UV-A waveband also did not compensate for the UV-B induced decrease of fresh weight, although the action spectrum for photorestoration of Ginkgo pollen growth peaks in the UV-A and blue light region (Klein 1963). One reason for these differences might be that the radiation was continuously employed and dark repair mechanisms were possibly necessary. Additionally, the reported photorestoration clearly depends on the dose/dose rate of photorestoring light (Klein 1963). Therefore the level of blue/UV-A radiation and white light might have been too low to make these repairing mechanisms effective. The daily peak PAR levels reported by Teramura et al. (1980) were in the range of 1500 µmol m^{-2} s^{-1}.

This study has shown an increase of anthocyanin and flavonoid biosynthesis in the presence of UV-B radiation. In addition, it has been shown that very small UV-B irradiance differences have large effects on the accumulation of anthocyanin and flavonoids. Furthermore, the effective waveband for anthocyanin biosynthesis induction must be below 300 nm, since UV-A radiation including small amounts down to about 300 nm had no effect. This is in accordance with the results of Wellmann (1982) who developed an action spectrum for anthocyanin synthesis in maize coleoptiles which peaked at about 295 nm. In contrast to anthocyanins, flavonoid accumulation was induced to some extent by UV-A radiation or the small amounts of UV-B radiation included in this radiation. However, at these low levels of UV-B radiation an irradiance dependence has not been observed. Unfiltered UV-A radiation, resulting in a slightly higher UV-B irradiance (Table 1), should have produced a greater amount of flavonoids in comparison to filtered radiation, and such an effect was not observed. Thus, the irradiance dependence requires relatively high levels of UV-B radiation. Flavonoid synthesis is much more sensitive to small amounts of UV-B radiation when compared to anthocyanin synthesis. This higher sensitivity might be of ecological importance since flavonoids strongly absorb in the UV-B waveband and in this way act as internal filters for the damaging UV-B wavelengths of radiation (Wellmann 1982; Caldwell et al. 1983). Anthocyanins, on the other hand, have only little absorption in the entire UV waveband.

Chlorophyll fluorescence induction kinetics, which can be taken as a relative measure of the activity of photosynthetic primary reactions (Papageorgiou 1975), was adversely affected by UV-B radiation or any combination where UV-B radiation was included. The initial level of fluorescence, F_O, was not changed, indicating that the antennae system was not affected by UV-B radiation after five days. But such an effect can occur after longer irradiation times as was shown by Tevini and Iwanzik (in this volume). The integrity of the pigment system can also be inferred from the nearly identical chlorophyll and carotenoid concentrations within any treatment. The reduction in variable fluorescence therefore is more likely the result of a direct intervention in the primary processes of photosynthesis, as was shown with isolated chloroplasts (Noorudeen and Kulandaivelu 1982; Iwanzik et al. 1983).

UV-A radiation did not decrease variable fluorescence or attenuate the UV-B-induced reduction. In contrast, Smillie (1983) found a strong inhibitory action of UV-A radiation on variable fluorescence. He must have use very high irradiances to produce an effect similar to photoinhibition by high light stress (Critchley 1981; Critchley and Smillie 1981; Lichtenthaler et al. 1981).

ACKNOWLEDGMENTS

This work was supported by Bundesministerium für Forschung und Technologie (KBF 48 and KBF 51). The author would like to thank Mrs. H. Hartmann for excellent technical assistance.

REFERENCES

Basiouny FM, Van TK, Biggs RH (1978) Some morphological and biochemical characteristics of C3 and C4 plants irradiated with UV-B. Physiol Plant 42:29-32

Brandle JR, Campbell WF, Sisson WB, Caldwell MM (1977) Net photosynthesis, electron transport capacity, and ultrastructure of Pisum sativum L. exposed to ultraviolet-B radiation. Plant Physiol 60:165-168

Caldwell MM (1971) Solar UV irradiation and the growth and development of higher plants. In: Giese AC (ed) Photophysiology, vol 6. Academic Press, New York, p 131

Caldwell MM, Robberecht R, Flint SD (1983) Internal Filters: prospects for UV-acclimation in higher plants. Physiol Plant 58:445-450

Chassagne M-H, Gaudillere J-P, Monties B (1981) Effets du rayonnement ultraviolet sur les plantes cultivées sous éclairement naturel et artificiel. I. Morphogenese et equippement polyphenolique de la tomate, la laitue et le poivron. Acta OEcol/OEcol Plant 2:267-282

Critchley C (1981) Studies on the mechanism of photoinhibition in higher plants. I. Effects of high light intensity on chloroplast activities in cucumber adapted to low light. Plant Physiol 67:1161-1165

Critchley C, Smillie RM (1981) Leaf chlorophyll fluorescence as an indicator of high light stress (photoinhibition) in Cucumis sativus L. Aust J Plant Physiol 8:133-141

Drumm-Herrel H, Mohr H (1981) A novel effect of UV-B in a higher plant (Sorghum vulgare). Photochem Photobiol 33:391-398

Forbush B, Kok B (1968) Reaction between primary and secondary electron acceptors of photosystem II of photosynthesis. Biochim Biophys Acta 162:243-153

Garrard LA, Van TK, West SH (1977) Plant response to middle ultraviolet (UV-B) radiation: carbohydrate levels and chloroplast reactions. Soil Crop Sci Soc Florida 36:184-188

Hashimoto T, Tajima M (1980) Effects of ultraviolet irradiation on growth and pigmentation in seedlings. Plant Cell Physiol 21:1559-1571

Iwanzik W, Tevini M (1982) Effect of enhanced UV-B irradiation on photosynthetic activity of barley seedlings and chloroplasts. In: Biological effects of UV-B radiation. Gesellschaft für Strahlen- und Umweltforschung mbH, München, p 121, ISSN 0721-1694

Iwanzik W, Tevini M, Dohnt G, Voss M, Weiss W, Gräber P, Renger G (1983) Action of UV-B radiation on photosynthetic primary reactions in spinach chloroplasts. Physiol Plant 58:401-407

Klein RM (1963) Interaction of ultraviolet and visible radiations on the growth of cell aggregates of Ginkgo pollen tissue. Physiol Plant 16:73-81

Lichtenthaler HK, Buschmann C, Döll M, Fietz H-J, Bach T, Kosel U, Meier D, Rahmsdorf U (1981) Photosynthetic activity, chloroplast ultrastructure, and leaf characteristics of high-light and low-light plants and of sun and shade leaves. Photosynthesis Research 2:115-141

Lindoo SJ, Caldwell MM (1978) UV-B radiation induced inhibition of leaf expansion and promotion of anthocyanin production. Plant Physiol 61:278-282

Lowry O, Rosebrough NT, Farr AL, Randall RJ (1951) Protein measurement with the folin phenol reagent. J Biol Chem 193:265-275

Malkin S, Kok B (1966) Fluorescence induction studies in isolated chloroplasts. I. Number of components involved in the reaction and quantum yield. Biochim Biophys Acta 126:413-432

Murata N, Hishimura M, Takamiya A (1966) Fluorescence of chlorophyll in photosynthetic systems. II. Induction of fluorescence in isolated spinach chloroplasts. Biochim Biophys Acta 120:23-33

Noorudeen AM, Kulandaivelu G (1982) On the possible site of inhibition of photosynthetic electron transport by ultraviolet-B (UV-B) radiation. Physiol Plant 55:161-166

Papageorgiou G (1975) Chlorophyll fluorescence: an intrinsic probe of photosynthesis. In: Govindjee (ed) Bioenergetics of photosynthesis. Academic Press, New York, London, p 319

Sisson WB, Caldwell MM (1976) Photosynthesis, dark respiration, and growth of Rumex patientia L. exposed to ultraviolet irradiance (288 to 315 nanometers) simulating a reduced atmospheric ozone column. Plant Physiol 58:563-568

Smillie RM (1983) Chlorophyll fluorescence in vivo as a probe for rapid measurement of tolerance to ultraviolet radiation. Plant Sci Lett 28:283-289

Teramura AH (1980) Effects of ultraviolet-B irradiances on soybean. I. Importance of photosynthetically active radiation in evaluating ultraviolet-B irradiance effects on soybean and wheat growth. Physiol Plant 48:333-339

Teramura AH, Caldwell MM (1981) Effects of ultraviolet-B irradiances on soybean. IV. Leaf ontogeny as a factor in evaluating ultraviolet-B irradiance effects on net photosynthesis. Am J Bot 68:934-941

Teramura AH, Biggs RH, Kossuth S (1980) Effects ultraviolet-B and photosynthetically active radiation on net photosynthesis, dark respiration, and transpiration. Plant Physiol 65:483-488

Tevini M, Iwanzik W (this volume) Effects of UV-B radiation on growth and development of cucumber seedlings. p 271

Tevini M, Iwanzik W, Thoma U (1981) Some effects of enhanced UV-B radiation on the growth and composition of plants. Planta 153:388-394

Tevini M, Iwanzik W, Thoma U (1982) The effects of UV-B irradiation on higher plants. In: Calkins J (ed) The role of solar ultraviolet radiation in marine ecosystems. Plenum Press, New York, London, p 581, ISBN 0-306-40909-7

Tevini M, Thoma U, Iwanzik W (1983a) Effects of enhanced UV-B radiation on germination, seedling growth, leaf anatomy and pigments of some crop plants. Z Pflanzenphysiol 109:435–488

Tevini M, Iwanzik W, Teramura AH (1983b) Effects of UV-B irradiation on plants during mild water stress. II. Effects on growth, protein and flavonoid content. Z Pflanzenphysiol 109:435–448

Van TK, Garrard LA, West SH (1977) Effects of 298 nm radiation on photosynthetic reactions of some crop species. Environ Exp Bot 17:107–112

Vu CV, Allen LH, Garrard LA (1979) Effects of supplemental ultraviolet radiation (UV-B) on growth of some agronomic crop plants. Soil Crop Sci Soc Florida, Proc 38:59–63

Vu CV, Allen LH, Garrard LA (1981) Effects of supplemental UV-B radiation on growth and leaf photosynthetic reactions of soybean (_Glycine max_). Physiol Plant 52:353–362

Wellmann E (1971) Phytochrome-mediated flavone glycoside synthesis in cell suspension cultures of _Petoselinum hortense_ after preirradiation with ultraviolet light. Planta 101:283–285

Wellmann E (1975) UV-dose dependent induction of enzymes related to flavonoid biosynthesis in cell suspension cultures of parsley. FEBS Lett 51:105–107

Wellmann E (1976) Specific ultraviolet effects in plant morphogenesis. Photochem Photobiol 24:659–660

Wellmann E (1982) Phenylpropanoid pigment synthesis and growth reduction as adaptive reactions to increased UV-B radiation. In: Biological effects of UV-B radiation. Gesellschaft für Strahlen- und Umweltforschung mbH, München, p 145, ISSN 0721-1694

Effects of Enhanced Ultraviolet-B Radiation on Yield, and Disease Incidence and Severity for Wheat Under Field Conditions

R. H. Biggs and P. G. Webb

University of Florida, Gainesville, Florida USA

ABSTRACT

The influence of enhanced UV-B radiation (280-320 nm) on wheat (Triticum aestivum cv. 'Florida 301') yield, and disease incidence and severity was investigated for two growing seasons under field conditions. Three levels of UV-B enhancement, simulating 8, 12, and 16% stratospheric ozone reduction, were employed during 1982, and two levels of UV-B radiation enhancement, simulating 12 and 16% ozone depletion, were used during 1983. At each level of UV-B enhancement during 1982, no significant direct effect upon yield was observed. However, in 1983, there was an indication that UV-B radiation was interacting with other parameters because there was a significant bimodal response on plant and seed dry weights. As determined by incidence survey, during 1982 the incidence of leaf rust (Puccinia recondita f. sp. tritici), leaf spot (Helminthosporium sativum), and glume blotch (Septoria nodorum) were not significantly affected by the enhanced levels of UV-B radiation. During 1983, specific small plots were used to test for leaf rust severity on a resistant and a susceptible cultivar. Leaf rust on the resistant cultivar 'Florida 301' decreased with plant age, but control and UV-treated plants were not different. With increased UV-B irradiation of the susceptible but tolerant cultivar 'Red Hart' severity increased with age, and all plots exposed to increased levels of UV-B radiation were significantly different from the control. However, these small plot tests did not allow us to determine if yield was influenced.

INTRODUCTION

Ultraviolet-B radiation (280-320 nm), especially high levels, could cause a variety of plant responses. These may be manifested in either subtle or dramatic changes in total biomass yield, or in damage to aerial plant parts (Bartholic et al. 1975; Biggs et al. 1975; Biggs et al. 1981; Biggs et al. 1982; Brandle et al. 1977; Van et al. 1976; Van et al. 1977; Webb and Biggs 1984). One result of increased incident solar UV-B radiation might be an enhancement of pathogenic microorganisms on a weakened or damaged host (Biggs and Basiouny 1975; Webb and Biggs 1982; Gold and Caldwell 1982; Webb and Biggs 1984).

Previous studies of UV-B-radiation effects on plants in growth chambers and greenhouses were conducted with light intensities less than ambient field levels, which accentuated the damage caused by UV-B radiation (Biggs and Kossuth 1978; Teramura et al. 1980; Teramura 1983). The present study was

NATO ASI Series, Vol. G8
Stratospheric Ozone Reduction, Solar Ultraviolet Radiation
and Plant Life. Edited by R. C. Worrest and M. M. Caldwell
© Springer-Verlag Berlin Heidelberg 1986

conducted under natural solar radiation levels at Gainesville, Florida, to examine the effects of enhanced UV-B radiation upon parameters of growth and development of wheat (<u>Triticum aestivum</u> L. cv. 'Florida 301'), particularly yield. Predisposition to, or enhancement of pathogenesis of, leaf rust (<u>Puccinia recondita</u> Rob. ex. Desm. f. sp. <u>tritici</u>) also was investigated under different UV-B radiation enhancement levels for two wheat cultivars, namely, 'Florida 301' (a resistant cultivar) and 'Red Hart' (a susceptible but tolerant cultivar).

MATERIALS AND METHODS

Yield Experiment -- 1982

Wheat (cv. 'Florida 301) seeds were planted on December 8, 1982, in 32 plots each 1.5 m wide and 2.4 m long for each of 3 UV-B enhancement levels equivalent to 8, 12, and 16% ozone reductions. There were 12 rows planted 7.5 cm apart per plot and emergent seedlings were thinned to 7.5 cm apart within rows. Plants were grown in constructed beds in a mixture of 40% peat, 40% perlite, and 20% sand with dolomite, Perk® and phosphate added. The plots were watered daily and a modified Hoaglands nutrient solution was supplied weekly.

The three levels of UV-B radiation enhancement plus a control, were obtained by using Westinghouse FS40 sunlamps suspended above the growing plants so that irradiances would simulate calculated levels for Gainesville, Florida (29°36'N latitude). Irradiance was determined with a UV spectroradiometer (Optronic Laboratories, Model 751) with the sensor positioned horizontally and directly below the center of the lamp array. The distance from lamps to plant terminal shoots was adjusted weekly, and as needed, so that the proper UV-B irradiances were maintained within ±10%. The control (0% enhancement) was achieved by encasing the lamp with a Mylar film (0.13 mm) to absorb UV-B radiation with wavelengths less than 320 nm. The simulated 8, 12, and 16% ozone reduction levels were achieved by using cellulose acetate filters of varying thickness: 0.12, 0.13, and 0.25 mm thick, respectively. Due to solarization, the filters were changed every 14 days to maintain the desired UV-B radiation enhancement levels.

For comparative purposes, UV-B spectral irradiances were weighted using a plant effective action spectrum (Caldwell 1971) normalized to 300 nm in calculating the effective dose rate throughout the day (Green et al. 1980). There was no attempt to compensate for cloudiness and other atmospheric UV-B radiation attenuators, so the error of enhancement is in the direction of greater than ambient, i.e., if the ground level measurement of UV-B radiation were based on ozone being the only attenuator and if the optical depth were 0.36 cm rather than 0.30 cm, then a 13.5% over-enhancement would occur. A sample solar irradiance measurement is presented in Table 1. Enhancement of solar ultraviolet-B irradiance was supplied in a 3-step symmetrical manner for 6 h daily, from 900 to 1500 EST.

Sample plants were removed for analysis at 51, 64, 77, and 91 days after planting with a final harvest at 129 days. Ten plants were collected randomly from the center 4 rows of each plot at 51 and 64 days, and 5 plants were collected at 77 and 91 days. Twenty plants were collected at the final harvest (129 days). Oven-dry total plant weight (g), total seed head dry weight (g), and number of tillers were recorded. The amount of chlorophyll (absorbance/g) was calculated at 66, 80, 94, and 108 days. The most recent fully expanded leaf was selected from 10 randomly located plants per bed for the chlorophyll analysis (Arnon 1949). Total nitrogen content (Nelson and Somers 1973), nitrate reductase (Deane-Drummond and Clarkson 1979), soluble protein content in the seed coat (Bradford 1976), and glutamine synthetase (Shapiro and Stadtman 1970) were evaluated to analyze the effect of UV-B radiation on nitrogen metabolism. Vigor tests were conducted on seeds from the final harvest by methods outlined in AOSA (1978). Accelerated aging tests were performed by placing 50 seeds in 4 replications on a wire screen suspended above 20 ml of deionized water in a covered 100 x 15 mm petri dish (Delouche and Baskin 1973). The petri dishes were exposed to a 41°C and 100% relative humidity environment for 96 hours. Aged seed were dried to 12% moisture after which germination was determined as described.

Table 1. Sample solar spectral irradiance measured at Gainesville, Florida, on June 5, 1981, at 12:31 EST.

Wavelength[a],	W m^{-2} nm^{-1} [b]
290 nm	3×10^{-6}
295	3.1×10^{-4}
300	8.3×10^{-3}
305	7.6×10^{-2}
310	9.8×10^{-2}
315	2.1×10^{-1}
320	3.0×10^{-1}

[a]1 nm resolution.

[b]Total UV-B (280-320 nm) = 3.090 W m^{-2}; DNA Eff. Radiation (280-320 nm) = 5.076 mW m^{-2}; Plant Eff. Radiation (280-320 nm) = 75.27 mW m^{-2}. A complete analysis of a solar day can be found in Kostkowski et al. (1980).

Yield Experiment -- 1983

Wheat (cv. 'Florida 301') seeds were planted in 8 beds each 1.5 m wide and 4.9 m long for each of 2 levels of UV-B enhancement simulating 12 and 16% ozone reduction and a control. There were 6 rows planted 15 cm apart per plot and emergent seedlings were thinned to 5 cm apart within rows. Forty-five plants were selected from the center 4 rows of each plot and this was replicated 8 times for each bed or 64 replications per treatment. The length of the growing season was 131 days and planting was on December 11, 1983. Oven-dry plant weight (g) and air-dry seed weight (g) were measured for each replications. Viability and vigor tests were conducted on seeds from the final harvest. After a cold treatment to remove seed dormancy, germination was determined by placing 50 seeds in three replications in standard rolled towel tests (AOSA) at 25°C for 4 days. "Total" germination was determined by the number of seed exhibiting visible radicle protrusion from the seed coat. The number of "normal" germinated seedlings (those expected to emerge from the soil) was determined in the rolled-towel test according to guidelines developed by seed testing rules (AOSA). The accelerated aging tests were performed as in 1982.

Disease Evaluation -- 1982

For the 1982 observations the fifth row of each bed used in the yield trials was selected for analysis of disease incidence. Disease incidence was based on Berger's (1980) description. An average of 32 plants in each bed were examined for incidence of wheat leaf rust (Puccinia recondita f. sp. tritici), glume blotch (Septoria nodorum (Berk.) Berk.) and leaf spot (Helminthosporium sativum P.K. and B.).

Disease Evaluation -- 1983

An experiment separate from the yield trials was designed and used to test for leaf rust severity for the resistant cultivar 'Florida 301' as compared to the disease severity for the susceptible cultivar 'Red Hart' at each of 3 levels of UV-B radiation enhancement simulating 8, 12, and 16% ozone reduction. Severity ratings were by methods described by Horsfall and Barratt (1945). Seeds of each variety were planted in pots containing 3 kg of a soil mix of 40% peat, 40% perlite, and 20% sand. The containers were buried to soil line in beds for exposure to each UV-B treatment. Upon emergence, each subplot (container) was thinned to 3 plants and 10 pots of each variety were randomly located in each bed planted with 'Florida 301' as border plants. A total of 30 plants in the pots of each variety were exposed to each of the 3 UV-B radiation enhancement levels plus a control. Plants were grown from December 1982 to April 1983, and disease severity measurements periodically made on 44, 57, 81, 116, and 159 days after planting. Chlorophyll analyses were made on samples 97 days after planting and consisted of determining the amount of chlorophyll in one randomly selected leaf per pot for the cultivar 'Florida 301' in each of the treatments. Germination and accelerated aging tests were conducted on seeds after final harvest.

RESULTS

Yield Experiments

During 1982 there were no significant differences using analysis of variance in total plant weight, total seed-head dry weight, number of tillers, and amount of chlorophyll among plants grown at the four UV-B enhancement levels. Total nitrogen content, nitrate reductase, soluble protein content, and glutamine synthetase was not significantly affected by the UV-B radiation enhancement treatments. Also, there were no differences found between treatments with the accelerated aging test on the seed. Average germination was 73% and the range was 71 to 75% germination across treatments.

During the 1983 season significantly greater plant dry weights (p = 0.03) and seed dry weights (p = 0.10) among plants grown in the middle rows under the higher UV-B radiation level (simulating 16% ozone reduction) and controls (0% ozone reduction) were observed when compared to the intermediate levels. A linear or a threshold response is the most likely indicator of a direct effect of UV-B radiation on yield. Based on the low r^2 values for a linear model, we have interpreted the bimodal response as an indirect effect. As in the 1982 tests, no differences between treatments were found with either the germination or accelerated aging tests on seeds.

Disease Evaluations

During 1982 there was no significant difference between treatments for the incidence of each disease (leaf rust, glume blotch, and leaf spot). During 1983, as the growing season progressed, leaf rust severity decreased significantly (p = 0.01) (Table 2) among the resistant cultivar 'Florida 301' plants at the control levels and at each of the three levels of UV-B radiation enhancement, but there were no differences between treatments. Leaf rust severity among the susceptible cultivar 'Red Hart' increased significantly (p = 0.01) (Table 3) during the growing season, with the highest incidence at the greatest UV-B enhancement levels. Among treatments of the cultivar 'Red Hart,' there was a significant (p = 0.05) treatment effect, and enhanced UV-B radiation seems to increase rust infection in approximately a linear manner. Number of seed heads, seed dry weights, tillering, chlorophyll content, total plant weight, seed germination, and seed aging were determined, but there were no significant differences between treatments.

Table 2. The average percentage infection by _Puccinia recondita_ f. sp. _tritici_ on wheat (cv. 'Florida 301') in 1983 at UV-B irradiances simulating four levels of ozone reduction. Figures in parentheses are standard errors.

Percentage ozone reduction	Days after planting				
	44	57	81	116	159
0	12.26 (2.74)	6.97 (1.27)	1.07 (0.14)	0.88 (0.15)	0.33 (0.15)
8	4.15 (1.01)	6.82 (1.28)	1.12 (0.06)	2.08 (0.14)	0.58 (0.16)
12	10.38 (1.23)	10.94 (2.02)	1.87 (0.19)	1.08 (0.08)	0.16 (0.06)
16	2.26 (0.46)	10.74 (0.52)	1.41 (0.09)	2.48 (0.19)	0.38 (0.16)

Table 3. The average percentage infection by <u>Puccinia recondita</u> f. sp. <u>tritici</u> on wheat (cv. 'Red Hart') in 1983 at UV-B irradiances simulating four levels of ozone reduction. Figures in parentheses are standard errors.

Percentage ozone reduction	Days after planting				
	44	57	81	116	159
0	13.00 (1.47)	9.96 (1.16)	4.14 (0.36)	15.93 (1.71)	46.44 (5.47)
8	4.03 (0.46)	12.56 (1.10)	4.93 (0.38)	23.41 (1.14)	55.11 (3.84)
12	12.48 (1.85)	12.40 (1.63)	8.23 (0.70)	27.15 (1.28)	63.19 (1.48)
16	5.06 (2.98)	14.95 (1.38)	9.77 (1.56)	34.09 (3.21)	69.66 (5.99)

DISCUSSION

Wheat (cv. 'Florida 301') plants when exposed to enhanced levels of UV-B radiation up to levels simulating 16% ozone reduction under field conditions for two growing seasons did not exhibit a significant reduction in yield or plant biomass. Total plant and seed dry weights were the same at the greatest enhancement level as at the control. This supports other studies which have reported wheat as resistant to UV-B-associated biomass loss either in growth chambers (Dumpert and Boscher 1982; Teramura et al. 1980; Bennett 1981; Biggs and Kossuth 1978) or field conditions (Becwar et al. 1982; Hart et al. 1975). No significant yield reduction has been observed among corn, soybean, and rice experiments under field conditions in similar experiments at Gainesville, Florida (Biggs and Webb, unpublished data). We feel that direct effects of increased UV-B radiation upon wheat yields should be minimal, at least until the UV-B enhancement level equivalent to a 0.252 cm ozone column in the atmosphere is surpassed.

The possibility of altered yields among wheat plants exposed to higher UV-B radiation levels remains a potential threat when the indirect effect of pathogenic predisposition is considered, as well as other possible pest competition, e.g., weeds (Gold and Caldwell 1983). Wheat seems to be tolerant to increased UV-B radiation and no yield reduction occurs. In this study no yield reduction was noted among plants inoculated with urediospores of P. recondita f. sp. tritici whether it was a resistant or susceptible cultivar. However, these were small plot tests and much larger plots will be needed to determine interactions on yield. Due to the daily watering regime, at no time during the growing season were the inoculated plants under moisture stress.

Since the majority of carbohydrates necessary for seed-head formation and seed fill comes from the flag leaf and since the flag leaves were not the most severely infected leaves by the organism under test, interactions between severity of infection and UV-B-radiation enhancement may be minimal with this organism. There was a bimodal treatment effect in the 1983 yield trials with 'Florida 301'. Either the expression in wheat biomass and seed yields at intermediate levels of UV-B radiation used is not real, or there is an indirect effect of UV-B radiation on some component(s) of the complex system that is integrated into a bimodal influence on biomass and yield.

We did observe a differential effect upon fungal infection between wheat cultivars exposed to enhanced UV-B radiation. Wheat rust is generally more severe on younger plants (Berger 1980). Thus, a potential reduction in yield would occur if there were an infection by a particular pathogen that colonized wheat seed heads or possibly the flag leaf. This idea warrants further study.

ACKNOWLEDGMENT

Although the research described in this chapter has been funded primarily by the U.S. Environmental Protection Agency through Cooperative Agreement CR808075 to the University of Florida, it has not been subjected to the Agency's required peer and policy review and therefore does not necessarily reflect the views of the Agency and no official endorsement should be inferred.

REFERENCES

AOSA (1978) Results for testing seed. J Seed Tech 3:1-126

Arnon DI (1949) Copper enzymes in isolated chloroplasts. Polyphenoloxidase in Beta vulgaris. Plant Physiol 24:1-15

Bartholic JF, Helsey LH, Garrard LA (1975) Field trials with filters to test for effect of UV radiation on agricultural productivity. In: Nachtwey DS, Caldwell MM, Biggs RH (eds) Impacts of climatic change on the biosphere, part I – ultraviolet radiation effects, monograph 5. Climatic Impact Assessment Program, US Dept Transportation Report No DOT-TST-75-55, NTIS, Springfield, Virginia, p (4)61

Becwar MR, Moore II FD, Burke MJ (1982) Effects of deletion and enhancement of ultraviolet-B (280-315 nm) radiation on plants grown at 3000 m elevation. J Amer Soc Hort Sci 107:771-774

Bennett JH (1981) Photosynthesis and gas diffusion in leaves of selected crop plants exposed to ultraviolet-B radiation. J Environ Qual 10:271-275

Berger RD (1980) Measuring disease intensity. Proc EC Stakman Commemorative Symposium on Crop Loss Assessment, Minn Agric Exp Stn Misc Pub 7, Minneapolis, Minnesota, p 28

Biggs RH, Basiouny FM (1975) Plant growth responses to elevated UV-B irradiation under growth chamber, greenhouse, and solarium condition. In: Nachtwey DS, Caldwell MM, Biggs RH (eds) Impacts of climatic change on the biosphere, part I, ultraviolet radiation effects, monograph 5. Climatic Impact Assessment Program, US Dept Transportation Report No DOT-TST-75-55, NTIS, Springfield, Virginia, p (4)127

Biggs RH, Kossuth SV (1978) Impact of solar UV-B radiation on crop productivity. In: Final report on UV-B biological and climate effects research, non-human organisms. US Environmental Protection Agency, Report No EPA-IAG-D6-0618, Washington, DC, p 11

Biggs RH, Kossuth SV, Teramura AH (1981) Response of 19 cultivars of soybeans to ultraviolet-B irradiance. Physiol Plant 53:19-26

Biggs RH, Sinclair TR, N'Diaye O, Garrard LA, West SH, Webb PG (1982) Field trials with soybeans, rice, and wheat under UV-B irradiances simulating O3 deletion levels. In: Biological effects of UV-B radiation. Gesellschaft für Strahlen- und Umweltforschung mbH, München, p 64, ISSN 0721-1694

Biggs RH, Sisson WB, Caldwell MM (1975) Responses of higher terrestrial plants to elevated UV-B irradiances. In: Nachtwey DS, Caldwell MM, Biggs RH (eds) Impacts of climatic change on the biosphere, part I, ultraviolet radiation effects, monograph 5. Climatic Impact Assessment Program, US Dept Transportation Report No DOT-TST-75-55, NTIS, Springfield, Virginia, p (4)34

Bradford M (1976) A rapid and sensitive method for the quantitation of microgram quantities of protein utilizing the principle of protein-dye binding. Anal Biochem 72:248

Brandle JR, Campbell WF, Sisson WB, Caldwell MM (1977) Net photosynthesis, electron transport capacity, and ultrastructure of Pisum sativum L. exposed to ultraviolet-B radiation. Plant Physiol 60:165-169

Caldwell MM (1971) Solar UV irradiance and the growth and development of higher plants. In: Giese AC (ed) Photophysiology, vol 6. Academic Press, New York, p 131

Deane-Drummond CE, Clarkson DT (1979) Effect of shoot removal and malate on the activity of nitrate reductase assayed in vivo in barley roots (Hordeum vulgare cv. Midas). Plant Physiol 64:660-662

Delouche JC, Baskin CC (1973) Accelerated aging techniques for predicting the relative storability of seed lots. Seed Sci and Tech 1:427-452

Dumpert K, Boscher J (1982) Response of different crop and vegetable cultivars to UV-B irradiance: Preliminary results. In: Biological effects of UV-B radiation. Gesellschaft für Strahlen- und Umweltforschung mbH, München, p 102, ISSN 0721-1694

Green AES, Cross KR, Smith LA (1980) Improved analytic characterization of ultraviolet skylight. Photochem Photobiol 31:59-65

Gold WG, Caldwell MM (1983) The effects of ultraviolet-B radiation on plant competition in terrestrial ecosystems. Physiol Plant 58:435-444

Hart RH, Carlson GE, Kleuter HH, Carns HR (1975) Response of economically valuable species to ultraviolet radiation. In: Nachtwey DS, Caldwell MM, Biggs RH (eds) Impacts of climatic change on the biosphere, part I, ultraviolet radiation effects, monograph 5. Climatic Impact Assessment Program, US Dept Transportation Report No DOT-TST-75-55, NTIS, Springfield, Virginia, p (4)263

Horsfall JG, Barratt RW (1945) An improved grading system for measuring plant diseases. Phytopathology 35:655

Kostkowski HJ, Saunders RD, Ward JF, Popenoe CH, Green AES (1980) New state of the art in solar terrestrial spectroradiometry below 300 nm. Opt Rad News 33:1-7

Nelson DW, Sommers LE (1973) Determination of total nitrogen in plant material. Agronomy J 65:109-112

Shapiro BM, Stadtman ER (1970) Glutamine synthetase (Escherichia coli). Methods of Enzymology 17A:910-922

Teramura AH, Biggs RH, Kossuth S (1980) Effects of ultraviolet-B irradiances on soybean. II. Interaction between ultraviolet-B and photosynthetically active radiation on net photosynthesis, dark respiration, and transpiration. Plant Physiol 65:483-488

Teramura AH (1983) Effects of ultraviolet-B radiation on the growth and yield of crop plants. Physiol Plant 58:415-527

Van TK, Garrard LA, West SH (1976) Effects of UV-B radiation on net photosynthesis of some crop plants. Crop Sci 16:715-718

Van TK, Garrard LA, West SH (1977) Effects of 298 nm radiation on photosynthetic reactions of leaf discs and chloroplast preparations of some crop species. Environ Exp Bot 17:107-112

Webb PG, Biggs RH (1982) Ultraviolet-B radiation influences Triticum aestivum growth, productivity, and microflora. Phytopathology 72:941 (abstract)

Webb PG, Biggs RH (1984) Ultraviolet-B irradiance effects on soybean productivity and incidence of fungi in seed. Phytopathology 74:822 (abstract)

Effects of Ultraviolet-B Radiation on the Growth and Productivity of Field Grown Soybean

J. Lydon, A. H. Teramura and E. G. Summers

University of Maryland, College Park, Maryland USA

ABSTRACT

The effects of enhanced UV-B radiation were studied on the growth and productivity of six field grown soybean cultivars. Control plants received ambient levels of UV-B radiation while experimental plants received daily doses equivalent to those expected on the summer solstice under clear sky conditions with a 16% and 25% ozone depletion at College Park, Maryland (39°N latitude). Plants were harvested at several reproductive growth stages. Enhanced UV-B radiation affected biomass allocation, total plant dry weight, average seed dry weight, and seed protein and lipid concentrations. Derived growth characteristics of UV-B irradiated plants (relative growth rate, net assimilation rate, leaf area duration and height growth rate) were also significantly different from controls. These effects were both cultivar and UV-B dose dependent.

INTRODUCTION

Soybean is an important crop that has been shown to be sensitive to moderate levels of UV-B radiation (Vu et al. 1978; Teramura 1980; Teramura et al. 1980; Teramura and Caldwell 1981). These studies have indicated that UV-B radiation affects vegetative growth (including total plant dry weight, leaf area and plant height) and physiological characteristics such as leaf net photosynthesis and stomatal conductance.

Biggs et al. (1981) have shown in controlled environment studies that large cultivar growth response differences to UV-B radiation exist in our modern soybean germplasm. Out of 19 cultivars tested, 4 were classified as extremely sensitive, 4 were moderately resistant, and 11 were intermediate. Quantitatively the responses reported in that study were undoubtedly much larger than might be expected under natural conditions due to the low photosynthetic photon flux densities (PPFD, between 400 and 700 nm) in the growth chambers (Warner and Caldwell 1983; Mirecki and Teramura 1984). However, the qualitative differences are indications of genetic variation in susceptibility. Therefore it is important to test different cultivars to gain an understanding of the full range of variability for UV-B sensitivity that exists in the soybean germplasm. This information could then be potentially useful in future breeding programs for UV resistance.

There were three primary objectives of the research: (1) To test the dose response relationship of UV-B radiation on the growth and productivity of

NATO ASI Series, Vol. G8
Stratospheric Ozone Reduction, Solar Ultraviolet Radiation
and Plant Life. Edited by R. C. Worrest and M. M. Caldwell
© Springer-Verlag Berlin Heidelberg 1986

field grown soybeans; (2) to determine whether particular reproductive phases of development are more susceptible to UV-B radiation than others, and (3) to determine qualitative and quantitative differences in UV-B radiation sensitivity among soybean cultivars grown under field conditions.

MATERIALS AND METHODS

The field site is located 5 km north of the University of Maryland. The experimental area covers approximately 2025 m^2 and has been continuously farmed since 1940. The soil texture is a silty loam with a pH of 6.6. The plots were fertilized with 0-100-100 (N-P-K) at a rate of 9 g m^{-2} before planting.

Prior to the field experiment, 24 soybean (<u>Glycine max</u> (L.) Merr.) cultivars were screened in the greenhouse for UV-B radiation sensitivity (unpublished data). Plants received either no UV-B radiation (Mylar-filtered controls) or a daily UV-B dose of 2479 J m^{-2} UV-B_{BE}. This was approximately equivalent to the daily dose should a 16% ozone reduction occur under clear sky conditions in College Park, Maryland (39°N latitude, 50 m elevation) during the summer solstice (Green et al. 1980). All plants were harvested at the full seed fill stage. A ranking computed by summing the percentage difference in total dry weight, leaf area, and height between nonirradiated control plants and UV-B irradiated plants

$$\frac{dw\ irrad - dw\ nonirrad}{dw\ nonirrad} + \frac{leaf\ area\ irrad - leaf\ area\ nonirrad}{leaf\ area\ nonirrad} +$$

$$\frac{ht\ irrad - ht\ nonirrad}{ht\ nonirrad} \times 100 \tag{1}$$

was used to assess cultivar sensitivity. Six cultivars were selected for field investigations based upon this ranking which included: three highly inhibited cultivars -- Essex, Williams (the two most commonly grown cultivars in the mid-Atlantic region of the USA) and Bay; an intermediately inhibited cultivar -- James; a moderately stimulated cultivar -- York; and a highly stimulated cultivar -- Forrest.

Seeds were planted in the field in early June in rows spaced 0.40 m apart at a seeding density of 33 seeds m^{-1}. Three border rows were planted on either side of the experimental rows. The overall experimental design was a completely randomized, split plot with 2 blocks, 6 cultivars, 3 treatments, 10 samples per treatment, and 3 sampling dates. The field treatments included 2 artificially enhanced UV-B radiation treatments and an ambient control (Fig. 1).

Supplemental UV-B radiation was supplied by filtered Westinghouse FS40 sunlamps suspended above and perpendicular to the plant rows. Lamps were set 0.30 m apart, fitted with 50-mm wide minireflectors, and held in racks supported by pulleys. Lamp ballasts were remotely mounted from the plants to minimize shading. All lamps were fitted with either presolarized 0.08-mm cellulose acetate (UV-B transmitting) or 0.13 mm Mylar Type S filters (control). Filters were routinely changed weekly to ensure transmission of the desired UV-B spectral distribution. Lamps were preburnt for at least 100 h prior to use (Teramura et al. 1980). The two experimental soybean rows were directly beneath the lamps, 0.20 m on either side of the lamp center, where UV-B irradiance varied by less than ± 5% (Teramura 1981).

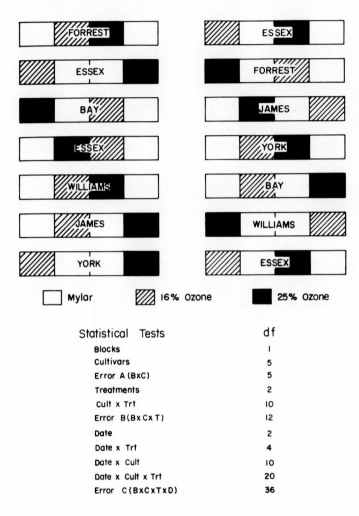

Fig. 1.   Field experimental design and statistical tests.

Spectral distribution of the lamps was determined with an Optronic Laboratories Inc. Model 742 Spectroradiometer. The spectral irradiance was weighted with a generalized plant action spectrum (UV-B$_{BE}$, Caldwell 1971) and a DNA action spectrum (UV-B$_{DNA}$, Setlow 1974). Supplemental UV-B radiation was provided as a two-step square wave for 3 h on either side of solar noon for a supplemental $UV\text{-}B_{BE}$ daily dose of 644 J m^{-2} $UV\text{-}B_{BE}$ (55 J m^{-2} $UV\text{-}B_{DNA}$) and 1137 J m^{-2} $UV\text{-}B_{BE}$ (103 J m^{-2} $UV\text{-}B_{DNA}$, normalized at 280 nm). This was approximately equivalent to the additional daily UV-B dose should a 16% and 25% ozone reduction occur under clear sky conditions in College Park, Maryland during the summer solstice. Supplemental UV-B radiation was supplied immediately after germination and continued throughout plant development. The two UV-B enhancements were achieved by varying the distance between the lamps and plants and using lamps of different age.

Subsamples of ten soybean plants were sequentially harvested during three reproductive stages of development. Previous studies (Teramura 1981) showed no significant UV-B effects on vegetative growth. These three reproductive stages of development were identified using the terminology of Fehr and Caveness (1977), and included full bloom (R2), full pod (R4), and full seed fill (R5-R6). At the time of harvest, a number of morphometric measurements were made on the plant material, including plant height, node number, leaf number, and when appropriate, flower and pod number. Total canopy leaf areas were determined with a Lambda Instruments Model LI-3100 area meter. Plants were then separated into leaves, petioles, stems, flowers, and pods and dried in a ventilated oven at 60°C before weighing. These data were then used to calculate dry matter allocation patterns (% stems, leaves, petioles, and reproductive tissue, i.e., flowers, pods, and seeds), and the mathematically derived growth parameters.

Overall cultivar performance under the various treatments was evaluated based on seed quality and quantity, and mathematical growth analysis of vegetative growth which included calculations for relative growth rate (RGR), net assimilation rate (NAR), leaf area duration (LAD), plant height growth (HTGROW), specific leaf weight (SLW), and leaf area ratio (LAR). RGR is the amount of dry matter produced per unit dry matter present per unit of time, g g^{-1} d^{-1}. This is calculated from the following relationship:

$$RGR = \frac{\ln(W_2) - \ln(W_1)}{T_2 - T_1} \qquad (2)$$

where W_1 and W_2 are the total plant biomass at the beginning and end of the harvest interval, and $T_2 - T_1$ is the length of the harvest interval, in days. The efficiency of the amount of dry matter produced per unit leaf area present per unit of time can be expressed as NAR in g m^{-2} d^{-1}. This is calculated from the following relationship:

$$NAR = \frac{W_2 - W_1}{A_2 - A_1} \times \frac{\ln(A_2) - \ln(A_1)}{T_2 - T_1} \qquad (3)$$

where A_1 and A_2 are the total leaf areas at the beginning and end of the harvest interval. LAD is a measure of how long a plant maintains its active assimilatory surface and is calculated by the following:

$$LAD = \frac{(A_2 - A_1) \times (T_2 - T_1)}{\ln(A_2) - \ln(A_1)} \tag{4}$$

HTGROW is an estimate of plant growth based upon plant height as a function of time:

$$HTGROW = \frac{H_2 - H_1}{T_2 - T_1} \tag{5}$$

where H_1 and H_2 are plant height measured during the first and second harvest intervals, respectively. SLW describes the amount of leaf dry weight per unit of leaf area in g dm^{-2} and LAR is the amount of leaf area per unit of plant dry weight in m^2 g^{-1}.

Seed protein and lipid content were determined with a Dickey-john Corp. Model GAC III-600 grain analysis computer. Seed from two plants were combined and twenty grams ground in an electric mill for 20 s, dried at 60°C for 24 h, and then allowed to equilibrate to room temperature before analyzing.

Statistical analysis was done using the Statistical Analysis System (SAS), incorporating the General Linear Model Procedure. The sample number for the morphometric and mathematically derived growth data was n=6, with 3 levels of irradiance, 2 plots per treatment, and 1 replicate per plot (an average of 10 plants). The sample number for the protein and lipid data was n=30, with 3 levels of irradiance, 2 plots per cultivar, and 5 replicates per plot. Linear regression analysis was conducted where inter-actions between UV-B irradiance and cultivar were found to be significant ($p < 0.05$) by analysis of variance (ANOVA). Only the regression models are presented.

RESULTS

Significant regression relationships between UV-B dose and morphometric parameters for all three harvests are shown in Table 1. By the first harvest when the plants were at full bloom, Williams was 48 days old, Essex, James and York 61 days old, and Bay and Forrest 68 days old. At this reproductive stage of development only two cultivars, Forrest and Williams, showed significant UV-B radiation effects. Both petiole dry weight and leaf number were quadratically related to UV-B dose in Forrest. Low levels of enhanced UV-B radiation equivalent to a 16% ozone reduction produced increases in these morphometric values compared to plants which received ambient levels of UV-B radiation. UV-B radiation was positively correlated with the percentage dry weight allocated to leaves and nega-tively correlated with the percentage dry weight allocated to stems and reproductive tissue in Williams.

At the second harvest, when the plants were in full pod, Williams was 70 day old, Essex and James 84 days old, York 91 days old, Bay 98 days old, and Forrest 103 days old. The UV-B effects on petiole dry weight and leaf number previously observed in Forrest were no longer evident at the second harvest. Nor did UV-B radiation continue to alter biomass allocation in Williams. Instead, a significant negative linear relationship between the level of UV-B radiation and leaf dry weight, petiole dry weight, and total plant dry weight was found in James. A quadratic relationship between leaf number and UV-B dose was also found in this cultivar. SLW and LAR were significantly affected by UV-B radiation in Bay, although morphological characteristics such as leaf dry weight, leaf area, and total plant weight (which are the parameters from which SLW and LAR are derived) were not significantly affected. SLW was negatively related, while LAR was positively related to UV-B dose.

Table 1. Regression relationships between UV-B$_{BE}$ dose and morphometric parameters. SLW = Specific Leaf Weight; LAR = Leaf Area Ratio. Daily supplemental UV-B$_{BE}$ dose: 0 J m^{-2}; 644 J m^{-2}; and 1137 J m^{-2}. All relationships are statistically significant at the 95% level.

Cultivar	Parameter	r^2	Equation
Harvest 1			
Forrest	Petiole Dry Wt	0.83	$Y = 2.93 + 3.62E{-}03X - 3.60E{-}06X^2$
Forrest	No. of Leaves	0.87	$Y = 16.1 + 2.01E{-}02X - 1.85E{-}05X^2$
Williams	% Leaves	0.83	$Y = 49.3 + 3.54E{-}03X$
Williams	% Stems	0.82	$Y = 29.0 - 2.75E{-}03X$
Williams	% Reprod Tissue	0.79	$Y = 0.63 - 2.38E{-}04X$
Harvest 2			
James	Leaf Dry Wt	0.80	$Y = 12.7 - 3.58E{-}03X$
James	Petiole Dry Wt	0.72	$Y = 5.96 - 1.80E{-}03X$
James	Total Dry Wt	0.74	$Y = 34.1 - 1.00E{-}02X$
James	No. of Leaves	0.94	$Y = 26.4 - 2.01E{-}02X + 1.30E{-}05X^2$
Bay	SLW	0.73	$Y = 4.65E{-}01 - 6.24E{-}05X$
Bay	LAR	0.85	$Y = 7.76E{-}03 + 1.05E{-}06X$
Harvest 3			
York	No. of Leaves	0.74	$Y = 6.95 - 4.78E{-}03X$
York	Flower Dry Wt	0.75	$Y = 7.77E{-}04 - 4.65E{-}07X$

All cultivars were harvested after 108 days because of impending frost. Plants were in full seed fill at this time. Flower number and dry weight were negatively correlated with UV-B dose in York at the time of the third harvest (Table 1). However, because of the low frequency of flowers during this stage and the small absolute changes in average flower dry weight (0.7 mg), these effects did not result in significant changes in the percentage dry weight allocated into reproductive tissue.

Table 2. Regression relationships between UV-B$_{BE}$ dose and the mathematically derived variables between the first and second harvest, and the second and third harvest: LAD = Leaf Area Duration; NAR = Net Assimilation Rate; and RGR = Relative Growth Rate. Daily supplemental UV-B$_{BE}$ dose: 0 J m^{-2}; 644 J m^{-2}; and 1137 J m^{-2}. All relationships are statistically significant at the 95% level.

Cultivar	Parameter	r^2	Equation
Harvest 1-2			
Forrest	LAD	0.41	$Y = 0.81 - 2.49E-03X + 2.13E-06X^2$
Forrest	NAR	0.73	$Y = 2.95 - 3.59E-03X + 2.92E-06X^2$
Forrest	RGR	0.56	$Y = 2.82E-02 - 3.45E-05X + 2.97E-08X^2$
Williams	LAD	0.41	$Y = -0.13 + 8.17E-04X - 8.07E-07X^2$
Williams	NAR	0.40	$Y = 2.73 + 3.56E-03X - 3.53E-06X^2$
Williams	RGR	0.37	$Y = 3.11E-02 + 4.03E-05X - 3.96E-08X^2$
Essex	HTGROW	0.55	$Y = 2.99E-03 + 2.29E-05X - 1.99E-08X^2$
Harvest 2-3			
Forrest	NAR	0.98	$Y = -4.19E-01 + 4.02E-03X - 2.37E-06X^2$
Forrest	RGR	0.95	$Y = -4.08E-03 + 3.38E-05X - 1.98E-08X^2$
York	NAR	0.76	$Y = 1.44 - 9.78E-03X + 8.14E-06X^2$
York	RGR	0.76	$Y = 1.30E-02 - 8.24E-05X + 6.79E-08X^2$

All of the mathematically derived growth variables that were significantly affected by UV-B radiation responded in a quadratic manner, regardless of cultivar. Both LAD and NAR in Forrest decreased with low UV enhancement levels and increased to values similar to that at ambient levels when exposed to the high enhanced UV-B levels between the first and second harvest (Table 2). LAD and NAR in Williams, and HTGROW in Essex increased

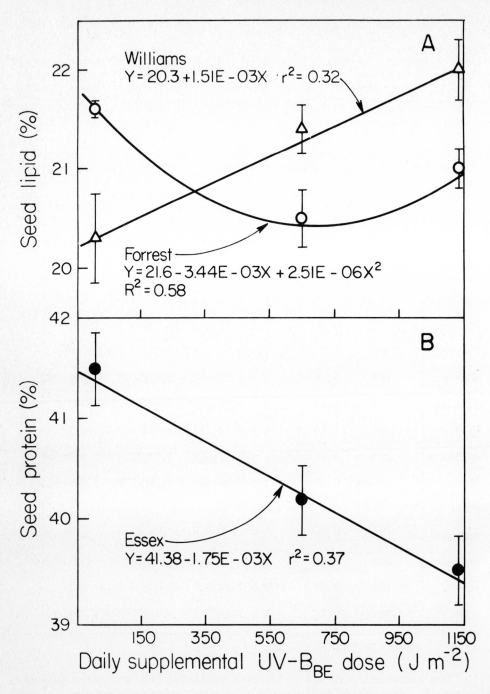

Fig. 2. Effects of UV-B radiation on: A. percentage seed lipid; and B. percentage seed protein. Daily supplemental UV-B$_{BE}$ dose: 0 J m^{-2}; 644 J m^{-2}; and 1137 J m^{-2}. Vertical bars represent $\pm$ 1 SE.

at low levels of enhanced UV-B radiation during the same growth period (Table 2). NAR in Forrest was again significantly affected by UV-B radiation during the last growth period (Table 2). However, unlike the previous growth period, the lower UV-B dose stimulated NAR in this cultivar. UV-B effects on NAR during this later growth period were also evident in York, but here the lower UV-B dose had a detrimental effect. Because LAR was not affected by UV-B radiation in these cultivars, and because RGR = NAR x LAR, the effect of UV-B dose on RGR was identical to its effect on NAR (Table 2).

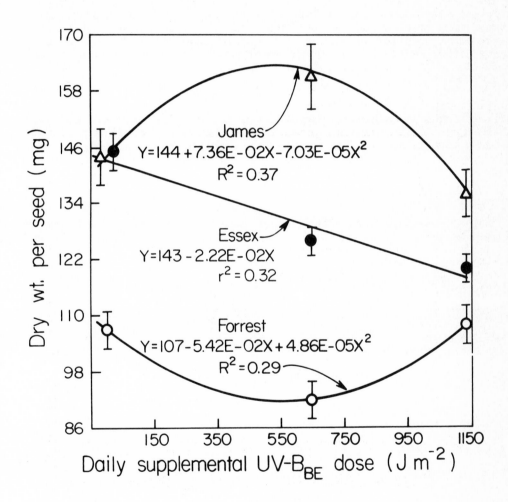

Fig. 3. Effects of UV-B radiation weight per seed. Daily supplemental UV-B$_{BE}$ dose: 0 J m^{-2}; 644 J m^{-2}; and 1137 J m^{-2}. Vertical bars represent $\pm$ 1 SE.

UV-B radiation affected the percentage of seed lipid in Forrest and Williams, but these cultivars responded differently (Fig. 2A). UV-B radiation and percentage seed protein were negatively related in Essex (Fig. 2B). Weight per seed varied quadratically in relation to UV-B dose in Forrest and James, but in a negative linear relationship in Essex (Fig. 3). Essex was the only cultivar tested that exhibited a significant relationship between UV-B dose and total seed weight; however, the r^2 value was less than 0.10 (data not shown).

DISCUSSION

The effectiveness of UV-B radiation on soybean growth varied with the developmental stage of the plant. For instance, a shift in biomass allocation from stems and reproductive tissue to leaves as a result of UV-B radiation occurred in Williams at the full bloom stage but not at later seed filling stages. Changes in sensitivity to UV-B radiation with plant development have been previously reported (Teramura 1981, Teramura et al. 1984).

Plant responses to UV-B radiation were also dose dependent. Surprisingly, lower levels of enhancement (equivalent to a 16% ozone reduction) were more effective in altering plant growth than higher levels (equivalent to a 25% ozone reduction). Similar results were reported for leaf, stem, and total dry weight of field grown corn exposed to a gradient of UV-B radiation (Biggs and Kossuth 1978).

Table 3. Relative ranking of cultivars based upon the percentage change in combined growth parameters (Eq. 1) between nonirradiated and UV-B irradiated (greenhouse) or ambient and UV-B supplemented (field) plants. Daily UV-B dose equivalent to that received given a 16% ozone reduction at College Park, Maryland (39°N latitude, 50 m elevation) during the summer solstice. Plants harvested at the full seed fill stage.

Greenhouse		Field	
Essex	−113	Essex	−75
Williams	−104	Williams	−41
Bay	−101	Bay	−37
James	−79	York	−27
York	+15	James	+6
Forrest	+253	Forrest	+29

There were distinctive cultivar differences in sensitivity to UV-B radiation. For example, the shift in biomass allocation as a result of UV-B radiation seen in Williams was not observed in any other cultivar. Interestingly, the relative ranking of cultivar sensitivity based upon Eq. (1) was very similar for the field and greenhouse studies, even though UV-B radiation produced a substantially larger effect in the greenhouse (Table 3). This implies that greenhouse studies may be useful in establishing relative differences in cultivar or species sensitivity to UV-B radiation. Nevertheless, they may not be appropriate for establishing absolute responses due to their inherently low PPFD's and due to a lack of important stress-factor interactions.

The six cultivars may be classified into two distinct categories of UV-B sensitivity based upon Eq. (1). Essex, Williams, Bay and York were detrimentally affected, while James and Forrest were stimulated by UV-B radiation. However, a sensitivity classification is more difficult to develop when only statistically significant differences are used in the determination. For example, Essex was the most UV-B sensitive cultivar based upon seed yield, yet vegetative growth was unaffected. In addition, seed yield components in York were unaffected by UV-B enhancements despite decreases in flower number and dry weight at the time of the third harvest. James was sensitive based upon morphological criteria, and tolerant based upon derived variables, yet weight per seed increased at the lower level of UV-B radiation.

Of major concern are the potential effects of UV-B radiation on seed yield. The data presented here indicate that quantitatively seed yield was relatively unaffected by UV-B radiation in all six cultivars tested. However, three of the cultivars did show changes in seed quality (protein and/or lipid content). In addition, two of those cultivars (Forrest and Essex) showed a concomitant decrease in weight per seed when exposed to a UV-B dose equivalent to a 16% ozone reduction at College Park, Maryland. These results are relevant not only in relation to crop plants but to natural plant populations as well, for alterations in seed reserves may substantially affect seedling survivorship.

In summary, the dose response relationship of UV-B radiation on the growth and productivity of field grown soybeans was specific to both the cultivar and the reproductive stage of soybean development. Thus, some of the detrimental effects of UV-B radiation on soybean may potentially be avoided by selecting tolerant cultivars or by breeding tolerant traits into the more sensitive yet popular varieties. UV-B radiation effects on vegetative characteristics are not necessarily reflected in reproductive characteristics. Therefore, emphasis should be specifically placed on yield, whether it is vegetative (as in leafy vegetables) or reproductive (as in cereals) when evaluating UV-B radiation effects on crop plants. The quality of the agricultural product should also be considered when estimating the impact of UV-B radiation on crop plants because seed protein and lipid content can be altered without quantitative changes in yield.

ACKNOWLEDGMENTS

The authors would like to thank Karen Teramura for illustrating the figures, and N. S. Murali for reviewing the manuscript. This work was supported by the United Stated Environmental Protection Agency (CR808035) and grants from the Graduate School and Provost to A. H. Teramura and J. Lydon. Scientific Article No. A-3739, Contribution No. 6715, of the Maryland Agricultural Experiment Station, Department of Botany. Although the work described in this chapter has been funded in part by the United States Environmental Protection Agency, it has not been subjected to the Agency's required peer and policy review and therefore does not necessarily reflect the views of the Agency and no official endorsement should be inferred.

REFERENCES

Biggs RH, Kossuth SV (1978) Impact of solar UV-B radiation enhancements under field conditions. In: Final report of UV-B biological and climate effects research, non-human organisms. US Environmental Protection Agency Report, No EPA-IAG-D6-0618, Washington, DC, p (IV)1-63

Biggs RH, Kossuth SV, Teramura AH (1981) Response of 19 cultivars of soybeans to ultraviolet-B irradiance. Physiol Plant 53:19-26

Caldwell MM (1971) Solar UV irradiance and the growth and development of higher plants. In: Giese AC (ed) Photophysiology, vol 6. Academic Press, New York, p 131

Green AES, Cross KR, Smith LA (1980)' Improved analytical characterization of ultraviolet skylight. Photochem Photobiol 31:59-65

Fehr WR, Caveness CE (1977) Stages of soybean development. Cooperative Extension Service, Agriculture and Home Economic Experiment Station, Iowa State Univ of Sci and Techn, p 1-11

Mirecki RM, Teramura AH (1984) Effects of ultraviolet-B irradiance on soybean. V. The dependence of plant sensitivity on the photosynthetic photon flux density during and after leaf expansion. Plant Physiol 74:475-480

Setlow RB (1974) The wavelengths in sunlight effective in producing skin cancer: A theoretical analysis. Proc Natl Acad Sci USA 71:3363-3366

Teramura AH (1980) Effects of ultraviolet-B irradiances on soybean. I. Importance of photosynthetically active radiation in evaluating ultraviolet-B irradiance effects on soybean and wheat growth. Physiol Plant 48:333-339

Teramura AH (1981) Cultivar differences in the effects of enhanced UV-B irradiation on the growth, productivity, and photosynthetic capacity of field grown soybeans. In: Annual report to US Environmental Protection Agency, No CR808-035-010, Washington, DC

Teramura AH, Caldwell MM (1981) Effects of ultraviolet-B irradiances on soybean. IV. Leaf ontogeny as a factor in evaluating ultraviolet-B irradiance effects on net photosynthesis. Am J Bot 68:934-941

Teramura AH, Biggs RH, Kossuth S (1980) Effects of ultraviolet-B irradiances on soybean. II. Interaction between ultraviolet-B and photosynthetically active radiation on net photosynthesis, dark respiration, and transpiration. Plant Physiol 65:483-488

Teramura AH, Perry MC, Lydon J, McIntosh MS, Summers EG (1984) Effects of ultraviolet-B radiation on plants during mild water stress. III. Effects on photosynthetic recovery and growth in soybean. Physiol Plant 60:484-492

Vu CV, Allen LH Jr, Garrard LA (1978) Effects of supplemental ultraviolet radiation (UV-B) on growth of some agronomic crop plants. Soil Crop Sci Soc Fla Proc 38:59-63

Warner CW, Caldwell MM (1983) Influence of photon flux density in the 400–700 nm waveband on inhibition of photosynthesis by UV-B (280–320 nm) irradiation in soybean leaves: separation of indirect and immediate effects. Photochem Photobiol 38:341–346

Interaction Between UV-B Radiation and Other Stresses in Plants

A. H. Teramura

University of Maryland, College Park, Maryland USA

ABSTRACT

Over 100 plant species and cultivars have been screened for sensitivity to UV-B radiation and these have displayed a wide range of responses. Yet despite this large body of information, we are still not at a point where we can make reliable quantitative or even qualitative estimates of the potential impact of enhanced levels of UV-B radiation resulting from a partial stratospheric ozone depletion. The primary reason for this dilemma is that nearly all of our biological effects information comes from controlled environment experiments (i.e., greenhouses and growth chambers). In such facilities UV-B radiation is supplied to plants while most other environmental variables such as temperature, water supply and nutrients are optimized for plant growth. Such ideal growth conditions rarely are found in nature. In fact, since most environmental factors co-vary, it would not be uncommon for plants in the field to experience several stresses concomitantly. The interaction between UV-B radiation and other stresses may produce a combined response which differs either antagonistically or synergistically from either stress independently. Such combined effects will be examined between UV-B radiation and plants grown under limiting conditions of visible radiation, water, and nutrients.

INTRODUCTION

Much of our knowledge regarding the effects of UV-B radiation is derived from studies conducted in controlled or partially controlled environments. This has resulted in a narrow and somewhat biased view of the effects of enhanced levels of UV-B radiation, since plants grown under such conditions generally look and respond very differently from field grown plants. As a direct result of this, we are yet unable to reliably assess the potential impacts of a partial depletion in stratospheric ozone upon global crop productivity. The primary reason for this dilemma is that environmental conditions within growth chambers are generally unlike those in the field.

One difficulty arising from growth chamber or greenhouse studies is that we generally manipulate only a single environmental component at a time, while optimizing all the others for ideal growth. This allows us to quantify the effects of a single factor stress on any number of morphometric and physiological responses. Unfortunately, such single factor stresses rarely, if ever, occur in nature. Under actual field conditions where environmental factors tend to co-vary, plants would commonly experience multiple stresses

NATO ASI Series, Vol. G8
Stratospheric Ozone Reduction, Solar Ultraviolet Radiation
and Plant Life. Edited by R. C. Worrest and M. M. Caldwell
© Springer-Verlag Berlin Heidelberg 1986

simultaneously. For example, periods of peak UV-B radiation coincide with periods of maximum temperature and evaporative demand (vpd) (Fig. 1). The net effect of two or more concomitant stresses on a given physiological or morphometric growth characteristic could be manifested in several different responses. In one case the response to one stress may compensate for the response to another stress. In this compensatory instance, a plant receiving two simultaneous stresses would be less affected than one receiving a single stress. Alternatively, the combined response may be identical to the response of each individual stress. Therefore, the net effect of two simultaneous stresses would show no differential effect. However, intuitively it is more likely that multiple stresses have synergistic negative effects, and plants experiencing several stresses concomitantly would be more adversely affected than plants subjected to only a single stress.

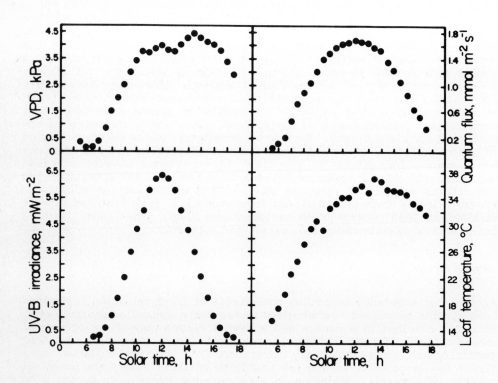

Fig. 1. Diurnal variation in quantum flux, leaf temperature, vapor pressure deficit, and UV-B irradiance at 39°N latitude on July 15, 1983, under clear sky conditions. UV-B irradiance was calculated from an analytical model of Green et al. (1980) and weighted with a DNA action spectrum (Setlow 1974). Other environmental measurements are from unpublished data of Teramura and Forseth.

Several studies have been expressly designed to test the effects of UV-B radiation on plants simultaneously experiencing other commonly encountered stresses. These were not merely attempts to simulate field conditions in growth chambers, but rather studies designed to test specific interactions in order to develop a better understanding of UV-B radiation responses in plants. This chapter will review those few studies which have specifically examined the interactions between UV-B radiation and other commonly experienced stresses including water stress, low photon flux density or shade stress, and mineral deficiency. These studies may help provide a link between the large data base of single-factor stresses from controlled experiments with ongoing and future field experiments.

INTERACTIONS BETWEEN UV-B RADIATION AND WATER STRESS

Plant water stress, developing from insufficient soil moisture or plant resistance to the uptake of water, is common throughout the world. Considering global vegetation types, plant productivity is limited more by available water than by any other single environmental factor (Salisbury and Ross 1978). Plant water stress occurs on both diurnal and seasonal bases and affects plant growth through several physiological processes (photosynthesis, dark respiration, etc.) and the reduction in turgor necessary for cell expansion (Hsiao 1973; Hanson and Hitz 1982; Kramer 1983).

Table 1. A summary of studies examining the interaction between UV-B and other environmental stresses

UV-B and water stress	UV-B and mineral deficiency	UV-B and low PPFD
Teramura et al. 1983	Bogenrieder & Douté 1982	Bartholic et al. 1974
Tevini et al. 1983		Van et al. 1976
Teramura et al. 1984		Sisson & Caldwell 1976
		Teramura 1980
		Teramura et al. 1980
		Warner & Caldwell 1983
		Mirecki & Teramura 1984

To date, only three studies have been reported which allow a direct assessment of the combined effects of water stress and UV-B radiation (Table 1). In the first two studies (Teramura et al. 1983; Tevini et al. 1983), cucumber (Cucumis sativus L. cv. Delikatess) and radish (Raphanus sativus L. cv. Saxa Treib) were grown in a factorial experiment with two UV-B irradiances and three levels of water stress. On a weighted, total daily dose basis, the UV-B irradiances were 3378 J m^{-2} d^{-1}, [weighted with the DNA action spectrum (Setlow 1974) and normalized to 300 nm] and 9709 J m^{-2} d^{-1}. The lower irradiance was equivalent to ambient levels at 49°N latitude during the beginning of the growing season. Water stress was induced by varying the frequency of watering and resulted in mean minimum midday tissue water potentials of -0.3, -0.7, and -1.5 MPa for the three treatments.

UV-B radiation had dramatic effects on the diurnal pattern of adaxial leaf diffusion resistance to water loss in cucumber, but not in radish (Figs. 2 and 3). During the first few days·of UV irradiation in cucumber, adaxial leaf resistance increased as much as three-fold. Immediately afterwards, stomatal function was apparently lost as leaf resistances sharply decreased and remained low throughout leaf senescense. Although the mechanism for this UV-B induced change in stomatal resistance is presently unknown, it is clear that UV-B radiation could potentially increase the susceptibility of cucumber to water stress.

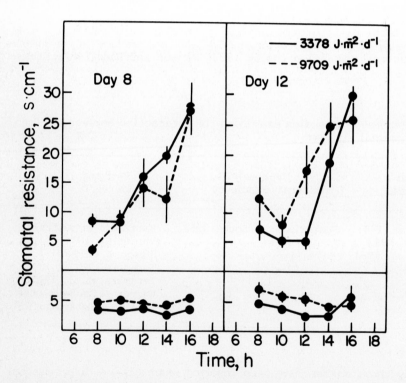

Fig. 2. Diurnal patterns of adaxial leaf resistance to water loss in Raphanus for plants with minimum midday leaf water potential of -0.3 MPa (lower figures) and -0.7 MPa (upper figures). Each point is the mean of 6 to 10 measurements and vertical lines represent ± 1 SE. Data from Teramura et al. (1983).

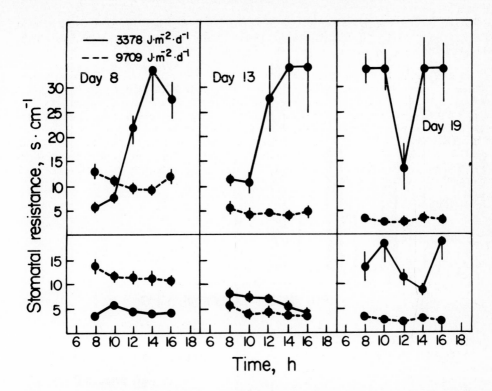

Fig. 3. Diurnal pattern of adaxial leaf resistance to water loss in
Cucumis for plants with minimum midday leaf water potential of -0.3 MPa
(lower figures) and -0.7 MPa (upper figures). Each point is the mean of 6
to 10 measurements and vertical lines represent ± 1 SE. Data from Teramura
et al. (1983).

The UV resistance seen in radish may be partially related to the effects
these stresses may have on flavonoid biosynthesis (Fig. 4). Flavonoids are
a large family of plant compounds, many of which have high absorption
coefficients in the UV region and are commonly found in plant epidermal
layers (Markham 1975). For these reasons, several investigators have
proposed flavonoids as potential UV screens (Wellmann 1982; Robberecht and
Caldwell 1978). Not only does UV stimulate flavonoid production in radish,
but the combination of UV and water stress results in a much greater
flavonoid concentration than would be predicted from exposure to either
stress alone (positive synergistic affect). In contrast, flavonoid biosynthesis decreased as UV dose increased in cucumber, possibly explaining why
cucumber is so UV sensitive.

Therefore, the results of these experiments not only suggest that UV-B
radiation might increase the susceptibility of some plants to water stress,
but also that water stress might alter plant sensitivity to UV.

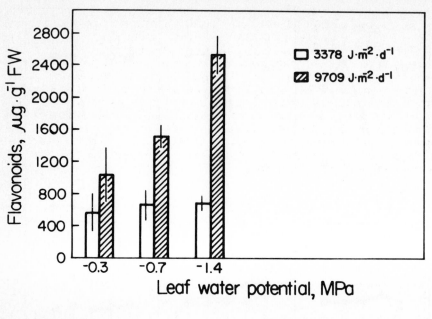

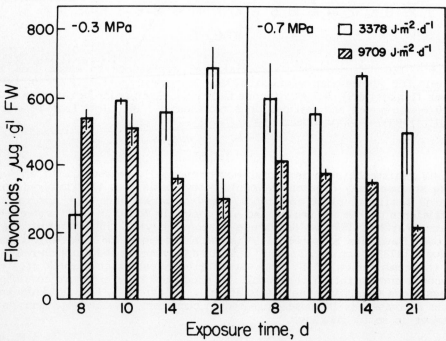

Fig. 4. Upper figure: Flavonoid content in Raphanus after 21 d irradiation based on fresh weight and cotyledon pair. Lower figure: Flavonoid content in Cucumis based on fresh weight after 8, 10, 14, and 21 d irradiation. Data from Tevini et al. (1983).

A third study examined the effects of UV-B radiation on the photosynthetic recovery from artificially induced water stress and their combined effects on plant growth (Teramura et al. 1984). In that study, soybeans (<u>Glycine max</u> (L.) Merr. cv. Essex) were grown under two levels of UV-B radiation. On a weighted, total daily dose basis, plants received either no UV or a dose equivalent to that incident at 39°N latitude at the beginning of the growing season in the event of a 23% ozone depletion [7078 J m^{-2} d^{-1}, weighted with DNA action spectrum (Setlow 1974) and normalized to 300 nm]. Water stress was achieved by withholding water until minimum midday leaf water potentials of −2.0 MPa were reached and photosynthetic recovery was monitored for several days after rewatering (Fig. 5). The entire experiment was repeated for plants in either early vegetative (30-d old) or reproductive (60-d old) development.

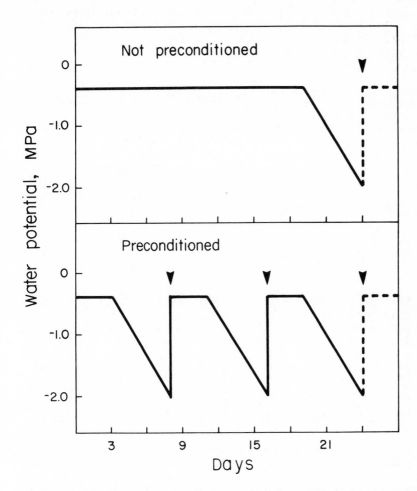

Fig. 5. Water stress was achieved by withholding water until a leaf water potential of −2.0 MPa developed. Plants were rewatered (arrows) and photosynthetic recovery from water stress was monitored for 3 d after rewatering (dashed lines). Data from Teramura et al. (1984).

On an absolute basis, UV-B radiation produced reductions in net photo-synthesis in young vegetative plants, and water stress resulted in further reductions of this already depressed rate (Fig. 6a). When this additional photosynthetic inhibition was integrated over time, the combination of UV and water stress resulted in larger reductions in total productivity (dry weight) than produced by either stress independently (Table 2). Older, reproductive plants appear to be more sensitive to water stress than to UV-B radiation (Fig. 6b). This coincided with a 2- to 3-fold greater accumulation of flavonoids in leaves of older, UV-B irradiated plants (Table 2) compared with vegetative plants.

Table 2. The effects of UV-B radiation and water stress on total plant productivity and flavonoid production (estimated from methanolic extract absorbance). -UV, 0 J m^{-2} d^{-1}; +UV, 7078 J m^{-2} d^{-1}; +W, one drying cycle to leaf Ψ of -2.0 MPa; -W, three drying cycles to leaf Ψ of -2.0 MPa. Values in rows followed by different letters are significantly different at the 5% level. Data from Teramura et al. (1984).

Age (days)	Plant characteristic	-UV +W	-UV -W	+UV +W	+UV -W
36	Total dry wt (g)	8.1 $\pm$ 1.5^a	6.4 $\pm$ 0.8^b	6.8 $\pm$ 0.8^b	4.8 $\pm$ 1.0^c
36	Absorbance (290 nm)	0.3 $\pm$ 0.1^b	0.3 $\pm$ 0.1^b	0.5 $\pm$ 0.1^a	0.4 $\pm$ 0.1^a
60	Total dry wt (g)	22.2 $\pm$ 2.3^{ab}	19.8 $\pm$ 2.0^b	24.4 $\pm$ 3.1^a	19.9 $\pm$ 4.5^b
60	Absorbance (290 nm)	0.8 $\pm$ 0.1^b	0.8 $\pm$ 0.1^b	0.9 $\pm$ 0.1^a	0.8 $\pm$ 0.2^b

The results of this study suggest that under natural field conditions, soybean plants would be more sensitive to UV-B radiation during the early stages of development, and especially sensitive during periods of concomitant water stress. Therefore, some of the deleterious effects of increased levels of UV-B radiation may be minimized during those years with favorable precipitation patterns early in the growing season. Alternatively, irrigation may provide a practical solution to minimize these effects.

INTERACTIONS BETWEEN UV-B RADIATION AND MINERAL DEFICIENCY

Plants require some 16 essential elements for completing their normal growth and development. These have various physiological roles including enzyme activation, development of osmotic potentials, regulation of membrane permeability, maintenance of electrostatic neutrality, etc. Inadequate supply of essential elements restricts these physiological functions and thus results in poor growth and development. In many soils, plants often experience mineral deficiencies due to low availability or inherently low elemental content. Therefore, we might intuitively expect that plants growing with mineral deficiencies would be even more susceptible to the deleterious effects of UV-B radiation. However, only one preliminary study has been published which directly tests this assumption (Table 1). In that study, Bogenrieder and Douté (1982) compared the effects of UV-B radiation on the growth and photosynthetic response of lettuce (<u>Lactuca</u> <u>sativa</u>) and alpine sorrel (<u>Rumex</u> <u>alpinus</u>) grown under four different mineral concentrations. Plants were grown hydroponically in nutrient solutions ranging in concentrations from 4 to 32 mmol l^{-1}. Unfortunately, neither the spectral UV distribution nor the weighted irradiances were reported in that study.

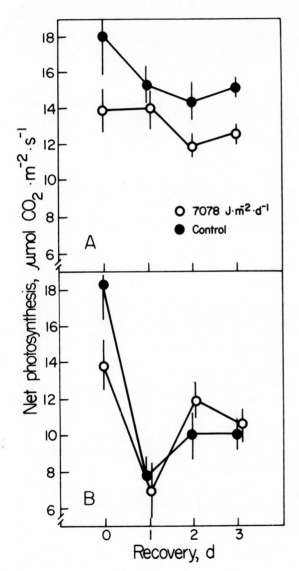

Fig. 6. Photosynthetic recovery from water stress in soybean plants during: A. Early vegetative (30-d old); B. Early reproductive (60-d old) development. Each symbol represents the mean of 10 measurements and vertical bars ± 1 SE. Data from Teramura et al. (1984).

Nevertheless, the results of that study indicated the sensitivity of net photosynthesis to UV radiation increased with increasing mineral concentration in lettuce, but not in alpine sorrel (Fig. 7). This increased sensitivity was coincident with an increase in total chlorophyll content. Interestingly, however, this greater photosynthetic inhibition did not result in an increase in the relative sensitivity of dry matter production. In fact, productivity (dry weight basis) became less sensitive to UV-B radiation as mineral supply increased (Table 3). Therefore, the increased photosynthetic inhibition resulting from short-term UV exposure was possibly compensated for by protective or repair processes stimulated by increased mineral supply over a longer time period. One example of this might be flavonoid biosynthesis, which has been shown to be sensitive to plant nutritional status (Harborne 1967).

Table 3. Relative reduction in total plant dry weight by UV-B radiation at four different nutrient levels for Lactuca and Rumex. Data from Bogenrieder and Douté (1982).

Species	Nutrient concentration (mmol l^{-1})	% reduction in dry wt per plant by UV radiation
Lactuca sativa	4	49
	8	31
	16	38
	32	24
Rumex alpinus	4	45
	8	42
	16	12

INTERACTION BETWEEN UV-B RADIATION AND LOW PPFD

Despite the many advantages of conducting experiments in growth chambers or greenhouses, these facilities have serious limitations when the conclusions drawn from such studies need to be extrapolated into the field. One reason, alluded to earlier, is that environmental conditions within growth chambers are quite unlike those found in the field. One of the obvious differences is in the level of visible radiation available to the plant for photosynthesis. Since most growth chambers supply visible radiation via a combination of fluorescent and incandescent lamps, photosynthetic photon flux density (PPFD, between 400 and 700 nm) is typically quite low, ranging from 10 to 33% of average midday irradiance.

The reason for concern over such low PPFD is that many of the deleterious effects of UV-B induced damage can be reversed by longer wavelength radiation (Caldwell 1971). If higher PPFD typically found in the field have an ameliorating effect on UV-B radiation damage, then conclusions extrapolated from growth chambers might substantially overestimate the impacts of enhanced levels of UV-B radiation.

Due to the obvious importance of this question, more work has been conducted on the interactions between UV-B radiation and PPFD than any of the other combined stresses (Table 1).   Table 4 summarizes the results of those published studies wherein the effects of UV-B radiation were tested under at least two contrasting PPFD levels.   Despite the use of a number of unrelated plant species in a diversity of growth environments, a clear trend emerges from these data: plants grown in higher PPFD tend to be less sensitive to the deleterious effects of UV-B radiation.   A corollary to this observation is plant sensitivity to UV-B radiation is highly dependent upon the PPFD available to plants during growth and that shaded conditions maximizes this sensitivity.

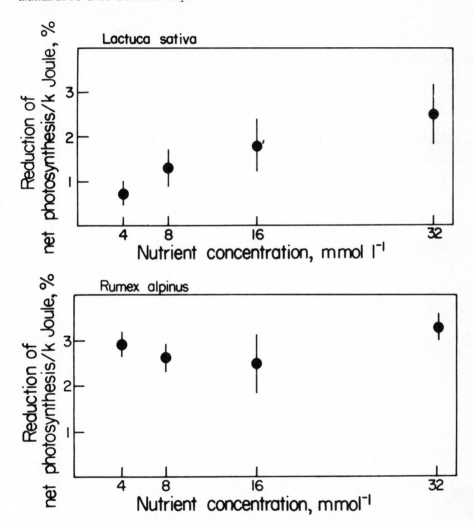

Fig. 7.   Sensitivity of photosynthetic inhibition to UV-B radiation in Lactuca (upper figure) and Rumex (lower figure) as a function of increasing nutrient concentration.   Reductions in photosynthesis are expressed per kJoule of UV provided by a 450-W xenon arc source filtered through an interference filter.   Data from Bogenrieder and Douté (1982).

Table 4. Percent change from Mylar controls in photosynthesis or growth by UV-B radiation at different photosynthetic photon flux density (PPFD). Values in parentheses are percent changes from controls resulting from concomitant PPFD.

Species	Peak PPFD (μmol m^{-2} s^{-1})						
	2000	1600	1400	800-900	400-500	200	70
Rumex patientia L.[a]	-38			-30	-38		
Phaseolus vulgaris L.[b]	+5						
Zea mays L.[b]	+7				-8		
Lycopersicon esculentum L.[b]	-22				-24		
Pisum sativum L.[b]					+4	-33	
Glycine max (L.) Merr.[c]		-6	-18	-58	-23		
Glycine max (L.) Merr.[d]		-14	-29	-65	-59	-52	
Pisum sativum L.[e]				-19	-35		
Lycopersicon esculentum L.[e]				-12	-33		
Brassica oleracea var. acephala L.[e]				-10	-27		
Brassica oleracea var. capitata L.[e]				-6	-33		
Zea mays L.[e]				+5	-16		
Avena sativa L.[e]				+5	-24		
Glycine max (L.) Merr.[f]			0	-20			
			0	(-20)			
Glycine max (L.) Merr.[g]				-5			-25
				(-17)			(-10)

a = Sisson and Caldwell (1976), b = Bartholic et al. (1974), c = Teramura et al. (1980), d = Teramura (1980), e = Van et al. (1976), f = Mirecki and Teramura (1984), g = Warner and Caldwell (1983)

Leaves which have developed under contrasting PPFD generally differ greatly
in morphology, anatomy, and physiology (Boardman 1977). For example,
leaves from high PPFD environments are smaller and thicker than those in
low PPFD. These and other differences make direct comparisons of UV-
induced damage difficult to ascertain. Recently, two studies (Mirecki and
Teramura 1984; Warner and Caldwell 1983) conducted to specifically separate
the effects of preconditioning or growth PPFD and concomitant PPFD.
Preconditioning PPFD is that prevailing during leaf development and
maturation, and primarily determines the anatomical and morphological
characteristics of the leaf. Concomitant PPFD is that supplied during UV-B
irradiation. In both studies concomitant PPFD did not alter leaf anatomy
and therefore allowed direct comparisons of UV-induced effects.

Table 5. Thickness of the upper epidermis, palisade layer, spongy
mesophyll, lower epidermis, and specific leaf weight (SLW) of soybean as
affected by preconditioning PPFD. Data from Mirecki and Teramura (1984).

Precon- ditioning PPFD (μmol m^{-2} s^{-1})	Upper Epidermis Thickness (μm)	Palisade Layer Thickness (μm)	Spongy Mesophyll Thickness (μm)	Lower Epidermis Thickness (μm)	SLW (g m^{-2})
1400	13.86 $\pm$0.34[a]	51.65 $\pm$0.92[a]	92.29 $\pm$1.96[a]	13.28 $\pm$0.22[a]	39.09 $\pm$0.58[a]
800	12.36 $\pm$0.37[b]	48.78 $\pm$0.64[b]	82.28 $\pm$1.32[b]	12.59 $\pm$0.23[b]	34.45 $\pm$0.46[b]

[a,b]Means in the same column followed by a different letter are significantly different at the 5% level.

In the study by Mirecki and Teramura (1984) soybean plants were initially
grown under two preconditioning PPFD levels (800 and 1400 μmol m^{-2} s^{-1}) for
10 days. These PPFD levels were the mean peak daily irradiances received
by plants growing in a greenhouse. After the first trifoliate leaf had
become fully expanded, plants from each treatment were then placed into a
factorial experiment with two levels of UV-B radiation, 0 and 80 mW m^{-2},
when weighted with the generalized plant response function of Caldwell
(1968) and two levels of concomitant PPFD, 800 and 1400 μmol m^{-2} s^{-1}. This
simultaneous irradiance was supplied for 9 h each day over a 10-day period.
The results of this study indicated that leaves preconditioned to high PPFD
were much less sensitive to UV-induced inhibition of photosynthesis than
those preconditioned to low PPFD. Similarly, leaves which were concomitantly irradiated with UV-B radiation and high PPFD were less sensitive to
photosynthetic inhibition than those concomitantly receiving low PPFD
(Fig. 8). The greater UV resistance exhibited by plants preconditioned to
high PPFD was at least partially due to greater flavonoid concentrations
produced in these leaves (Fig. 9) as well as the development of thicker
leaves, which effectively places a greater proportion of potentially

sensitive organelles into deeper, more protected tissue layers (Table 5). However, these changes in anatomical properties and flavonoid biosynthesis do not account for the reduced UV sensitivity of plants concomitantly receiving high PPFD. Therefore, this study indicates that not only are anatomical/morphological factors important in mitigating UV-induced damage, but also that longer wavelength radiation is involved in compensatory or repair process during UV irradiation.

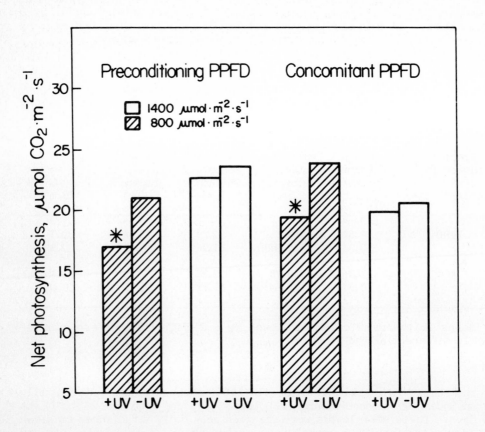

Fig. 8. The effects of UV-B radiation on plants grown in high and low preconditioning (expansion PPFD) or concomitant (irradiation PPFD) PPFD's. An asterisk denotes a significant difference from the control at the 5% level. Data from Mirecki and Teramura (1984).

The study by Warner and Caldwell (1983) was similar in design to the previous experiment, the major difference being in the UV and PPFD level employed and the duration of the concomitant PPFD. In this study, soybean were grown in constant preconditioning PPFD of 70 and 750 $\mu mol\ m^{-2}\ s^{-1}$ in growth chambers. UV-B radiation and concomitant PPFD (150 and 850 $\mu mol\ m^{-2}\ s^{-1}$) were supplied during a single 5-h period. As in the previous experiment, high preconditioning PPFD nearly eliminated any UV-induced photosynthetic inhibition through the production of thicker leaves with greater flavonoid concentration. In contrast, however, plants receiving high PPFD concomitantly with UV-B radiation did not show an amelioration of UV effects. Instead, a small but significant increase in photosynthetic inhibition was observed and this was independent of the previous preconditioning PPFD. Presently, this response is not well understood.

Despite this discrepancy, both studies concluded that plants preconditioned to PPFD more typical of field levels showed greater UV resistance. Plants grown under high PPFD are somewhat preadapted or more tolerant to enhanced levels of UV-B radiation by virtue of anatomical and physiological modifications resulting from the longer-wavelength radiation.

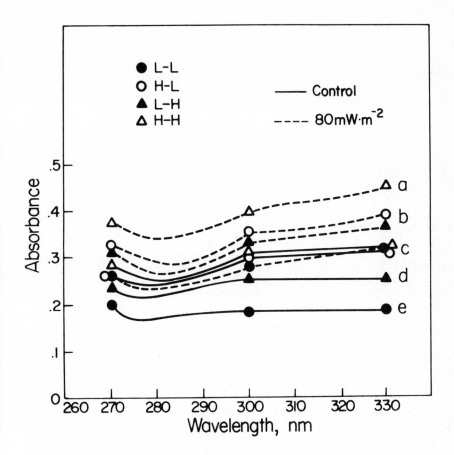

Fig. 9. Absorbance of methanolic extracts at 330 nm as determined by interactions among preconditioning PPFD, concomitant PPFD, and UV-B irradiation level. Values followed by the same letter are not significantly different at the 5% level. Data from Mirecki and Teramura (1984).

In conclusion, it appears that studies of the interactions between UV-B radiation and other commonly experienced stresses are fruitful approaches since they elicit plant responses which would otherwise be undetected in single stress experiments. However, since these studies were also conducted under artificial (partially controlled) conditions, their results may not be truly indicative of the complicated response surfaces found in nature. Nevertheless, they may bring us one step closer to an understanding of the potential impacts of enhanced levels of UV-B radiation on plants.

ACKNOWLEDGMENTS

The author would like to thank Karen Teramura for illustrating the figures and Inez Miller and Leslie Posey for their careful typing. This work was supported by the United States Environmental Protection Agency through Cooperative Agreement No. CR808035 to the University of Maryland and grants from the Graduate School and Provost to the author. Scientific Article No. A-3700, Contribution No. 6676 of the Maryland Agricultural Experiment Station, Department of Botany. Although the work described in this chapter has been funded in part by the United States Environmental Protection Agency, it has not been subjected to the Agency's required peer and policy review and therefore does not necessarily reflect the view of the Agency and no official endorsement should be inferred.

REFERENCES

Bartholic JF, Halsey LH, Biggs RH (1974) Effects of UV radiation on agricultural productivity. In Third Conference on CIAP (February 1974), US Dept Trans, Washington DC. p 498

Boardman NK (1977) Comparative photosynthesis of sun and shade plants. Ann Rev Plant Physiol 28:355-377

Bogenrieder A, Douté Y (1982) The effect of UV on photosynthesis and growth in dependence of mineral nutrition (Lactuca sativa L. and Rumex alpinus L.). In: Bauer H, Caldwell MM, Tevini M, Worrest RC (eds) Biological effects of UV-B radiation. Gesellschaft für Strahlen- und Umweltforschung MbH, München, p 164, ISSN 0721-1694

Caldwell MM (1968) Solar ultraviolet radiation as an ecological factor for alpine plants. Ecol Monogr 38:243-267

Caldwell MM (1971) Solar UV irradiation and the growth and development of higher plants. In: Giese AC (ed) Photophysiology, vol 6. Academic Press, New York, p 131

Green AES, Cross KR, Smith LA (1980) Improved analytical characterization of ultraviolet skylight. Photochem Photobiol 31:59-65

Hanson AD, Hitz WD (1982) Metabolic responses of mesophytes to plant water deficits. Ann Rev Plant Physiol 33:163-203

Harborne JB (1967) Comparative biochemistry of the flavonoids. Academic Press, New York, p 383

Hsiao TC (1973) Plant responses to water stress. Ann Rev Plant Physiol 24:519-570

Kramer PJ (1983) Water relations of plants. Academic Press, New York, p 489, ISBN 0-12-425040-8

Markham KR (1975) Isolation techniques for flavonoids. In: Harborne JB, Mabry TJ, Mabry H (eds) The flavonoids, part 1. Academic Press, New York, p 1, ISBN 0-12-364601-6

Mirecki RM, Teramura AH (1984) Effects of ultraviolet-B irradiance on soybean. V. The dependence of plant sensitivity on the photosynthetic photon flux density during and after leaf expansion. Plant Physiol 74:475–480

Robberecht R, Caldwell MM (1978) Leaf epidermal transmittance of ultraviolet radiation and its implication for plant sensitivity to ultraviolet-radiation induced injury. Oecologia (Berl) 32:277–287

Salisbury FB, Ross CW (1978) Plant Physiology, 2nd ed. Wadsworth Co, Belmont, California, p 361, ISBN 0-534-00562-4.

Setlow RB (1974) The wavelengths in sunlight effective in producing skin cancer: a theoretical analysis. Proc Natl Acad Sci USA 71:3363–3366

Sisson WB, Caldwell MM (1976) Photosynthesis, dark respiration, and growth of Rumex patientia L. exposed to ultraviolet irradiance (288 to 315 nm) simulating a reduced atmospheric ozone column. Plant Physiol 58:563–568

Teramura AH (1980) Effects of ultraviolet-B irradiances on soybean. I. Importance of photosynthetically active radiation in evaluating ultraviolet-B irradiance effects on soybean and wheat growth. Physiol Plant 48:333–339

Teramura AH, Biggs RH, Kossuth S (1980) Effects of ultraviolet-B irradiances on soybean. II. Interactions between ultraviolet-B and photosynthetically active radiation on net photosynthesis, dark respiration, and transpiration. Plant Physiol 65:483–488

Teramura AH, Tevini M, Iwanzik W (1983) Effects of ultraviolet-B irradiation on plant during mild water stress. I. Effects on diurnal stomatal resistance. Physiol Plant 57:175–180

Teramura AH, Perry MC, Lydon J, McIntosh MS, Summers EG (1984) Effects of ultraviolet-B irradiation on plants during mild water stress. III. Effects on photosynthetic recovery and growth in soybean. Physiol Plant 60:484–492

Tevini M, Iwanzik W, Teramura AH (1983) Effects on UV-B radiation on plants during mild water stress. II. Effects on growth, protein and flavonoid content. Z Pflanzenphysiol 110:459–467

Van TK, Garrard LA, West SH (1976) Effects of UV-B radiation on net photosynthesis of some crop pants. Crop Sci 16:715–718

Warner CW, Caldwell MM (1983) Influence of photon flux density in the 400–700 nm waveband on inhibition of photosynthesis by UV-B (280–320 nm) irradiation in soybean leaves: separation of indirect and immediate effects. Photochem Photobiol 38:341–346

Wellmann E (1982) Phenylpropanoid pigment synthesis and growth reduction as adaptive reactions to increased UV-B radiation. In: Bauer H, Caldwell MM, Tevini M, Worrest RC (eds) Biological effects of UV-B radiation. Gesellschaft für Strahlen- und Umweltforschung MbH, München, p 145, ISSN 0721-1694

Models and Data Requirements for Measuring the Economic Consequences of UV-B Radiation on Agriculture

R. M. Adams

Oregon State University, Corvallis, Oregon USA

ABSTRACT

Any environmental stress which either reduces potential agricultural productivity or increases its variability may have substantial economic consequences in both national and world markets. UV-B radiation is a potential environmental stress for vegetation, including agricultural crops. Concern over the economic consequences of such stresses on agriculture in the United States has motivated several economic assessments to aid policy makers in setting air quality standards. The purpose of this paper is to review the biological and economic data needs and procedures for performing an assessment of the effect of UV-B stress on agriculture. Conducting an economic assessment of air pollution or UV-B stress on agriculture first requires biological or production response data linking stress levels to yield or productivity. Then, consideration must be given to the structure of the economic problem being addressed. The economic assessment methodology must consider not only changes in yields, but also associated adjustments in commodity prices, and distributional consequences attendant to these price changes. The important dimensions of an economic assessment, and general economic methodologies capable of addressing them, are discussed using examples from air pollution. Data required to implement such methodologies are examined with emphasis on the integration of biological and economic information into a conceptually valid framework. While the tractability of such frameworks have been directly established for air pollution economic assessments, application to UV-B assessments will ultimately depend on the plant scientists' ability to discriminate the yield effects of this stress from other environmental stresses.

INTRODUCTION

The effects of airborne pollutants and other environmental stresses on vegetation are well documented in the plant science literature (see, for example, U.S. Environmental Protection Agency [U.S. EPA] 1984). One important class of vegetative effects relevant to policy makers is agricultural production, given the role played by food and fiber in any society's well-being. Environmental stress which either reduces potential agricultural productivity or increases its variability may have substantial economic consequences in both national and world markets. Concern over the economic impact of airborne pollutants on U.S. agriculture has recently motivated several economic assessments to aid policy makers in setting air quality standards.

NATO ASI Series, Vol. G 8
Stratospheric Ozone Reduction, Solar Ultraviolet Radiation
and Plant Life. Edited by R. C. Worrest and M. M. Caldwell
© Springer-Verlag Berlin Heidelberg 1986

More is presently known about the effect of air pollutants on agricultural productivity than about the effects of many other environmental stresses, such as UV-B radiation. However, it is likely that an increase in the amount of UV-B radiation has a similar type of effect on crop yield as would air pollutants. Thus, from an analytical point of view, measuring the economic effect to consumers and producers of changes in agricultural yields due to UV-B radiation can follow the same procedures used in economic studies of air pollutants.

It is the overall purpose of this paper to review the biological and economic data needs and procedures for performing an assessment of the effects of UV-B radiation on agriculture. While the utility of these assessment methods have been directly established for other stress problems, their application to a UV-B assessment will ultimately depend on the plant scientists' ability to discriminate the yield effects of this stress from other environmental stresses. Therefore, the specific objectives are to:

1. define the role of biological and meteorological data in economic assessments of ecosystem effects;

2. discuss previous integrated bioeconomic assessment programs of analytically similar environmental effects;

3. discuss qualitative and quantitative economic issues and procedures as they apply to this type of assessment, drawing from recent economic assessments which are empirical analogues to the hypothesized UV-B radiation problem;

4. present a heuristic bioeconomical model for assessing UV-B radiation effects, borrowing from recent benefits assessments funded by the U.S. EPA and others; and

5. discuss implementation of such a bioeconomical model for UV-B radiation.

PROBLEM SETTING: MEASURING AGRICULTURAL EFFECTS

The economic implications of reducing agricultural productivity over the long run can be established by examining the current magnitude of the U.S. agricultural sector. The importance of selected agricultural crops to the U.S. economy is portrayed in Table 1 for the 1982 crop year. As the table indicates, the farm-level value of these six crops amounts to nearly $45 billion. When these intermediate commodities are processed or fed through the livestock system, their value increases considerably. Further, several of these commodities (e.g., maize, soybeans) are major U.S. export items, amounting to approximately 20% of total U.S. exports in recent years. Dramatic short-term changes in expected yields and production, such as those associated with the U.S. drought of 1983, have a pronounced effect on crop prices and hence consumers' welfare, nationally and internationally. Even subtle or moderate chronic yield effects, such as due to air pollution, may have substantial economic consequences.

Conducting an economic assessment of environmental stress on agriculture, or on nonagricultural ecosystems, requires biological or production response data linking stress levels, e.g., pollutant level, and performance parameters of the ecosystem in question. For agriculture this response is typically represented by a relationship linking yield and pollutant dose levels. The relationship may be quantified directly using data generated from biological experimentation, indirectly from observed data on producer input, usage, and output levels or from some combination of data sources. From the standpoint of economic analysis, response estimates that reflect observed data drawn from producer decisions concerning actual input and output patterns are preferred (Crocker et al. 1981).

Table 1. Major U.S. commodities: 1982 acreage, production, and value.[a]

Crop	Acreage Harvested (million hectares)	Production (million bushels)	Value (million dollars)
Cotton	5.601	16.646 (bales)	4,109.069
Maize	34.247	8,200.951	20,000.805
Barley	3.703	478.301	1.168
All Hay	24.367	143.105	9.212
Winter Wheat	23.711	2,098.719	7,692.431
Soybeans	26.988	2,030.452	12,943.174
Total	118.617		44,755.859

[a]Source: U.S. Department of Agriculture (1983). Preliminary data.

A relationship that expresses production, or yield, as a function of both economic and biological factors is preferable because it is more likely to capture the true effect of a given change than a controlled experiment. Such relationships are typically referred to as production functions and their estimation and use is quite common in addressing certain economic issues. While some success has been achieved in applying these production function procedures to measure yield response in environmental change at the local level (e.g., see Mjelde et al. 1984), assessors in general have had little success in directly applying such techniques across large geographical areas, due to both data and statistical difficulties (e.g., see Adams et al. 1982; Manuel et al. 1981). Even for those regional studies where plausible response estimates are obtained, some experimental data are required to formulate testable hypotheses as well as to establish the credibility of any estimates derived from producer production data. As a result, practical, policy-oriented economic assessments of environmental stress to agriculture must rely on supplemental data from controlled biological experimentation to define the relationship between the stress and plant response. The more closely these experiments simulate actual production conditions and decisions faced by producers, the greater the utility of the estimates in subsequent economic assessments.

The U.S. Environmental Protection Agency, through its National Crop Loss Assessment Network (NCLAN), has recently attempted to improve on the state of knowledge in the area of crop response to air pollutants and associated economic consequences. In this multi-site, multi-year interdisciplinary program field experiments on major agricultural crops will provide information on crop response to tropospheric ozone pollution, the primary air pollutant in terms of reduced agricultural productivity (Heck et al. 1983). The output from these experiments is intended to support U.S. EPA's standard-setting responsibilities as well as provide policy makers with regulatory information on the economic consequences (benefits and costs) of ozone pollution on U.S. agriculture. In addition, the experience gained from NCLAN over the past three years can provide guidance to other environmental effects research programs directed at plant response. Another major goal of NCLAN is to correct some of the limitations inherent in past economic assessments. The results of this research can be used to improve other interdisciplinary studies aimed at quantifying the economic effects of ecosystem damage.

GENERATION AND USE OF YIELD RESPONSE DATA
IN ECONOMIC ASSESSMENTS

In many economic decisions pertaining to agricultural productions, some type of biological information is typically needed before an economic analysis of that decision can be performed. A common example is the calculation of the most profitable level of fertilizer application for a given crop, which requires information on the yield-fertilizer response. However, for this biological information to ultimately be translated into economic considerations, it should be consistent with the needs of the economist. These needs are well documented in the economic literature (e.g., Anderson and Dillon 1968; Adams and Crocker 1982). Two general requirements concerning such information are important. First, biological experiments structured to provide broad coverage of possible input levels (e.g., the use of a wide range of fertilizer levels or pollution exposure levels) typically are of greater value to the economist than studies or experiments primarily designed to minimize variance within plots. Second, as noted above, experiments performed under field conditions in which the experiment input levels correspond to those observed in commercial production provide a more realistic basis for measuring economic effects.

In studying the economic consequences of alternative levels of UV-B radiation an estimated response function is required to describe the physical relationship between yield and UV-B levels. In most economic analyses of environmental stress, a generalized or averaged (averaged across sites or regions) response relationship between relative crop yields and pollutant concentrations levels is used. Such an average response follows from the diverse nature of certain environmental stress levels across rural areas and the lack of systematic data on yields and UV-B radiation and other pollution levels for often widely dispersed locations. When using biological data from controlled experiments, effects of other covariates on response are not usually treated in the experiment. Thus, one typically is required to estimate a response function which expresses average yield as a function of the pollutant concentration, abstracting from other explanatory variables that may vary with location.

Given such a response function, it is then possible to determine the proportionate changes in yield which may occur with varying levels of UV-B radiation. By keying on proportionate or relative yield changes, some of the difficulties of extrapolating from site-specific data to regional effects can be minimized. Similar functions which describe the yield responses for ozone pollution and other air pollutants have already been estimated from NCLAN data and are used to predict relative yield changes used in economic assessments (e.g., Adams et al. 1984).

Estimating Response Functions

Operationally, response functions are fit using the classical methods of regression analysis, where yields for different crops and varieties per land area depend on various experimental levels of the pollutant in question and other controlled variables, e.g., fertilizer, moisture levels, other interactive pollutants. Typically, a number of statistical and practical issues must be addressed in the estimation process, such as the approximate definition and measurement of variables, and the specification of a functional form.

In performing an economic analysis of alternative pollution scenarios, one needs to estimate the changes in expected crop yield corresponding to a certain pollutant level or standard. This expected or mean value of yield associated with a given pollution level can be portrayed by the following simple response function relationship:

$$E(Y|X = x) = \alpha - \beta x \tag{1}$$

$$\text{with } var(Y|X = x) = \sigma^2 \tag{2}$$

where X denotes the level of pollutant experienced by the crop and Y denotes the corresponding observed crop yield. The α and β parameters measure the effect of the pollutant on yields and are hence the means by which one can predict yields at varying levels of X. This specification explicitly abstracts from other inputs or factors known to affect plant yields, but these could be captured in the response parameters.

The utility of the estimated response functions depends on the "reason-ableness" of the estimated parameters with respect to standard statistical diagnostic procedures. For example, β is typically assumed to be negative, based on prior experimental information. That is, an increase in the level of pollutant experienced by a crop translates into a decrease in the expected crop yield. The variance of the estimate of β ($\hat{\beta}$) reflects some of the uncertainty in the data and the given response model (linear, quadratic, etc.), uncertainty which may also affect resultant economic estimates. Additional experiments on response for a given crop or the introduction of prior information provide supplemental evidence on the effect of the pollutant on crop yields and reduce the variance of the estimate ($\hat{\beta}$).

Within the NCLAN program, a number of response functions have been estimated for major agricultural crops following similar procedures (Heck et al. 1983). The yield estimates for any given crop vary across experiments and across different expressions (linear, etc.) of the relationship between yields and dose. These variations are attributable to a wide range of uncertainties, including natural uncertainty, model uncertainty, and statistical uncertainty. In the context of the NCLAN experiments, natural uncertainty arises from pollutant and other stress interactions, alternative cultivar selection, and temporal and spatial variations across experiments. Model and statistical uncertainties are associated with the use of response model forms and data that yield estimated empirical models which differ from the true underlying physical relationship. Each of these sources of uncertainty can lead to different quantitative estimates of yield response and associated economic impacts. While there may be no prior statistical or theoretical reason to choose between the alternative response functions, differences in predicted yield for the same ozone pollution level could have implications for any analysis of the economic gains and losses due to changes in air pollution standards.

The effects of different response assumptions and associated predicted yields on economic estimates of the benefits of pollution control for the Corn Belt region of the U.S. has been tested by Adams and McCarl (1984). Measurement of the contribution of the response data to economic benefit estimates is also examined in Adams et al. (1984). The results of these analyses indicate that some biological information on ozone pollution response of the type generated in the NCLAN program has a high payoff in terms of redefining such economic estimates. However, the payoff of additional response information quickly diminishes. This is particularly true when the economist and biologist can draw on prior information such as the effect of other stresses on yields to restrict some of the uncertainties in the response estimates. The implication is that in estimating the economic gains from controlling air pollution, it is possible to obtain acceptable economic estimates using response data and associated response functions which are not perfect characterizations of the true response.

ANALYTICS OF ECONOMIC ASSESSMENTS

Usable dose-response information from biological experiments is clearly necessary to ensure a credible economic assessment. However, consideration must also be given to the structure of the economic problem being addressed. It is apparent from data on historical supply, consumption, and price that U.S. and international agricultural commodity markets function according to conventional supply and demand concepts. There are complex economic linkages between farmers (producers), their input suppliers and the ultimate retail consumer. This implies that the imposition of an air pollutant or other environmental stress at the farm level may have far-reaching and often overlooked effects.

Previous economic assessments have typically ignored the complexity of the agricultural economy, abstracting from such important dimensions of the problem as the impact of pollution on prices and ultimately on consumers as well as the possibility of adjustments by producers that may partially mitigate the effect of pollution. These limitations have been discussed elsewhere (Leung et al. 1978; Adams and Crocker 1981). The NCLAN and other recent economic assessment efforts attempt to deal with those dimensions that have important impacts on economic estimates (e.g., Adams et al. 1984; Kopp et al. 1983). The relationships between crop yield, farm production and price, and their relevance in an economic assessment are discussed subsequently.

Yield, Production, and Price Relationships

Crops display different tolerance levels to environmental stress and specifically UV-B radiation (Teramura 1983). Within a given crop group, if a substantial proportion of the crop acreage experiences reduced yields due to the effects of UV-B radiation, then increased prices for these and perhaps other crops may be expected. The net effects of these price increases on consumers and producers are not obvious. Thus, what the economist must consider in assessing any economic effects of UV-B radiation are changes in yields, associated adjustments in commodity prices, and distributional consequences attendant to these price changes.

In an economic context, both magnitudes and rates of change of yields affect the individual producer's decisions. Slow or minor rates of change may be mitigated by switching to more tolerant crop varieties. This may tend to offset some or all effects of air pollution on agriculture, and have minimal economic effects as long as the input price structure is not altered by the substitution. However, substantial changes in environmental conditions may cause major alterations in cropping patterns and locations, and potentially in total production of sensitive crops. Such changes in total production then introduce the prospect of changes in prices of those crops.

A simple analytical model of price determination frequently used by agricultural economists is adopted here. Specifically, the equilibrium price of agricultural commodities, taken in the aggregate or individually, is derived from the intersection of the relevant supply and demand curves, as depicted in Fig. 1. The effects of UV-B radiation or other stress that reduces yields may be viewed as a supply phenomenon, shifting the supply curve S_0 to S_1. Conversely, reductions in UV-B incidence may increase yields and hence production, resulting in an outward shift in the supply function (to S_2). The demand for U.S. agricultural commodities is generally inelastic, indicating that small shifts in the supply curve for an agricultural commodity will translate into rather large shifts in the equilibrium price of that commodity. In addition, this may trigger changes in prices of other food items. Thus, one may hypothesize changes in a variety of commodity prices if UV-B radiation affects the position of the supply curve for a single commodity.

These changes in agricultural prices not only affect producers, they also work their way through the system, affecting the welfare of consumers, producers, input suppliers, resource owners, and other parties. For example, reductions in commodity supply in the face of inelastic demand may actually increase farmers' total net revenue, as the attendant price rise (from P_0 to P_1) may be greater than the percentage reduction in quantity supplied or produced. However, the increase in prices from a supply reduction will reduce consumers' welfare so that they are not as well off as they were before the price change. Conversely, increases in supply will lower prices (from P_0 to P_2), reducing farmers' income but improving the well-being of consumers. Thus, if UV-B radiation alters yield of a substantial portion of the acreage for a given crop or causes a reduction

in harvested acreage of that crop, then the overall change in supply may
result in changes in the price at the farm level which will ultimately be
felt at the consumer level.   Alternately, if farmers employ mitigative
measures to adjust for UV-B radiation, then any additional costs of such
measures may also affect consumers through increases in producers' costs.

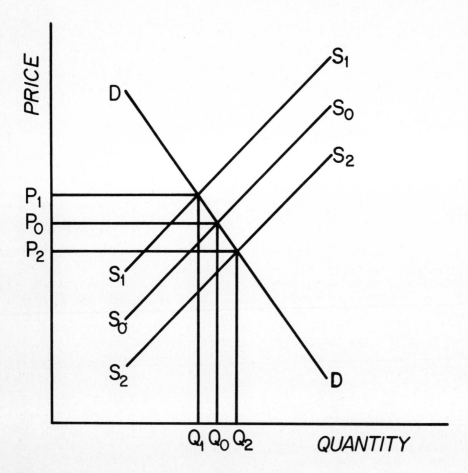

Fig. 1.   Price changes due to shifts in supply.

ASSESSING ECONOMIC EFFECTS ON AGRICULTURE: INTERFACING MODELS
AND DATA

Conceptually, the measurement or assessment problem may be viewed as one of
capturing the supply adjustments and corresponding price effects portrayed
in Fig. 1.   A standard approach used in most past assessments (Benedict et
al. 1971; Shriner et al. 1982) may be described as the damage function or
survey method.   This largely descriptive procedure calculates or specifies
some percentage reduction in crop production at current input levels and
then multiplies the relevant reduction by the assumed constant market

price. The advantage of such an approach is that the data requirements are relatively modest, allowing the assessor to arrive at a quantitative measure of damages quickly and inexpensively. However, assuming constant price ignores the price changes in Fig. 1 which ultimately determine production and consumption levels. By ignoring these economic realities, this approach places greater emphasis on the biological data, thus increasing the likelihood of diverse estimates of economic damage. Further, given different crop mixes, air pollution or UV-B radiation levels, weather patterns, and soil types within and across major production areas, some representation of the environmental and agronomic conditions and other aspects of regional agricultural production are required. Also, the quantification of the distributional effects, e.g., between consumers and producers, of these damages is useful information to policy makers. The damage function procedure is not capable of addressing distributional concerns, as it at best only measures effects on producers. These and other criticisms of this procedure are well documented in the economic assessment literature (Freeman 1979; Crocker 1982; Adams et al. 1984).

Fortunately, there are a number of defensible economic procedures or methodologies that are applicable to this type of assessment. For example, one way biological and economic relationships can be integrated in a defensible economic model is by specifying the economic problem as a mathematical programming problem. While this approach may require extensive economic and physical data needed to accurately "model" the agricultural environment and some means of solving the problem such as large-scale computer algorithm, it has been widely used in agricultural and environmental economic analyses because of its ability to project the economic consequences of as yet unrealized environmental effects. Alternatively, economic models based on actual or current producer and consumer behavior may also be used to reflect changes in yield and agricultural production due to environmental stress. In these latter models, yield effects may be addressed exogenously through experimental dose-response data or endogenously through estimation of production functions as mentioned earlier. Still other economic assessment methods are available that have a narrower focus than the programming or simulation methods but which do provide economically consistent estimates on some subset of the agricultural sector, e.g., the methods used by Kopp et al. (1983) or Mjelde et al. (1984). All of these approaches to measuring economic effects are capable of capturing the important economic aspects of the problem, unlike the traditional damage function approach. The intent of this discussion is only to note that defensible and tractable economic assessment methods exist and that policy makers should not have to base decisions on incorrect economic methods. The interested reader is referred to Freeman (1979) for a detailed discussion of economic methodologies. To demonstrate how one may apply a comprehensive economic assessment procedure, including data needs, a programming approach as used to assess ozone pollution effects in the NCLAN program is discussed subsequently.

An Economic Model in the Mathematical Programming Framework

One type of mathematical programming used to assess the effects of alternative ozone pollution levels on the agricultural sector is a price-endogenous model such as used by Adams et al. (1982), Howitt et al. (1984), and Adams et al. (1984b) in which price is a function of the supply of a given commodity. This and other types of programming models have been used by agricultural economists to simulate the effect of alternative agricultural policies or technological change (Heady and Srivastava 1975; Duloy and Norton 1973).

The model applied in the NCLAN assessment consists of both a producer or farm level component and a more aggregate or sector component to assess total changes in ozone pollution. The features of this particular model are discussed briefly here as an example of how this may be applied to the UV-B assessment problem. Specifically the producer level is portrayed by production "activities," which represent the alternative ways or techniques of producing various agricultural commodities. The production side is modeled with considerable detail on cropping activities (twelve annual crops plus hay and beef, pork, and milk production), input use, and environmental and other fixed constraints. Such detail is needed to adequately model potential producer mitigative behavior in the face of environmental change. As noted by Crocker (1982), failure to account for these adaptive opportunities tends to overstate potential environmental damages and understate benefits. In the aggregate, these production activities determine the total supply of the commodities in the model. The supply of each commodity is then used in combination with demand relationships for each crop or livestock product to determine the market price, following the logic of Fig. 1.

The solution procedure underlying the model is one that solves for the set of market prices that maximizes a given "objective function". The objective function in this, or any programming model, is simply a mathematical expression or equation that takes on economic significance when the variables in that equation are defined to represent prices and quantities of commodities. The values that the objective function can take are constrained by other relationships, such as total farm land available or total processing capacity. By specifying the variables in the model to conform to real economic value as defined by supporting historical data, the model solution approximates actual conditions for a given time period.

The objective function in the NCLAN assessment may be interpreted as a measure of "economic surplus," a somewhat abstract concept that is frequently used by economists to measure in dollars the social benefits of alternative policies or actions (see Just et al. 1982, for more discussion). Analytically, this is defined as the area between the demand and supply curves to the left of their intersection. This area may also be disaggregated into components representing effects on consumers and on producers. Such information is of particular interest to policy makers who must consider compromises between those who stand to gain or lose by such policies.

Since the demand functions and aggregate production responses in the model are specified at the national level, the solution of the agricultural model provides objective function values at the national level. By then imposing alternative ozone pollution levels on the model as manifested in yields predicted by NCLAN dose-response functions, changes in crop production and prices and ultimately the objective function may be measured. Comparisons of changes in the model objective function value between these hypothetical ozone situations and current ambient ozone pollution levels provide economic estimates of the benefits or costs of these alternative levels of ozone pollution. It is this kind of information that can then be used by policy makers to assess the economic efficiency of changing current ozone pollution standards.

IMPLEMENTING A UV-B ASSESSMENT: DATA NEEDS

A flow chart representing a generalized national model for assessing UV-B
radiation, and the critical inputs for implementing such a model is
provided in Fig. 2. The blocks on the left-hand side of the figure clearly
show the essential link between meteorological, biological, and economic
data needs. The key inputs for the model are given in the bold-face blocks
which deal with environmental quality, crop dose-response, and yield
adjustment data. These three input boxes then serve to drive the economic
analysis portion of the model represented in the right-hand side of Fig. 2.
The focus of this discussion is on the inputs in the three bold-face
blocks.

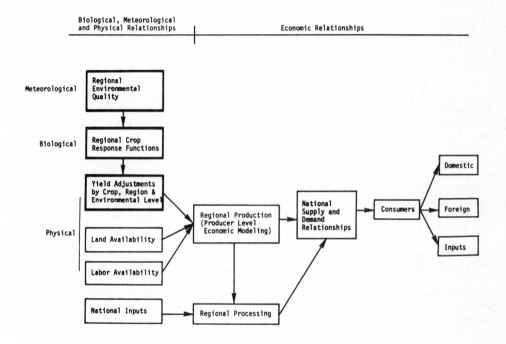

Fig. 2. Flow chart of a generalized economic model of agriculture.

The data on regional environmental quality serves two purposes. First,
when using actual ambient conditions, the model can be solved for a given
year for which results are already available. For example, 1980 was used
as a base for validation of the NCLAN model. This base solution then
serves as the benchmark or base situation against which to compare hypothetical environmental conditions associated with proposed standards or
regulations. Second, if alternative standards or assumptions on environmental quality are to be considered, these assumed conditions for each
region of the economic model need to be defined. In the case of UV-B
radiation, the base case situation would reflect actual incidence of UV-B
radiation for each region under meteorological conditions representing that

base year. The assumed changes in UV-B radiation due to differing assumptions concerning changes in stratospheric ozone would then need to be projected for these same regions. These projections would likely result from an appropriate meteorological model of UV-B incidence.

The projected regional changes in UV-B radiation are then fed into the dose-response block of the figure. Ideally, this box would contain regional dose-response functions for each major crop grown in that region, adjusted for the effects of other stresses. In practice, such detailed data probably would not be available. By assuming responses to be somewhat homogeneous across regions, a smaller number of response functions may be acceptable. These functions are typically estimated using response data generated from experiments on the environmental stress or stresses in question, as in the NCLAN program. In the case of UV-B radiation, some of these data could be provided by current research underway at the University of Maryland and the University of Florida (e.g., Teramura 1983).

By adjusting the dose parameters or UV-B incidence in each response function, alternative yield changes relative to ambient conditions are estimated. The changes in dose would be determined by the regional environmental quality block. The projected regional yields arising from the interplay of dose and response functions, become the driving force in the economic analysis. These data are needed for the third block. For tractability, these yield changes should be measured as relative changes; that is, the dose-response functions project proportionate or percentage yields, not actual. The use of proportionate adjustments also allows for potential extrapolation of site-specific data to regional response if the proportionate responses are statistically homogeneous (Rawlings and Cure 1984). These proportionate yield adjustments are then used to modify the base yields in the economic model. It is this resultant difference in yields between base or actual yields and those under the hypothetical environmental situation which is at the heart of the economic analysis.

Depending on the scope and extent of the environmental assessment, a number of yield scenarios may be required. In each case, the differences in yields trigger changes in the economic model which ultimately translate into changes in the model solution or objective function values, i.e., the measure of economic surplus. Comparison of the differences in objective function values, as measured in dollars, is then an estimate of the benefits or costs of each environmental scenario. Thus, the interface between the meteorological, biological, and economic analyses is essentially captured in the linkages in the three boxes described in Fig. 2.

At present, the key boxes in projecting UV-B radiation levels and the biological link between those levels and yields is ill defined. The literature does not appear to have fully resolved the direction and magnitude of changes in stratospheric ozone levels and hence in UV-B radiation. Further, the relationship between UV-B radiation and crop yield shows much less impact than, say, ambient ozone pollution on crop yields. However, there is the suggestion of possible interactive effects between UV-B radiation and other environmental stresses, e.g., drought, ambient ozone pollution, increased CO_2 levels. An ideal assessment of the benefits of environmental regulation should account for such major interactive effects, given that environmental regulation and control typically affects anthropogenic sources of several pollutants simultaneously.

Thus, even though the direct effects of UV-B radiation on agriculture may seem small when studied in isolation, they may offer possible explanations for the substantial variability found in other, more well defined, agricultural stress assessments.

SUMMARY AND CONCLUSIONS

The models, procedures, and data needs outlined in this paper are meant to be illustrative of techniques available for assessing the agricultural benefits of controlling air pollution or other environmental stresses. The treatment here has been general, defining major conceptual concerns and data needs for performing benefits analyses on alternative environmental situations. The analysis draws upon other bioeconomic assessments to highlight the potential data needs in a UV-B radiation assessment.

As the biological and meteorological data become available, an assessment of economic benefits to agriculture from different UV-B levels can be performed with existing economic methodologies. Properly structured estimates can provide decision-makers with the compromises associated with environmental regulation. How such benefit estimates are ultimately used in setting policy is outside the purview of the economist. However, for the benefit estimates to be meaningful, the economic assessment must capture the principal mechanisms involved in the agricultural economy including supply and demand phenomena. Failure to adequately account for both the physical or biological and economic aspects of the problem will result in possibly misleading and even contradictory assessment estimates.

ACKNOWLEDGMENTS

Although the research described in this chapter has been partially funded by the United States Environmental Protection Agency through cooperative agreements with the University of Wyoming and Oregon State University, it has not been subjected to the Agency's required peer review and therefore does not necessarily reflect the views of the Agency and no official endorsement should be inferred. This research was performed while the author was Associate Professor of Agricultural Economics, University of Wyoming, Laramie.

REFERENCES

Adams RM, Crocker TD (1981) Analytical issues in economic assessments of vegetation damages. In: Teng PS, Krupa SV (eds) Crop loss assessment. Miscellaneous publication no 7. University of Minnesota Agricultural Experiment Station, p 198

Adams RM, Crocker TD (1982) Dose-response information and environmental damage assessments: An economic perspective. Air Poll Cont Assoc J 32:1062-1067

Adams RM, Crocker TD, Thanavibulchai N (1982) An economic assessment of air pollution damages to selected annual crops in Southern California. J Environ Econ Manag 9:42-58

Adams RM, McCarl BA (1985) Assessing the benefits of alternative oxidant standards on agriculture: The role of response information. J Envir Econ Manag 12:264–276

Adams RM, Hamilton SA, McCarl BA (1984) The economic effects of ozone on agriculture. EPA-600/3-84-090. U.S. Environmental Protection Agency, Environmental Research Laboratory, Corvallis, Oregon 97333

Anderson JR, Dillon JL (1968) Economic considerations in response research. Amer J Agric Econ 50:130–142

Benedict NM, Miller CJ, Smith JS (1971) Assessment of economic impact of air pollutants on vegetation in the United States: 1969–1971. EPA-650/5-78-002. Stanford Research Institute, Menlo Park, California

Crocker TD (1982) Economic assessments of pollution damage to managed ecosystems. In: Jacobson, JS, Miller AA (eds) Effects of air pollution on farm commodities. Izaak Walton League of America, Washington, D.C.

Crocker TD, Dixon BL, Howitt R, Oliveria R (1981) A program assessing the economic benefits of preventing air pollution damages to U.S. agriculture. Report to Corvallis Environmental Research Laboratory, June

Duloy J, Norton R (1973) CHAC: A programming model of Mexican agriculture. In: Goreux L, Manne A (eds) Multilevel planning: Case studies in Mexico. North Holland Publishing, LC 72–93090

Freeman AM, III (1979) The benefits of environmental improvement. The Johns Hopkins University Press, Baltimore, Maryland, ISBN 0-8018-2163-0

Heady E, Srivastava U (1975) Spatial sector programming models in agriculture. Iowa State University Press, Ames, ISBN 0-8138-1575-4

Heck WW, Adams RM, Cure WW, Kohut R, Kress L, Temple P (1983) A reassessment of crop loss from ozone. Environ Sci Tech 17:572A–581A

Howitt RE, Gassard TE, Adams RM (1984) Effects of alternative ozone levels and response data on economic assessments. J Air Poll Cont Assoc 34:1122–1127

Just RE, Hueth DL, Schmitz A (1982) Applied welfare economics and public policy. Prentice-Hall, New York, New York, ISBN 0-13-043398-5

Kopp RJ, Vaughn WT, Hazilla M (1983) Agricultural benefits analysis: alternative ozone and photochemical oxidant standards. Final Report to Economic Analysis Branch, U.S. Environmental Protection Agency, Research Triangle Park, North Carolina.

Leung SK, Reed W, Cauchius S, Howitt R (1978) Methodologies for evaluation of agricultural crop yield changes: a review. EPA-600/5-78-018. Corvallis Environmental Research Laboratory

Manuel EH, Horst RL, Brennan KM, Laner WN, Duff MC, Tapiero JK (1981) Benefits analysis of alternative secondary national ambient air quality standards for sulfur dioxide and total suspended particulates, vol 4. EPA-68-D2-3392. Final report by Math Tech, Inc. to U.S. Environmental Protection Agency, Research Triangle Park

Mjelde JW, Adams RM, Dixon BL, Garcia P (1984) Using farmers actions to measure the cost of air pollution. J Air Poll Cont Assoc 31:360–364

Rawlings JD, Cure WW (1984) The Weibull function as a dose-response model for studying air pollution effects. Crop Sci (in press)

Shriner DS, Cure WW, Heagle AS, Heck WW, Johnson DW, Olson RJ, Skelly JM (1982) An analysis of potential agriculture and forestry impacts of long-range transport of air pollutants. ORNL-5910, Oak Ridge National Laboratory, Tennessee

Teramura AH (1983) Effects of ultraviolet-B radiation on the growth and yield of crop plants. Physiol Plant 58:415–427

U.S. Department of Agriculture (1983) Agricultural statistics, 1982. U.S. Government Printing Office. Washington, DC

U.S. Environmental Protection Agency (1984) Air quality criteria for ozone and other photochemical oxidants. EPA-600/8-84-020A, ECAO, Research Triangle Park, North Carolina

APPENDIX 1

SUBROUTINE FOR SCHIPPNICK AND GREEN MODEL

R. RUNDEL

```
1000 REM *********************************************************
1001 REM *                                                       *
1002 REM * SCHIPPNICK AND GREEN UV SPECTRAL IRRADIANCE MODEL     *
1003 REM * (1982) PHOTOCHEMISTRY AND PHOTOBIOLOGY 35:89-101      *
1004 REM * (HEREAFTER REFERRED TO AS "SG")                       *
1005 REM * REFERENCE IS ALSO MADE TO GREEN, CROSS, AND SMITH     *
1006 REM * (1980) PHOTOCHEMISTRY AND PHOTOBIOLOGY 31:59-65       *
1007 REM * (HEREAFTER REFERRED TO AS "GCS")                      *
1008 REM *                                                       *
1009 REM * INPUT PARAMETERS TO THIS ROUTINE ARE:                 *
1010 REM *     L = WAVELENGTH IN NANOMETERS                      *
1011 REM *     Y1 = ALTITUDE IN KILOMETERS                       *
1012 REM *     Z = SOLAR ZENITH ANGLE IN DEGREES                 *
1013 REM *     M = COS(Z)                                        *
1014 REM *     W = OZONE COLUMN THICKNESS IN ATMOS-CM            *
1015 REM *     B = SURFACE ALBEDO                                *
1016 REM *                                                       *
1017 REM * THIS SUBROUTINE WRITTEN IN WANG BASIC-2               *
1018 REM *                                                       *
1019 REM *********************************************************

1020 B1=0
   : B2=0
   : REM SET INITIAL ALBEDO CONTRIBUTIONS TO ZERO
1030 IF B=0 THEN 1060
1040 Y=0
   : GOSUB 1100
   : GOSUB 1170
   : GOSUB 1220
   : GOSUB 1310
   : GOSUB 1480
   : REM PREPARE TO CALCULATE TOTAL IRRADIANCE AT GROUND LEVEL
1050 G=D+S*M*D0
   : REM TOTAL IRRADIANCE AT GROUND LEVEL
1060 Y=Y1
   : GOSUB 1100
   : GOSUB 1170
   : GOSUB 1220
   : GOSUB 1310
   : GOSUB 1480
   : GOSUB 1590
   : REM PREPARE TO CALCULATE TOTAL IRRADIANCE AT ALTITUDE Y1
1070 G1=D+S1*M1*D0+B1
   : REM TOTAL DOWNWARD IRRADIANCE
1080 G2=S2*M2*D0+B2
   : REM TOTAL UPWARD IRRADIANCE
1090 G=G1
   : REM FINAL TOTAL UPWARD IRRADIANCE RETURNED BY SUBROUTINE
```

NATO ASI Series, Vol. G 8
Stratospheric Ozone Reduction, Solar Ultraviolet Radiation
and Plant Life. Edited by R. C. Worrest and M. M. Caldwell
© Springer-Verlag Berlin Heidelberg 1986

```
1100 REM CALCULATES OPTICAL DEPTHS VERSUS WAVELENGTH L.
   : REM 1 = RAYLEIGH SCATTERING
   : REM 2 = AEROSOL SCATTERING
   : REM 3 = OZONE ABSORPTION
   : REM 4 = AEROSOL ABSORPTION
1110 T1=1.221*(300/L)^4.27
   : REM GCS EQ. 18
1120 REM K3=9.9405/(0.0445+EXP((L-300)/7.294))
   : REM T3=W*K3
   : REM GCS EQ. 19 (USE THIS OPTION TO COMPARE WITH GCS MODEL)
1130 T2=(0.08052/0.204)*(0.205+(L-302.5)*1.75E-4)
   : REM PERSONAL COMMUNICATION FROM A.E.S. GREEN (1983)
1140 K3=9.788*1.0556/(0.0556+EXP((L-300)/6.978))
   : T3=W*K3
   : REM PERSONAL COMMUNICATION FROM A.E.S.GREEN (1983)
1150 T4=(0.034/0.204)*(0.08052/0.034)*(0.034-(L-302.5)*5E-5)
   : REM PERSONAL COMMUNICATION FROM A.E.S. GREEN (1983)
1160 RETURN

1170 REM CALCULATES RELATIVE SPECIES CONCENTRATION VS. HEIGHT
   : REM 1 = RAYLEIGH SCATTERING
   : REM 2 = AEROSOL SCATTERING
   : REM 3 = OZONE ABSORPTION
   : REM 4 = AEROSOL ABSORPTION
1180 N1=1.437/(0.437+EXP(Y/6.35))
1190 N2=0.8208/(-0.145+EXP(Y/0.952))+0.04*(1+EXP(-16.33/3.09))
       /(1+EXP((Y-16.33)/3.09))
1200 N3=0.13065/(2.35+EXP(Y/2.66))+0.961*(1+EXP(-22.51/4.92))
       /(1+EXP((Y-22.51)/4.92))
   : REM N1, N2, N3 FROM GCS EQ. 9 AND TABLE 1
1210 RETURN

1220 REM CALCULATES S(L,Z,Y) RATIO OF IRRADIANCE AT ZENTIH
   : REM ANGLE Z TO IRRADIANCE AT ZENITH ANGLE 0
   : REM        L = WAVELENGTH
   : REM        Z = SOLAR ZENITH ANGLE
   : REM        Y = HEIGHT (KM)
1230 REM        1 = DOWN
   : REM        2 = UP
1240 P1=(1.0226/(M^2+0.0226))^0.5-1
1250 P2=(1.0112/(M^2+0.0112))^0.5-1
   : REM SG EQ. 11
1260 F1=1/(1+84.37*(T3+T4)^0.6776)
1270 F5=1/(1+28.8*(T3+T4)^1.325)
   : REM SG EQ.10
1280 S1=(F1+(1-F1)*EXP(-T3*N3^2.392*P1))
       *EXP(-(0.5346*T1*N1^0.3475+0.6077*T2*N2^0.3445)*P1)
1290 S2=(F5+(1-F5)*EXP(-T3*P2))
       *EXP(-(0.6440*T1*N1^0.0795+0.1020*T2)*P2)
   : REM SG EQ. 9
1300 RETURN
```

```
1310 REM CALCULATES M(L,Y) RATIO OF DIFFUSE IRRADIANCE AT
   : REM HEIGHT Y TO DIRECT IRRADIANCE AT Y=0
1320 REM        1 = DOWN
   : REM        2 = UP
1330 REM N6=7.389/(6.389+EXP(0.921*Y/6.35))
   : REM SG EQ. 17 FOR N1 BAR
1340 A1=1.735-0.346*N1^5
1350 A2=0.8041*T1^A1*N6
   : REM F1D(L,Y) FROM SG EQ. 15
1360 A3=1/(1+0.3264*W^1.223*N3^(1+1.7*T3)*K3^0.7555)
   : REM G3D(L,Y) FOR DH MODEL FROM SG EQ. 13 (PREFERRED OPTION)
1370 REM A3=1/(1+0.3747*W3^1.223*N3^(1+1.5*T3)*K3^0.7555
   : REM G3D(L,Y) FOR BD MODEL FROM SG EQ. 13
1380 A4=1/(1+1.554*T4^0.88*N2^0.49)
   : REM G4D(L,Y) FROM SG EQ. 13
1390 A5=(A2*A3+N2^0.564)*1.437*(T2^1.12)/A3
   : REM F2D(L,Y) FROM SG EQ. 13
1400 M1=(A2+A5)*A3*A4
   : REM SG EQ. 12
1410 A6=1.1032*T1^1.735*(1-N1)^0.921
   : REM F1U(L,Y) FROM SG EQ. 14
1420 A7=2.027*T1^1.735*T2^1.12*(1-N1)^0.921+0.4373*T2^1.12
   *(1-N2)^0.564
   : REM F2U(L,Y) FROM SG EQ. 19
1430 A8=1/(1+0.1983*W^1.1181*K3^0.7555*N3)
   : REM G3U(L,Y) FOR DH FROM SG EQ. 13 (PREFERRED OPTION)
1440 REM A8=1/(1+0.2248*W3^1.1181*K3^0.7555*N3)
   : REM G3U(L,Y) FOR DB FROM SG EQ. 13
1450 A9=1/(1+6.2*T4^0.88*N2^0.356)
   : REM G4U(L,Y) FROM SG EQ. 13
1460 M2=(A6+A7)*A8*A9
  :REM SG EQ. 12
1470 RETURN

1480 REM CALCULATES DIRECT IRRADIANCE
1490 H=0.582*(300/L)^5*(EXP(9.102)-1)/(EXP(9.102*300/L)-1)
   : REM UNITS OF W NM^-1 M^-2
   : REM EXTRATERRESTRIAL IRRADIANCE FROM SG TABLE 1
1500 REM H=1.095*(1-EXP(-0.6902*EXP((L-300)/23.74)))
   : REM UNITS OF W NM^-1 M^-2
   : EXTRATERRESTRIAL IRRADIANCE FROM GCS EQ. 17
1510 H=H*(1-0.738*EXP(-(L-279.5)^2/2/2.96^2)-0.485
       *EXP(-(L-286.1)^2/2/1.57^2)-0.243
       *EXP(-(L-300.4)^2/2/1.8^2)+0.192
       *EXP(-(L-333.2)^2/2/4.26^2-0.167
       *EXP(-(L-358.5)^2/2/2.01^2)+.097
       *EXP(-(L-368)^2/2/2.43^2))
1520 REM ADD SMOOTHED FRAUNHOFER STRUCTURE
1530 S5=SQR((M*M+1.8E-3)/(1+1.8E-3))
1540 S6=SQR((M*M+3E-4)/(1+3E-4))
1550 S7=SQR((M*M+7.4E-3)/(1+7.4E-3))
1560 A=T1*N1/S5+T2*N2/S6+T3*N3/S7+T4*N2/S6
   : REM TOTAL OPTICAL DEPTH
   : A0=T1+T2+T3+T4
   : REM TOTAL OPTICAL DEPTH FOR Z = 0 AND Y = 0
```

```
1570 D=M*H*EXP(-A)
   : REM DIRECT IRRADIANCE FROM CGS EQ. 1
   : D0=H*EXP(-A0)
   : REM DIRECT IRRADIANCE FOR Z = 0 AND Y = 0
1580 RETURN

1590 REM CALCULATES EFFECT OF NON-ZERO ALBEDO
1600 REM        1 = DOWN
   :            2 = UP
1610 IF B=0 THEN 1670
   : REM        B = ALBEDO
1620 R=(0.4424*T1^0.5626/(1+0.2797*W^1.0132*K3^0.8404)
       +0.1*T2^0.88)/(1+3.7*T4)
   : REM SG EQ. 28, 29
1630 E1+N1*EXP(-(0.4638*T2*(1-N2)+3.5*T3*(1-N3)+0.4638*T4(1-N2)))
1640 E2=EXP(-(0.4909*T1*(1-N1)+0.5658*T2*(1-N2)+2.417*T3*(1-N3)
       +0.5658*T4*(1-N2)))
   : REM SG EQ. 32, 33
1650 B1=R*B/(1-R*B)*E1*G
1660 B2=B/(1-R*B)*E2*G
   : REM ALBEDO CONTRIBUTION FROM SG EQ. 25,27,31
1670 RETURN
```

LIST OF ATTENDEES

R. M. Adams
Division of Agricultural Economics
University of Wyoming
Laramie, Wyoming 82071
USA

H. Bauer
Gesellschaft für Strahlen-
   und Umweltforschung
Bereich Projektträgerschaften
Josephspitalstraße 15
D-8000 München 2
FEDERAL REPUBLIC OF GERMANY

R. H. Biggs
Fruit Crops Department
1143 Fifield Hall
University of Florida
Gainesville, Florida 32611
USA

L. O. Björn
Department of Plant Physiology
University of Lund
P.O. Box 7007
S-220 07 Lund
SWEDEN

J. F. Bornman
Department of Plant Physiology
University of Lund
P.O. Box 7007
S-220 07 Lund
SWEDEN

G. Brasseur
Institut d'Aéronomie Spatiale de Belgique
3, Avenue Circulaire
B-1180 Bruxelles
BELGIUM

M. M. Caldwell
Department of Range Science
Utah State University (UMC 52)
Logan, Utah 84322
USA

B. L. Diffey
Regional Medical Physics Department
Dryburn Hospital
Durham DH1 5TW
UNITED KINGDOM

G. Döhler
Botanisches Institut der J. W. Goethe-Universität
Siesmayerstraße 70
D-6000 Frankfurt a. M.
FEDERAL REPUBLIC OF GERMANY

S. A. W. Gerstl
Theoretical Division, MS-B279
Los Alamos National Laboratory
P.O. Box 1663
Los Alamos, New Mexico 87545
USA

D.-P. Häder
Fachbereich Biologie-Botanik (Lahnberge)
Philipps-Universität Marburg
D-3550 Marburg
FEDERAL REPUBLIC OF GERMANY

W. Iwanzik
Botanisches Institut II
Universität Karlsruhe
Kaiserstraße 12
D-7500 Karlsruhe
FEDERAL REPUBLIC OF GERMANY

J. Kiefer
Strahlenzentrum der
    Justus-Liebig-Universität
Leihgesterner Weg 217
D-6300 Giessen
FEDERAL REPUBLIC OF GERMANY

O. Klais
Angewandte Physik
Hoechst Aktiengesellschaft
Postfach 80 03 20
D-6230 Frankfurt a. M. 80
FEDERAL REPUBLIC OF GERMANY

J. C. van der Leun
State University of Utrecht
Institute of Dermatology
Catharijnesingel 101
3511 GV Utrecht
THE NETHERLANDS

J. Lydon
Department of Botany
University of Maryland
College Park, Maryland 20742
USA

H. D. Mennigmann
Institut für Mikrobiologie
Fachbereich Biologie der J. W. Goethe-Universität
Theodor-Stern-Kai 7, Haus 75 A
D-6000 Frankfurt a. M.
FEDERAL REPUBLIC OF GERMANY

A. Mookerjee
School of Environmental Sciences
Jawaharlal Nehru University
New Delhi-110067
INDIA

W. Rau
Botanisches Institut der Universität München
Menzinger-Straße 67
D-8000 München 19
FEDERAL REPUBLIC OF GERMANY

A. Rau-Hund
Botanisches Institut der Universität München
Menzinger-Straße 67
D-8000 München 19
FEDERAL REPUBLIC OF GERMANY

G. Renger
Max-Volmer-Institut für Biophysikalische
   und Physikalische Chemie
Technische Universität Berlin, PC-14
Straße des 17. Juni 135
D-1000 Berlin 12
FEDERAL REPUBLIC OF GERMANY

R. Robberecht
Range Resources
University of Idaho
Moscow, Idaho 83843
USA

C. S. Rupert
Biology Programs -- FO 3.1
University of Texas at Dallas
P.O. Box 688
Richardson, Texas 75080
USA

R. D. Rundel
Physics Department
Mississippi State University
Mississippi State, Mississippi 39762
USA

U. Schmailzl
Chemie der Atmosphäre
Max-Planck-Institut für Chemie
Postfach 3060
D-6500 Mainz
FEDERAL REPUBLIC OF GERMANY

E. Schrott
Botanisches Institut der Universität
Menzingerstraße 67
D-8000 München 19
FEDERAL REPUBLIC OF GERMANY

W. B. Sisson
New Mexico State University
Department of Animal and Range Science
P.O. Box 3I
Las Cruces, New Mexico 88003
USA

D. Steinmüller
Botanisches Institut II
Universität Karlsruhe
Kaiserstraße 12
D-7500 Karlsruhe
FEDERAL REPUBLIC OF GERMANY

M. D. Steven
Environmental Physics Unit
School of Agriculture
University of Nottingham
Sutton Bonington
Loughborough LE12 5RD
UNITED KINGDOM

A. H. Teramura
Department of Botany
University of Maryland
College Park, Maryland 20742
USA

M. Tevini
Botanisches Institut II
Universität Karlsruhe
Kaiserstraße 12
D-7500 Karlsruhe
FEDERAL REPUBLIC OF GERMANY

R. M. Tyrrell
Institut Suisse Recherches
   Expérimentales sur le Cancer
CH-1066 Epalinges s./Lausanne
SWITZERLAND

M. Voss
Max-Volmer-Institut PC14
TU Berlin
Straße des 17. Juni 135
D-1000 Berlin 12
FEDERAL REPUBLIC OF GERMANY

G. Wagner
Botanisches Institut I
   der Justus-Liebig-Universität
Senckenbergstraße 17-21
D-6300 Giessen
FEDERAL REPUBLIC OG GERMANY

A. Webb
Environmental Physics Unit
School of Agriculture
University of Nottingham
Sutton Bonington
Loughborough LE12 5RD
UNITED KINGDOM

E. Wellmann
Biologisches Institut II der Universität
Schänzlestraße 1
D-7800 Freiburg
FEDERAL REPUBLIC OF GERMANY

R. C. Worrest
Department of General Science
Oregon State University
Corvallis, Oregon 97331
USA

V. L. Worrest
Student Health Service
Oregon State University
Corvallis, Oregon 97331
USA

F. Zölzer
Strahlenzentrum der
   Justus-Liebig-Universität Giessen
Leihgesterner Weg 217
D-6300 Giessen
FEDERAL REPUBLIC OF GERMANY

NATO ASI Series G